Basics of Matrix Algebra for Statistics with R

Chapman & Hall/CRC
The R Series

Series Editors

Aims and Scope

This book series reflects the recent rapid growth in the development and application of R, the programming language and software environment for statistical computing and graphics. R is now widely used in academic research, education, and industry. It is constantly growing, with new versions of the core software released regularly and more than 6,000 packages available. It is difficult for the documentation to keep pace with the expansion of the software, and this vital book series provides a forum for the publication of books covering many aspects of the development and application of R.

The scope of the series is wide, covering three main threads:

- Applications of R to specific disciplines such as biology, epidemiology, genetics, engineering, finance, and the social sciences.
- Using R for the study of topics of statistical methodology, such as linear and mixed modeling, time series, Bayesian methods, and missing data.
- The development of R, including programming, building packages, and graphics.

The books will appeal to programmers and developers of R software, as well as applied statisticians and data analysts in many fields. The books will feature detailed worked examples and R code fully integrated into the text, ensuring their usefulness to researchers, practitioners and students.

Published Titles

Stated Preference Methods Using R, *Hideo Aizaki, Tomoaki Nakatani, and Kazuo Sato*

Using R for Numerical Analysis in Science and Engineering, *Victor A. Bloomfield*

Event History Analysis with R, *Göran Broström*

Computational Actuarial Science with R, *Arthur Charpentier*

Statistical Computing in C++ and R, *Randall L. Eubank and Ana Kupresanin*

Basics of Matrix Algebra for Statistics with R, *Nick Fieller*

Reproducible Research with R and RStudio, Second Edition, *Christopher Gandrud*

R and MATLAB® *David E. Hiebeler*

Nonparametric Statistical Methods Using R, *John Kloke and Joseph McKean*

Displaying Time Series, Spatial, and Space-Time Data with R, *Oscar Perpiñán Lamigueiro*

Programming Graphical User Interfaces with R, *Michael F. Lawrence and John Verzani*

Analyzing Sensory Data with R, *Sébastien Lê and Theirry Worch*

Parallel Computing for Data Science: With Examples in R, C++ and CUDA, *Norman Matloff*

Analyzing Baseball Data with R, *Max Marchi and Jim Albert*

Growth Curve Analysis and Visualization Using R, *Daniel Mirman*

R Graphics, Second Edition, *Paul Murrell*

Data Science in R: A Case Studies Approach to Computational Reasoning and Problem Solving, *Deborah Nolan and Duncan Temple Lang*

Multiple Factor Analysis by Example Using R, *Jérôme Pagès*

Customer and Business Analytics: Applied Data Mining for Business Decision Making Using R, *Daniel S. Putler and Robert E. Krider*

Implementing Reproducible Research, *Victoria Stodden, Friedrich Leisch, and Roger D. Peng*

Graphical Data Analysis with R, *Antony Unwin*

Using R for Introductory Statistics, Second Edition, *John Verzani*

Advanced R, *Hadley Wickham*

Dynamic Documents with R and knitr, Second Edition, *Yihui Xie*

Basics of Matrix Algebra for Statistics with R

Nick Fieller

University of Sheffield, UK

CRC Press
Taylor & Francis Group
Boca Raton London New York

CRC Press is an imprint of the
Taylor & Francis Group, an **informa** business
A CHAPMAN & HALL BOOK

CRC Press
Taylor & Francis Group
6000 Broken Sound Parkway NW, Suite 300
Boca Raton, FL 33487-2742

Printed on acid-free paper
Version Date: 20150602

International Standard Book Number-13: 978-1-4987-1236-1 (Hardback)

Visit the Taylor & Francis Web site at
http://www.taylorandfrancis.com

and the CRC Press Web site at
http://www.crcpress.com

To Rafe and Hal

Contents

Preface

The subject matter of this book is basic matrix algebra with a focus on those parts of the theory that relate to practical statistical applications (so attention is especially focused on real symmetric matrices) together with a guide to the implementation of the calculations in **R**. The aim is to provide a guide to elementary matrix algebra sufficient for undertaking specialist courses such as multivariate data analysis and linear models encountered in final year undergraduate and master's courses. In addition the material goes a little further than this, providing an initial guide to more advanced topics such as generalized inverses of singular matrices and manipulation of partitioned matrices. This is to provide a second step for those who need to go a little further than just courses such as these, for example, when embarking on a master's dissertation or the first year of doctoral research and also to satisfy those who develop an interest in the subject in its own right. It is not intended to be a comprehensive text on matrix theory so topics not of direct relevance to these needs are not included

In addition to the algebraic manipulation of matrices, the objective is to give numerical examples since it is necessary for a practicing statistician to be able to do numerical calculations, both by hand and by using **R**. Much of the exposition is by presenting examples with detailed solutions.

People undertaking specialist statistics courses are typically in one or the other of two broad groups. First are those who have arrived through a general mathematics degree programme or at least the first two years of such a programme. Second are those arriving after experience in applied sciences such as biomedical or engineering or sociological sciences, perhaps with a brief introductory course covering essentials of mathematics and elementary statistics. Both groups will have some knowledge of basic statistics, including common univariate distributions, estimation, hypothesis testing and simple linear regression on at least one variable. Both groups will be familiar with basic [differential] calculus (polynomial and exponential functions, products, ratios and the chain rule). Typically those in the first group will have encountered matrices through a study of linear or vector spaces while those in the second will have escaped such material and most probably have encountered matrices through solving systems of simultaneous equations or maybe even only via unmotivated definitions and initially strange rules for matrix multiplication. Generally, neither group is prepared for the heavy reliance on matrix manipulation that is required by subjects such as multivariate analysis or linear models. This text is intended to provide that experience. Keeping in mind the backgrounds of those coming to statistics through non-mathematical routes (a growing and crucially important constituency for the discipline of statistics), the presentation given here does not rely on notions of linear spaces. This is undeniably a little restrictive when providing detailed theoretical proofs of some results (for example, the equivalence of row and column ranks of a rectangular matrix) but examples, illustrations and intuitive explanations should compensate for the lack of formal proofs.

Arithmetical calculations "by hand" involving matrices are unquestionably confusing and tedious. Whilst it may be important to know the principles of such calculations, it is now so easy to use a computer package such as S-PLUS or **R** or MATLAB® or FREEMAT to perform everything from simple addition and multiplication to the most sophisticated of techniques so that the numerical complexities of matrix calculations should not be a hindrance to performing them. This is an aspect which is largely ignored in mathematical courses on linear spaces and unlikely to be covered in any detail in courses in most other disciplines. Because **R** is the computer language of choice for statistics (not least since it is free and open source), the presentation here is focused on the use of **R** rather than any of its rivals or competitors. This entails some investment in learning **R** itself but as the intended primary readership is those embarking on further statistical work, this should be an investment that provides dividends beyond just the implementation of matrix algebra. Even though **R** provides ready-made functions for executing standard statistical techniques such as estimating linear models and performing principal component analysis with all the matrix calculations involved hidden from the user there will always be a need to go beyond the routine everyday calculations provided by the statistical package.

Chapter 7 provides a brief introduction to differentiation involving vectors and matrices. This chapter concerns problems involving maximum likelihood in mind, not just calculation of maximum likelihood estimates but construction of likelihood ratio tests of hypotheses which only arise in multivariate problems. **R**'s symbolic differentiation capabilities are limited and cumbersome, especially in the context of vectors and matrices, but **R** is certainly invaluable in numerical implementation of such tests. Also included is a section (§7.6) on constrained optimisation of quadratic forms where introduction of a Lagrange multiplier converts the problem to solving an eigenequation. Some illustrations of these are given in the final Chapter 9 of the text.

A further comment concerns the exercises. These are provided as an integral part of the exposition and as a vehicle for self assessment of understanding of the material. Outline solutions are provided to enable checking of numerical answers and a guide to the techniques required, but deliberately are not very detailed.

Finally, there are many people to thank for this book, not least several reviewers (most of whom remain anonymous) whose suggestions and comments have greatly improved the content and presentation of the book and the help and rapid response over recent months from Rob Calver, Amber Donley and Saf Khan among others at Chapman & Hall/CRC Press and especially to Daniel Fieller for the cover photograph *Baobabs in Tete, Mozambique*.

Nick Fieller, Sheffield, UK.
nick.fieller@sheffield.ac.uk

1

Introduction

1.1 Objectives

The aim of this book is to provide a guide to elementary matrix algebra sufficient for undertaking intermediate and advanced statistical courses such as multivariate data analysis and linear models. Starting from a definition of a matrix and covering the basic rules of addition, subtraction, multiplication and inversion, the later topics include determinants, calculation of eigenvectors and eigenvalues and differentiation of linear and quadratic forms with respect to vectors. These later topics are sometimes not included in basic courses on linear algebra but are virtually essential for full discussion of statistical topics in multivariate analysis and linear models. The notes go a little beyond meeting just this need, providing an initial guide to more advanced topics such as generalized inverses of singular and rectangular matrices and manipulation of partitioned matrices. This is to provide a second step for those who need to go a little further than standard lecture courses on advanced statistics, for example, when embarking on a dissertation. As well as describing the basics of matrix algebra, including numerical calculations "by hand", for example, of matrix multiplication and inversion, the notes give guidance on how to do numerical calculations in **R** (R Core Team 2014). **R** is broadly similar in operation to the package MATLAB® but oriented specifically towards statistical applications rather than more general areas of applied mathematics. **R** is an open source system and is available free. It is closely similar to the commercial package S-PLUS: the prime difference is that **R** is command-line driven without the standard menus and dialogue boxes for statistical operations in S-PLUS. Otherwise, most code written for the two systems is interchangeable. There are however a few differences, for example there may be differences in the available optional arguments for some functions. Generally, all options available in S-PLUS are also available in **R** but not necessarily *vice versa*. These are quickly verified by use of the help system. The sites from which **R** and associated software (extensions and libraries) and manuals can be found are listed at http://www.ci.tuwien.ac.at/R/mirrors.html.

There are many available user interfaces for **R**, such as RStudio (Verzani, 2011), some of which provide a menu-driven facility, but this text assumes use of only the standard version of **R** without any such special facility.

Free versions of full manuals for **R** (mostly in PDF format) can be found at any of these mirror sites. There is a wealth of contributed documentation.

Some sections are considerably more advanced in concept and technical requirements than others, for example, those on partitioned matrices (§2.6.1) and generalized inverses (§8.3), but these have not been starred or marked since what may seem technically complex to one may seem straightforward to another with a different background. It is left to the readers to glean what they can from them on initial readings and then return later if a complete understanding is wanted. Some topics often covered in texts such as this one are not included (for example linear spaces) which means that there are a few results (e.g. equality of row and column rank, §3.1) which will need to be taken on trust or else checked in one of the more specialist texts referenced in §1.2

Since the applications envisaged for the majority of users of these notes are within statistics, most emphasis is given to real matrices and mostly to real symmetric matrices. Unless specifically indicated, all matrices are presumed to be real.

In addition to the algebraic manipulation of matrices the notes give numerical examples since it is necessary to be able to do numerical calculations, both by hand and by using **R**. Much of the exposition is by presenting examples with solutions. There are additionally some examples for self study and solutions are available separately (for the desperate) to these. For numerical questions, the calculations are given only in **R** and in some cases only the numerical answers are given. A brief guide to installing and running **R** is included.

1.2 Further Reading

1.2.1 Matrix algebra

There are many books on matrix algebra and a search will reveal hundreds if not thousands aimed at readers in the later years at school, throughout university and at advanced levels of mathematical research. A particular one consulted in the preparation of this text is Abadir and Magnus (2005) which consists largely of worked exercises. This will complete many of the details omitted from this book (in particular complex matrices) but, as is common with all post-school level texts, it does require a good understanding of linear spaces. It is not specifically written for any particular discipline but does contain several sections on statistical topics such as least squares and maximum likelihood estimation. A rather more theoretical book by Magnus and Neudecker (1988) goes into further details on vector and matrix calculus. Many statistical texts on topics such as linear models and multivariate analysis, such as Draper and Smith (1998), Seber and Lee (2012), Anderson (2003), Cox (2005) and Mardia et al. (1979), contain brief sections or appendices on matrix algebra but there are also many books aimed at the statistical reader. Three such are Searle (1982), Basilevsky (2013) and Harville (2008). A comprehensive account of linear algebra and matrix analysis in statistics is given by Banerjee and Roy (2014). This contains much advanced material and provides a full guide to the topic but is designed for experienced readers. A more specialist book that concentrates on applications to linear statistical models is Puntanen et al. (2011). This also is aimed at experienced readers. All have been consulted for ideas on exposition and subjects of examples. None of these gives any guidance on the computational aspects other than by hand and none mentions **R**. A recent book by Vinod (2011) does give a comprehensive guide to implementation of matrix calculations in **R** and includes a guide to getting started in **R**. The book is particularly good for those wanting to know about more specialist packages such as `fBasics` (Wuertz, D. and Rmetrics Core Team Members, 2013) which contains routines for determining the rank of a matrix (§3.1). This book has avoided extensive discussion of **R** packages since many readers will be encountering **R** for the first time but their use cannot be avoided for long and so some details are given of the most frequently used ones such as the `MASS` library (Venables and Ripley, 2002) as well as a few with useful additional facilities such as `Matrix` (Bates and Maechler, 2014) and `CCA` (Gonzlez and Djean, 2012).

1.2.2 Elementary R

There is a wealth of information on **R** readily available and easily accessible free of charge on the CRAN webpage at http://cran.r-project.org/. The menu on the left hand side under documentation quickly leads to a basic introduction to **R** (Venables et al., 2014) in a pdf file of about a hundred pages. This provides all that is needed on the **R** language for this text. For those who want to go further, Zuur et al. (2009) provides an introduction to the **R** language and for an introduction to the use of **R** in statistics, a starting point is Dalgaard (2008) or Crawley (2005). Crawley (2012) and Adler (2010) provide comprehensive references to further facilities.

1.3 Guide to Notation

- Generally, matrices are denoted by upper case letters at the beginning and end of the alphabet: $\boldsymbol{A}, \boldsymbol{B}, \boldsymbol{C}, \boldsymbol{D}, \boldsymbol{U}, \boldsymbol{V}, \boldsymbol{W}, \boldsymbol{X}, \boldsymbol{Y}, \boldsymbol{Z}$.
- Generally, [column] vectors are denoted by lower case letters at the beginning and end of the alphabet: $\boldsymbol{a}, \boldsymbol{b}, \boldsymbol{c}, \boldsymbol{d}, \boldsymbol{u}, \boldsymbol{v}, \boldsymbol{w}, \boldsymbol{x}, \boldsymbol{y}, \boldsymbol{z}$.
- Generally, *elements* of a vector $\boldsymbol{x}$ are denoted by $x_1, x_2, \ldots, x_p$.
- Generally, *elements* of a matrix $\boldsymbol{X}$ are denoted by $x_{11}, x_{12}, \ldots, x_{1n}, x_{21}, \ldots, \ldots, x_{pn}$.
- Sometimes the *columns* of a matrix $\boldsymbol{X}$ are denoted by $\boldsymbol{x}_1, \boldsymbol{x}_2, \ldots, \boldsymbol{x}_n$.
- Lower case letters in the middle of the alphabet $i, j, k, l, m, n, p, q, r, s, t$ generally indicate integers. Often i, j, k, l are used for dummy or indexing integers (e.g. in summation signs) whilst m, n, p, q, r, s, t are usually used for fixed integers, e.g. for $i = 1, 2, \ldots, n$.

- The upper case letters **H**, **I** and **J** are usually reserved for special matrices.
- The transpose of a matrix X is indicated by X' . Some texts may use an alternative X^T and this may be used in some circumstances here.
- The inverse of a [non-singular] matrix X is indicated as X^{-1}.
- The [Moore–Penrose] generalized inverse of a [not necessarily non-singular or square] matrix is indicated as A^+ and the generalized inverse by A^- (A^+ is a restricted form of A^-).
- **R** commands, functions and output are printed in a Courier font `like this.` Extensive **R** output is printed in two (or occasionally three) columns separated by a vertical rule. The length of the rule indicates the length of the column.
- Brackets. There are three types of brackets: `( )`, `[ ]` and `{ }` in **R**. The first, parentheses `( )`, are used to enclose the arguments of **R** functions. The second, brackets `[ ]`, are used to extract the elements of arrays and the third, braces `{ }`, are used to combine several **R** expressions into one.

1.4 An Outline Guide to R

1.4.1 What is R?

R is a powerful interactive computer package that is oriented towards statistical applications. It will run on the most commonly used platforms (or operating systems), Windows, Linux and Mac. The notes here are oriented towards use on a Windows platform. **R** consists of a base system that can be downloaded without charge together with many contributed packages for specialist analyses. It offers:

- An extensive and coherent set of tools for statistics and data analysis
- A language for expressing statistical models and tools for using linear and non-linear statistical models
- Comprehensive facilities for performing matrix calculations and manipulations, enabling concise efficient analyses of many applications in multivariate data analysis and linear models
- Graphical facilities for interactive data analysis and display
- An object-oriented programming language that can easily be extended
- An expanding set of publicly available libraries or packages of routines for special analyses
- Libraries or packages available from the official Contributed Packages webpages are thoroughly tested by the **R** Core Development Team
- Packages have manuals, help systems and usually include illustrative datasets

1.4.2 Installing R

Full instructions for installing **R** are given on the **R** Project home page at http://www.r-project.org/. The first step is to choose a site close to you from which to download the package. Next choose the appropriate operating system, select the option base and download the system. Accepting the option to run the download file will install the package after downloading it. Accepting the default options for locations is simplest but these can be customized. By default an icon ℝ (with version number) is placed on the desktop. Clicking on the icon will open an **R** session (i.e. start the **R** program). The **R** graphical user interface (RGui) opens and you are presented with a screen like this:

```
R version 3.1.1 (2014-07-10) -- "Sock it to Me"
Copyright (C) 2014 The R Foundation for Statistical Computing
Platform: x86_64-w64-mingw32/x64 (64-bit)

R is free software and comes with ABSOLUTELY NO WARRANTY.
You are welcome to redistribute it under certain conditions.
Type 'license()' or 'licence()' for distribution details.

  Natural language support but running in an English locale

R is a collaborative project with many contributors.
Type 'contributors()' for more information and
'citation()' on how to cite R or R packages in publications.

Type 'demo()' for some demos, 'help()' for on-line help, or
'help.start()' for an HTML browser interface to help.
Type 'q()' to quit R.
>
```

Above this is a menu bar with several drop-down menus and a row of icons. The Windows and Mac versions are broadly equivalent but differ in detail. Throughout this text, specific sequences of choices from the drop-down menus refer to the Windows version.

1.4.3 R is an interactive program

The symbol > is the command line prompt symbol; typing a command or instruction will cause it to be executed or performed immediately. If you press RETURN before completing the command (e.g., if the command is very long), the prompt changes to + indicating that more is needed. This is first illustrated in §2.1.2 on Page 23. Sometimes you do not know what else is expected and the + prompt keeps coming. Pressing Esc will kill the command and return to the > prompt. If you want to issue several commands on one line, they should be separated by a semicolon (;) and they will be executed sequentially. This is first illustrated in §2.3.2 on Page 28.

Along the top of the window is a limited set of menus. The `Packages` menu allows you to install specific packages (which needs to be done only once) and then load them into the session. Each time a new session is started, you have to load the packages which you will need. This can be done from the `Packages>Load Packages...` menu or by the command `library(packagename)`. Some of the commands needed for matrix manipulations are within the `MASS` library which is automatically installed (together with a few others such as `stats, Matrix, graphics, ...`) when **R** is first installed, i.e., it does not need to be installed from the `Packages` menu but it does need to be loaded if needed during each **R** session. [`MASS` is *Modern Applied Statistics with S* by Venables and Ripley, 2002]. Some packages are automatically loaded during each **R** session (e.g., `stats` and `graphics` but not `Matrix` and `MASS`). To discover which packages are loaded during a session issue the command `search()`.

A convenient way of running several commands in sequence is to open a *script window* using the `File>New script` menu which opens a simple text editing window for typing the commands. Highlighting a selection and then clicking on an icon in the menu bar will run the commands in the selection. Lines in the script window can be edited and run again. A script can be saved in a file (with default extension .R) and opened in a later session via the menus.

Pressing the up arrow key ↑ retrieves the last command. This can be edited (using the left and right arrow cursor keys and not the mouse to move the cursor) and then run again (by pressing RETURN). Pressing the up arrow key repeatedly will cycle back through the previous commands until you find the one you want.

1.4.4 Obtaining help in R

Obtaining help on specific functions during an **R** session can by done by using either `help(functionname)` or `?functionname`. This will give you the list of possible arguments and the list of possible values produced. There may also be examples of their use, including script files which can be cut and pasted into a script window for you to run. Typing `example(functionname)` may run one or more examples. This of course requires you to know the name of the function.

There are certain "reserved" words in **R** such as "for", "while", "if", "break", "next" and "if" where typing `help(for)` generates an error message. Instead it is necessary to type `help("for")`, `help("while")` etc. to access the **R** help system for these control words.

Typing `library(help=libraryname)` will give summary description of the library together with an index of all functions and datasets supplied with the library. Having found the name of a function or a dataset then use `help(.)`, `?` and `example(.)` to find out more about it. For example `library(help=stats)` lists all the functions in library stats; these are mostly statistical functions such as `t.test` and then `help(t.test)` shows exactly how to perform a Student's t-test.

To find out what various packages can do, look at the CRAN website and click on packages. This has a basic search facility with CTRL+F (for the Windows Find dialogue box) which will search for an exact match for the character string entered in the dialogue box. For example, to find packages which have functions for imputation of missing values, go to the Packages page on the CRAN project page and scroll down to the list headed `Available Bundles and Packages`, press CTRL+F and enter `impute` in the dialogue box. This will list in turn `arrayImpute`, `impute`, `imputeMDR`, and `yaImpute`. This technique will only find strings which appear on the very brief summary descriptions of the packages. A more thorough search can be performed within an **R** session with `help.search` or `??` For example `help.search(characterstring)` or equivalently, `??characterstring`, will search all installed packages for an approximate match in the summary description of each function in the package to the `characterstring`. The default is to use fuzzy matching which may have unexpected consequences. For example using `help.search(impute)` or equivalently `??impute` will also find all such occurrences of `compute`. To find out how to avoid this and instead use exact matching, try `help(help.search)`.

To find out more about **R**, the **R** website has links (under Documentation) to various manuals and other contributed documentation. Following the link from the CRAN page under `Other` and **R-related projects** transfers the user to **The R Wiki** at **http://wiki.r-project.org/rwiki/doku.php**.

1.4.5 R is a function language

All commands in **R** are regarded as functions; they operate on arguments, e.g., `plot(x, y)` plots the vector `x` against the vector `y`, that is, it produces a scatter plot of x vs. y. Even `help` is regarded as a function: to obtain help on the function matrix use `help(matrix)`. To end a session in **R** use `quit()`, or `q()`, i.e. the function `quit()` or `q()` with a null argument. In fact the function `quit` can take optional arguments; type `help(quit)` to find out what the possibilities are.

1.4.6 R is an object-oriented language

All entities (or things) in **R** are objects. This includes vectors, matrices, data arrays, graphs, functions, and **the results of an analysis**. For example, the set of results from performing a two-sample *t*-test is regarded as a complete single object. The object can be displayed by typing its name or it can be summarized by the function `summary()`. Even the results of `help` are objects, e.g. of `help(matrix)`. If you want to store the object created by this for later examination (though the need for this may be rare), give it, say, the name `matrixhelp` then input `matrixhelp<-help(matrix)`. Typing `matrixhelp` will print the help information on the screen (or it can be exported). Note here the use of the *assignment symbol* `<-` (a left pointing arrow) which assigns the object with name `matrixhelp` the results of the function `help`. `<-` is the most commonly used symbol for this but `_` and `=` will work in exactly the same way and are included in the **R** language for compatibility with other programming languages. Thus `matrixhelp_help(matrix)` and `matrixhelp=help(matrix)` will produce exactly the same results as `matrixhelp<-help(matrix)`. Generally `<-` will be used throughout this text.

Some functions create an object consisting of just a single value, e.g. the function `sqrt()` will consist of a single value, the square root of its argument. Other functions such as `t.test()` create objects containing many different values. In the case of `t.test()`, these include the value of the t-statistic, the degrees of freedom, the p-value and a confidence interval for the mean appropriate to the specified alternative hypothesis (and several others). These individual values are stored in simple objects with names including a $ symbol. So, if the command `testresults<-t.test(x)` is used to test the hypothesis that the mean of the vector $\boldsymbol{x}$ is 0 and store the complete results of the test in an object `testresults`, the value of the t-statistic, the degrees of freedom and the p-value are stored in objects `testresults$statistic`, `testresults$parameter` and `testresults$p.value`, respectively. The first time this is illustrated is with the function `eigen()` in §1.7.12 on Page 18 used for performing an eigenanalysis (see Chapter 6). The values produced by this function are the eigenvalues and eigenvectors stored in `objectname$values` and `objectname$vectors`.

Of course, some functions such as `help()` and `plot()` produce objects containing no specific values. To find out what values (if any) and names are produced by a function, enter `help(functionname)`.

1.4.6.1 The class of an object

Every object in **R** has a class (or type of object) associated with it. Examples of classes are `"numeric"`, `"matrix"` and `"character"`. The function `t.test()` produces objects of class `"htest"`. The importance of class is that some functions require that their arguments are of a particular class or even may operate differently depending upon which class their argument happens to be. This phenomenon is illustrated in §2.1.2 on Page 22. The class of an object can be changed by functions such as `as.matrix(.)` which will convert a `"numeric"` class vector into a `"matrix"` class. This is only critical in a few cases since **R** usually takes care of this matter internally.

1.4.7 Saving objects and workspaces

Objects such as a matrices and vectors (see below) created during an **R** session can be saved in an **R** workspace file through the `File>Save Workspace...` menu or via the icon. They can be loaded into the **R** session by the menu or (if the default .RData file extension is accepted) the file can be located in Windows Explorer and clicking initiates an **R** session with this workspace open and making the working directory that containing the file. Issuing the command `objects(.)` (or equivalently `ls(.)` or using the menu under `Misc`) will list the objects created (or retrieved from a workspace) during the current session.

When you close down **R** you are prompted as to whether you want to save the workspace image. If you have loaded a workspace during the session, this will be overwritten by the current one. When you next run **R**, you will start by loading the most recently saved workspace, i.e. it will retrieve all the objects in the workspace you last saved. If you have started **R** by clicking on a saved workspace with the .Rdata extension, the two workspaces will be merged, perhaps with unexpected consequences.

1.4.7.1 Mistakenly saved workspace

BEWARE: If you have created all the objects in the workspace during the current session (i.e., you have not loaded or opened a previously saved workspace) when you accept *Save workspace image?* at the end of a session it will be saved somewhere on your drive. ***When you next start R, this workspace will be automatically restored.*** This can have unexpected consequences and can cause mysterious problems because you will have objects in the workspace that you might not have intended to be there. To cure the problem, you can remove all objects by `rm(list=ls(all=T))` or use the menu under `Misc` and choose `Remove all objects`. This should give a response `character(0)` to the command `ls(.)`.

1.4.8 R is a case-sensitive language

Note that **R** treats lower case and upper case letters as different, for example, inverting a matrix is performed using the function `solve()` but **R** does not recognize `Solve()`, nor `SOLVE()`, nor .... The objects `x` and `X` are distinct (and easy to confuse). The function `matrix` and the library `Matrix` are distinct.

1.5 Inputting Data to R

1.5.1 Reading data from the keyboard

Small amounts of data can be typed directly from the keyboard. For example, to create a vector `x` of length 4 containing the four numbers 1.37, 1.63, 1.73, 1.36 do `x<-c(1.37, 1.63, 1.73, 1.36)` and to enter numbers into a matrix, see the next section. The function `scan(.)` can be used to enter data and will stop when a complete blank line is read. For example:

```
> x<-scan()
1: 1.37
2: 1.63 1.73
4: 1.36
5:
Read 4 items
> x
```

```
[1] 1.37 1.63 1.73 1.36
```

`scan()` is a very flexible function with facilities for entering tables of numbers and other data; to find out more type `help(scan)`. Note that here and elsewhere lines starting with > (the prompt symbol) have been entered from the console and other lines are the responses from **R**. Thus `x<-scan()` followed by the data and `x` are entered from the console.

1.5.2 Reading data from files

The three main functions for reading tabular data from files are, `read.table()`, `read.csv()` and `read.delim()`. The first is used primarily for plain text files (i.e., with extension .txt or .dat), the second for comma separated values (e.g., as produced by Excel) and the third for tab-separated values. The default format of the data in `read.table()`is that the first row should contain the column names (i.e., variable names) and the first item of each row is the row name, so the first row contains one fewer item than the other rows. If the data are not in such a standard form, look at the help system to find out how to use the additional arguments `header`, `row.names` and `col.names` to read the data correctly.

If a data file has been saved during an **R** session (using the `save(.)` function (see `help(save)`), the data can just be retrieved by `load(filename)`. The `source(filename)` function will execute all the **R** commands in the specified file and this can be a convenient method of delivering a data file. The library `foreign` can be used for reading data files created by Minitab, S, SAS, SPSS, Stata, Systat, dBase, ..., (but not S-PLUS). There are commercially available packages for reading S-PLUS other files with substantial discounts for academic use.

1.6 Summary of Matrix Operators in R

The operations given below assume that the orders of the matrices involved allow the operation to be evaluated. More details of these operations will be given in the notes later.

- **R** is case sensitive so `A` and `a` denote distinct objects.
- To control the number of digits printed to 3 `options(digits=3)`.
- To create a vector $\boldsymbol{x}$ $\mathtt{x} < -\mathtt{c}(\mathtt{x}_1, \mathtt{x}_2, ..., \mathtt{x}_\mathtt{p})$.
- To access an individual element in a vector $\boldsymbol{x}$, the ith, `x[i]`.
- To create a matrix $\boldsymbol{A}$, `A<-matrix(data, nrow=m, ncol=n, byrow=F)`.
- To access an individual element in a matrix $\boldsymbol{A}$, the $(i,j)^{th}$, `A[i,j]`.
- To access an individual row in a matrix $\boldsymbol{A}$, the ith, `A[i,]`

- To access an individual column in a matrix $\boldsymbol{A}$, the jth, `A[,j]`.
- To access a subset of rows in a matrix $\boldsymbol{A}$, $\texttt{A}[\texttt{i}_1:\texttt{i}_2,]$.
- To access a subset of columns in a matrix $\boldsymbol{A}$, $\texttt{A}[,\texttt{j}_1:\texttt{j}_2]$.
- To access a sub-matrix of $\boldsymbol{A}$, $\texttt{A}[\texttt{i}_1:\texttt{i}_2,\texttt{j}_1:\texttt{j}_2]$.
- Addition $\boldsymbol{A}+\boldsymbol{B}$, `A+B`.
- Subtraction $\boldsymbol{A}-\boldsymbol{B}$, `A-B`.
- Multiplication $\boldsymbol{AB}$, `A%*%B`.
- Hadamard multiplication, $\boldsymbol{A}\odot\boldsymbol{B}$, `A*B`.
- Kronecker multiplication $\boldsymbol{A}\otimes\boldsymbol{B}$, `A%x%B`.
- Transpose $\boldsymbol{A}'$, `t(A)`.
- Matrix cross-product $\boldsymbol{A}'\boldsymbol{B}$, `crossprod(A,B)`.
- Inversion $\boldsymbol{A}^{-1}$, `solve(A)`.
- Moore–Penrose generalized inverse $\boldsymbol{A}^{+}$, `ginv(A)` (in `MASS` library) [or `MPinv(A)` (in gnm library)].
- Note: `ginv(.` will work with almost any matrix but it is safer to use `solve(.)` if you expect the matrix to be non-singular since `solve(.)` will give an error message if the matrix is singular or non-square but `ginv(.)` will not.
- Determinant $det(\boldsymbol{A})$ or $|\boldsymbol{A}|$, `det(A)`.
- Eigenanalysis, `eigen(A)`.
- To extract a diagonal of a matrix $\boldsymbol{A}$ as a vector, `diag(A)`.
- Trace of a matrix $\boldsymbol{A}$, `sum(diag(A))`.
- To create a diagonal matrix, $\texttt{diag}(\texttt{c}(\texttt{x}_{11},\texttt{x}_{22},\ldots,\texttt{x}_{\texttt{pp}}))$.
- To create a diagonal matrix from another matrix, `diag(diag(A))`.
- To change a dataframe into a matrix, `data.matrix(dataframe)`.
- To change some other object into a matrix, `as.matrix(object)`.
- To join vectors into a matrix as columns, $\texttt{cbind}(\texttt{vec}_1,\texttt{vec}_2,\ldots,\texttt{vec}_\texttt{n})$.
- To join vectors into a matrix as rows, $\texttt{rbind}(\texttt{vec}_1,\texttt{vec}_2,\ldots,\texttt{vec}_\texttt{n})$.
- To join matrices $\boldsymbol{A}$ and $\boldsymbol{B}$ together side by side, `cbind(A,B)`.
- To stack $\boldsymbol{A}$ and $\boldsymbol{B}$ together on top of each other, `rbind(A,B)` .
- To find the length of a vector $\boldsymbol{x}$, `length(x)`.
- To find the dimensions of a matrix $\boldsymbol{A}$, `dim(A)`.

1.7 Examples of R Commands

This section is for quick reference; details are explained later in the text.

1.7.1 Expressions

$x'(XX')^{-1}x$: `t(x)%*%solve(X%*%t(X))%*%x`

1.7.2 Inputting data

```
> A<-matrix(c(1,2,3,4,5,6),nrow=2,ncol=3,byrow=F)
> B<-matrix(c(1,2,3,4,5,6),nrow=2,ncol=3,byrow=T)
```

```
> A
     [,1] [,2] [,3]
[1,]    1    3    5
[2,]    2    4    6
> A[1,2]
[1] 3
```

```
>
> A[1,]
[1] 1 3 5
> A[,2]
[1] 3 4
```

```
> B
     [,1] [,2] [,3]
[1,]    1    2    3
[2,]    4    5    6
> B[2,2]
[1] 5
```

```
>
> B[2,]
[1] 4 5 6
> B[,3]
[1] 3 6
```

```
> C<-matrix(c(1,2,3,4,5,6),2,3)
> D<-matrix(c(1,2,3,4,5,6),2,3,byrow=T)
```

```
> C
     [,1] [,2] [,3]
[1,]    1    3    5
[2,]    2    4    6
```

```
> D
     [,1] [,2] [,3]
[1,]    1    2    3
[2,]    4    5    6
```

1.7.3 Calculations

```
> A+B
     [,1] [,2] [,3]
[1,]    2    5    8
[2,]    6    9   12
```

```
> A-B
     [,1] [,2] [,3]
[1,]    0    1    2
[2,]   -2   -1    0
```

```
> 2*A
     [,1] [,2] [,3]
[1,]    2    6   10
[2,]    4    8   12
```

```
> A*2
     [,1] [,2] [,3]
[1,]    2    6   10
[2,]    4    8   12
```

Beware:

```
> A%*%B
Error in A %*% B :
 non-conformable arguments
```

```
> t(A)
     [,1] [,2]
[1,]    1    4
[2,]    2    5
[3,]    3    6
```

```
> t(B)
     [,1] [,2]
[1,]    1    4
[2,]    2    5
[3,]    3    6
```

```
> t(A)%*%B
     [,1] [,2] [,3]
[1,]    9   12   15
[2,]   19   26   33
[3,]   29   40   51
```

```
> A%*%t(B)
     [,1] [,2]
[1,]   22   49
[2,]   28   64
```

BEWARE: ***A⋆B*** gives element-by-element multiplication (the Hadamard or Schur product) which is rarely required (but see §6.3.0.1):

```
> A*B
     [,1] [,2] [,3]
[1,]    1    6   15
[2,]    8   20   36
```

Kronecker products

```
> A%x%B
     [,1] [,2] [,3] [,4] [,5] [,6] [,7] [,8] [,9]
[1,]    1    2    3    3    6    9    5   10   15
[2,]    4    5    6   12   15   18   20   25   30
[3,]    2    4    6    4    8   12    6   12   18
[4,]    8   10   12   16   20   24   24   30   36
>
> B%x%A
     [,1] [,2] [,3] [,4] [,5] [,6] [,7] [,8] [,9]
[1,]    1    3    5    2    6   10    3    9   15
[2,]    2    4    6    4    8   12    6   12   18
[3,]    4   12   20    5   15   25    6   18   30
[4,]    8   16   24   10   20   30   12   24   36
>
```

1.7.4 Dimensions and lengths of matrices of vectors

```
> C<-matrix(c(1,2,3,4,5,6),2,3)
> dim(C)
[1] 2 3
> dim(t(C))
[1] 3 2
> length(C)
[1] 6
```

```
>
> x<-c(1,2,3,4)
> length(x)
[1] 4
> dim(x)
NULL
```

Beware:

```
> x
[1] 1 2 3 4
> t(x)
     [,1] [,2] [,3] [,4]
[1,]    1    2    3    4
> dim(x)
NULL
```

```
> dim(t(x))
[1] 1 4
> matrix(x)
     [,1]
[1,]    1
[2,]    2
[3,]    3
[4,]    4
```

```
> dim(matrix(x))
[1] 4 1
```

```
> t(matrix(x))
     [,1] [,2] [,3] [,4]
[1,]    1    2    3    4
> dim(t(matrix(x)))
[1] 1 4
```

Here, it seems that creating a vector $\boldsymbol{x}$ with `x<-c(1,2,3,4)` creates a string which has length 4 but is not regarded by **R** as a matrix until it is involved in some operation that only applies to matrices, such as taking the transpose. The fact that the transpose has dimensions 1 and 4 indicates that $\boldsymbol{x}$ is assumed to be a column vector. However, the following shows that **R** will interpret $\boldsymbol{x}$ as either a column vector or as a row vector according to context, e.g. in pre- or post-multiplication by a square matrix of conformable dimension. This is demonstrated in the following example:

```
> x<-c(1,2)
> Z<-matrix(c(1,2,3,4),2,2)
```

```
> x
[1] 1 2
```

```
> Z
     [,1] [,2]
[1,]    1    3
[2,]    2    4
```

```
> Z%*%x
      [,1]
[1,]     7
[2,]    10
```

```
> x%*%Z
     [,1] [,2]
[1,]    5   11
>
```

If x is forced to be a matrix (i.e. of class `matrix` in **R**), one of these multiplications will give an error because of non-comformability.

```
> class(x)
[1] "numeric"
> x<-matrix(x)
> class(x)
[1] "matrix"
```

```
> Z%*%x
     [,1]
[1,]    7
[2,]   10
> x%*%Z
Error in x %*% Z :
 non-conformable arguments
```

1.7.5 Joining matrices together

```
> A
     [,1] [,2] [,3]
[1,]    1    3    5
[2,]    2    4    6
```

```
> C
     [,1] [,2] [,3]
[1,]    1    3    5
[2,]    2    4    6
```

```
> cbind(A,C)
  [,1] [,2] [,3] [,4] [,5] [,6]
[1,] 1    3    5    1    3    5
[2,] 2    4    6    2    4    6
```

```
> rbind(A,C)
     [,1] [,2] [,3]
[1,]    1    3    5
[2,]    2    4    6
[3,]    1    3    5
[4,]    2    4    6
```

```
> t(rbind(t(A),t(B)))
  [,1] [,2] [,3] [,4] [,5] [,6]
[1,] 1    3    5    1    2    3
[2,] 2    4    6    4    5    6
```

```
> t(cbind(t(A),t(B)))
     [,1] [,2] [,3]
[1,]    1    3    5
[2,]    2    4    6
[3,]    1    2    3
[4,]    4    5    6
```

Beware: Joining non-conformable matrices will generate error messages

1.7.6 Diagonals and trace

```
> E<-matrix(c(1,2,3,4,5,6,7,8,9),3,3,byrow=T)
```

```
> E
     [,1] [,2] [,3]
[1,]    1    2    3
[2,]    4    5    6
[3,]    7    8    9

> diag(E)
[1] 1 5 9

> sum(diag(E))
[1] 15
```

```
> diag(c(1,5,9))
     [,1] [,2] [,3]
[1,]    1    0    0
[2,]    0    5    0
[3,]    0    0    9

> diag(diag(E))
     [,1] [,2] [,3]
[1,]    1    0    0
[2,]    0    5    0
[3,]    0    0    9
```

1.7.7 Trace of products

```
> F<-matrix(c(1,2,3,4,5,6,7,8,9),3,3,)
```

```
> F
     [,1] [,2] [,3]
[1,]    1    4    7
[2,]    2    5    8
[3,]    3    6    9
> F%*%E
     [,1] [,2] [,3]
[1,]   66   78   90
[2,]   78   93  108
[3,]   90  108  126
```

```
> E%*%F
     [,1] [,2] [,3]
[1,]   14   32   50
[2,]   32   77  122
[3,]   50  122  194
> sum(diag(E%*%F))
[1] 285
> sum(diag(F%*%E))
[1] 285
```

1.7.8 Transpose of products

```
> t(E%*%F)
     [,1] [,2] [,3]
[1,]   14   32   50
[2,]   32   77  122
[3,]   50  122  194
```

Beware:

```
> t(E)%*%t(F)
     [,1] [,2] [,3]
[1,]   66   78   90
[2,]   78   93  108
[3,]   90  108  126
```

```
> t(F)%*%t(E)
     [,1] [,2] [,3]
[1,]   14   32   50
[2,]   32   77  122
[3,]   50  122  194
```

Note $\boldsymbol{EF}$ and $\boldsymbol{FE}$ are symmetric but neither $\boldsymbol{E}$ nor $\boldsymbol{F}$ is symmetric. Also $\boldsymbol{E'F'} \neq (\boldsymbol{EF})'$.

1.7.9 Determinants

```
> G<-matrix(c(1,-2,2,2,0,1,1,1,-2),3,3,byrow=T)
> G
     [,1] [,2] [,3]
[1,]    1   -2    2
[2,]    2    0    1
[3,]    1    1   -2
> det(G)
[1] -7
```

1.7.10 Diagonal matrices

```
> G<-matrix(c(1,-2,2,2,0,1,1,1,-2),3,3,byrow=T)
```

```
> G
     [,1] [,2] [,3]
[1,]    1   -2    2
[2,]    2    0    1
[3,]    1    1   -2
> diag(G)
```

```
[1]  1  0 -2
> diag(diag(G))
     [,1] [,2] [,3]
[1,]    1    0    0
[2,]    0    0    0
[3,]    0    0   -2
```

1.7.11 Inverses

```
> options(digits=3)
> solve(A%*%t(B))
       [,1]   [,2]
[1,]  1.778 -1.361
[2,] -0.778  0.611
```

```
> library(MASS)
> ginv(A%*%t(B))
       [,1]   [,2]
[1,]  1.778 -1.361
[2,] -0.778  0.611
```

Beware:

```
> ginv(t(A)%*%B)
        [,1]    [,2]   [,3]
[1,]  1.741  0.4630 -0.815
[2,]  0.269  0.0741 -0.120
[3,] -1.204 -0.3148  0.574
```

BUT:

```
> solve(t(A)%*%B)
Error in
solve.default(t(A) %*% B):
system is computationally
singular: reciprocal
condition number=1.794e-17
```

1.7.12 Eigenanalyses

```
> eigen(A%*%t(B)
$values
[1] 85.579  0.421
$vectors
       [,1]   [,2]
[1,] -0.610 -0.915
[2,] -0.792  0.403
```

```
> eigen(t(A)%*%B)
$values
[1] 8.6e+01 4.2e-01 1.5e-15
$vectors
       [,1]   [,2]   [,3]
[1,] -0.242 -0.815  0.408
[2,] -0.528 -0.125 -0.816
[3,] -0.814  0.566  0.408
```

1.7.13 Singular value decomposition

```
> X<-matrix(c(1,2,3,4,5,6,7,8,9),3,3,byrow=T); X
     [,1] [,2] [,3]
[1,]    1    2    3
[2,]    4    5    6
[3,]    7    8    9
> svd(X)
$d
[1] 1.68e+01 1.07e+00 4.42e-16
```

```
$u
       [,1]   [,2]   [,3]
[1,] -0.215  0.887  0.408
[2,] -0.521  0.250 -0.816
[3,] -0.826 -0.388  0.408
```

```
$v
       [,1]    [,2]   [,3]
[1,] -0.480 -0.7767 -0.408
[2,] -0.572 -0.0757  0.816
[3,] -0.665  0.6253 -0.408
```

1.8 Exercises

*(These exercises are intended only for those completely new to **R**)*

(1) Install **R** on your computer; see §1.4.2.

(2) Go to the CRAN home page and download (but don't print unless really desperate to do so) the manual by Venables et al. (2014); see §1.2.2.

(3) Still on the CRAN home page, look at the software under Packages from the link in the menu on the left, browsing through the packages sorted by name. Find the description of the package CCA (used in §9.5) and look briefly at the reference manual.

(4) Try typing direct into the **R** console window some of the examples given in §1.7.

(5) Open a new script file (§1.4.3 on Page 6) using the menu under File on the top left of the window.

(6) Type a few **R** commands into this script window and then highlight them with the mouse and then click on the middle icon in the top row to run them. (Note that this icon appears only when the **R** Editor window is the *active* window.)

(7) Click on the **R** console window and then using the menu under File change the working directory to somewhere convenient. Usually **R** starts with the working directory (where it looks for and saves files by default) at the very top level.

(8) Making the **R** editor window the active window (by clicking the mouse when the cursor is in it) and using the File on the top left, save the script file in the new working directory, using a name with extension `.R` (so that it can be opened in a a later **R** session with `File > Open script ...`, provided the working directory is the same).

2

Vectors and Matrices

2.1 Vectors

2.1.1 Definitions

A vector $\boldsymbol{x}$ of *order* p (or *dimension* p) is a column of p numbers:

$$\boldsymbol{x} = \begin{pmatrix} x_1 \\ x_2 \\ \vdots \\ x_p \end{pmatrix}.$$

Technically $\boldsymbol{x}$ is an element of p-dimensional Euclidean space $\Re^p$ but this will not be central to the material below. The numbers x_i are the ***components*** (or ***elements***) of $\boldsymbol{x}$. $\boldsymbol{x}$ may be referred to as a **column** vector for emphasis. The transpose of $\boldsymbol{x}$, $\boldsymbol{x}' = (x_1, x_2, \ldots, x_p)$ is a **row** vector. Vectors are presumed to be column vectors unless specified otherwise. Addition and subtraction of vectors of the same order is performed element by element

$$\boldsymbol{x}+\boldsymbol{y} = \begin{pmatrix} x_1+y_1 \\ x_2+y_2 \\ \vdots \\ x_p+y_p \end{pmatrix}.$$

It is not possible to add or subtract vectors which are not ***conformable***, i.e., which do not have the same dimension or order.

Scalar multiplication of a vector is element by element:

$$\lambda\boldsymbol{x} = \begin{pmatrix} \lambda x_1 \\ \lambda x_2 \\ \vdots \\ \lambda x_p \end{pmatrix}$$

for any scalar (real number) λ.

The results of addition and subtraction of two vectors are vectors of the same order. The results of adding to, subtracting from or multiplying a vector by a scalar

are vectors of the same order. Two vectors $\boldsymbol{x}$ and $\boldsymbol{y}$ are equal if they are of the same order and each pair of corresponding elements is equal, i.e., $x_i = y_i$, for $i = 1,2,\ldots,p$. A vector with all elements 0 is denoted by $\boldsymbol{0}$, i.e.,if $x_i = 0$ for $i = 1,2,\ldots,p$, $\boldsymbol{x} = \boldsymbol{0}$. A vector $\boldsymbol{e}_i$ with i^{th} element 1 and all others 0 is the ith unit vector. A vector with all elements 1 is and is denoted by $\iota_{\mathbf{p}}$, i.e.,if $x_i = 1$ for $i = 1,2,\ldots,p$ then $\boldsymbol{x} = \iota_{\mathbf{p}}$. Clearly $\sum_{i=1}^{p} \boldsymbol{e}_i = \iota_{\mathbf{p}}$.
$\iota_{\mathbf{p}}$ is referred to as the ***sum vector*** [because $\boldsymbol{x}'\iota_{\mathbf{p}} = \iota_{\mathbf{p}}'\boldsymbol{x} = \sum_{i=1}^{p} x_i$, see below].
Note that $\boldsymbol{x}+\boldsymbol{y} = \boldsymbol{y}+\boldsymbol{x}$ (commutativity) and $(\boldsymbol{x}+\boldsymbol{y})+\boldsymbol{z} = \boldsymbol{x}+(\boldsymbol{y}+\boldsymbol{z})$ (associativity).
Vectors cannot be multiplied together in a usual way but a useful scalar function of two vectors is the ***scalar product*** (or ***inner product***). Mathematicians denote this by $\langle \boldsymbol{x},\boldsymbol{y} \rangle$ and it is defined by $\langle \boldsymbol{x},\boldsymbol{y} \rangle = \sum_{i=1}^{p} x_i y_i$. In statistics the inner product is more usually denoted by $\boldsymbol{x}'\boldsymbol{y}$ (see later under matrix multiplication). The ***outer product*** of $\boldsymbol{x}$ and $\boldsymbol{y}$ is defined as $\boldsymbol{x}\boldsymbol{y}'$ (see §2.3.1 on Page 25.
Note that $\langle \boldsymbol{x},\boldsymbol{y} \rangle$ is a scalar (i.e.,an ordinary number) and we shall see that $\boldsymbol{x}\boldsymbol{y}'$ is a $p \times p$ matrix (§2.3.1).

Example 2.1:
If $\boldsymbol{x} = (1,2,3)'$ and $\boldsymbol{y} = (4,5,6)'$ then $\boldsymbol{x}'\boldsymbol{y} == 1 \times 4 + 2 \times 5 + 3 \times 6 = 32$ Two vectors are said to be orthogonal if $\boldsymbol{x}'\boldsymbol{y} = \langle \boldsymbol{x},\boldsymbol{y} \rangle = 0$. For example if $\boldsymbol{x} = (1,2,0)'$, $\boldsymbol{y} = (-6,3,0)'$, $\boldsymbol{z} = (0,0,7)'$ then $\boldsymbol{x}$, $\boldsymbol{y}$ and $\boldsymbol{z}$ are mutually orthogonal. Note that $\langle \boldsymbol{x},\boldsymbol{x} \rangle = \boldsymbol{x}'\boldsymbol{x} = \sum x_i^2$ = sum of the squared elements of x.

2.1.2 Creating vectors in R

The easiest way to create a vector is to use the function $\mathtt{c}(\mathtt{x}_1,\mathtt{x}_2,\ldots,\mathtt{x}_\mathtt{p})$:

```
> a<-c(1,2,3)
> a
[1] 1 2 3
> b<-c(4,5,6)
```

and this works in some cases but $\boldsymbol{a}$ and $\boldsymbol{b}$ will be interpreted as **either** a column **or** a row vector according to context. It is better to remove ambiguity and ensure that the vector is of the right class and force this by making using of the `matrix(.,.,.,.)` function with 1 column:

```
> c<-matrix(c(3,2,1),3,1,byrow=T)
> d<-matrix(c(6,5,4),nrow=3,ncol=1,byrow=T)
```

```
> c
     [,1]
[1,]    3
[2,]    2
[3,]    1
```

```
> d
     [,1]
[1,]    6
[2,]    5
[3,]    4
```

Without the `matrix` function, the result has class `"numeric"`, not `"matrix"`. Note that using the `matrix` function ensures that **R** prints the result as a column vector.

An equivalent way of coercing a string of numbers to be of class "matrix" is to use the function as.matrix(.):

```
> b<-c(4,5,6)
> b
[1] 4 5 6
> class(b)
[1] "numeric"
> b<-as.matrix(b)
```

```
> b
     [,1]
[1,]    4
[2,]    5
[3,]    6
> class(b)
[1] "matrix"
```

Note also the use of nrow and ncol and the default order that matrix assumes if they are omitted:

```
> u<-matrix(c(3,2,1),1,3,
+ byrow=T)
> u
     [,1] [,2] [,3]
[1,]    3    2    1
```

```
> v<-matrix(c(6,5,4),
+ ncol=1,nrow=3,byrow=T)
> v
     [,1]
[1,]    6
[2,]    5
[3,]    4
```

Note the occurrence of the continuation symbol +.
When entering column vectors or row vectors the byrow argument has no effect, i.e.,byrow=T and byrow=F give the same result but this is not the case when entering matrices (see below).

2.2 Matrices

2.2.1 Definitions

An $m \times n$ matrix $\boldsymbol{X}$ is a rectangular array of scalar numbers:

$$\boldsymbol{X} = \begin{pmatrix} x_{11} & x_{12} & \dots & x_{1n} \\ x_{21} & x_{22} & \dots & x_{2n} \\ \vdots & \vdots & \ddots & \vdots \\ x_{m1} & x_{m2} & \dots & x_{mn} \end{pmatrix}.$$

This matrix has m rows and n columns; it is an $m \times n$ matrix (m **by** n matrix), it is a matrix of ***order*** $m \times n$. $\boldsymbol{X}$ has dimensions m and n. Technically $\boldsymbol{X}$ is an element of $m \times n$-dimensional Euclidean space $\Re^{m \times n}$ but this will not be central to the material below. Sometimes we may write $\boldsymbol{X} = (x_{ij})$.

For example $\boldsymbol{A} = \begin{pmatrix} 1 & 2 & 3 \\ 4 & 4 & 6 \end{pmatrix}$ is a 2×3 matrix. A [column] vector is a matrix with one column, it is a $n \times 1$ matrix where n is the order of the vector. A row vector is a $1 \times n$ matrix.

The numbers x_{ij} are the ***components*** (or ***elements***) of $\boldsymbol{X}$. A matrix is a ***square matrix*** if $m = n$, i.e.,if it has the same number of rows and columns, i.e.,an $n \times n$ matrix is a square matrix. The transpose of the $m \times n$ matrix $\boldsymbol{X}$ is the $n \times m$ matrix $\boldsymbol{X}'$:

$$\boldsymbol{X} = \begin{pmatrix} x_{11} & x_{12} & \dots & x_{1m} \\ x_{21} & x_{22} & \dots & x_{2m} \\ \vdots & \vdots & \ddots & \vdots \\ x_{n1} & x_{m2} & \dots & x_{mn} \end{pmatrix}.$$

A square matrix $\boldsymbol{X}$ is ***symmetric*** if $\boldsymbol{X}' = \boldsymbol{X}$. Two matrices $\boldsymbol{X}$ and $\boldsymbol{Y}$ are equal if they are of the same order and each pair of corresponding elements are equal, i.e.,$x_{ij} = y_{ij}$, for $i = 1, 2, \dots, m$ and $j = 1, 2, \dots, n$.

A matrix with all elements 0 is denoted by $\mathbf{0}$, i.e.,if $x_{ij} = 0$ for $i = 1, 2, \dots, m$ and $j = 1, 2, \dots, n$ then $\boldsymbol{X} = \mathbf{0}$. A square matrix with all elements not on the diagonal equal to 0 is a ***diagonal matrix***, i.e.,if $x_{ij} = 0$ for all $i \neq j$ (and $x_{ii} \neq 0$ for at least one i).

A diagonal matrix with all diagonal elements 1 (and all others not on the diagonal 0) is denoted by $\mathbf{I}_n$, i.e.,if $x_{ii} = 1$ for $i = 1, 2, \dots, n$ and $x_{ij} = 0$ for for all $i \neq j$ for $i, j = 1, 2, \dots, n$ then $\boldsymbol{X} = \mathbf{I_n}$. It is referred to as the identity matrix.

$$\mathbf{I_n} = \begin{pmatrix} 1 & 0 & \dots & 0 \\ 0 & 1 & \dots & 0 \\ \vdots & \vdots & \ddots & 0 \\ 0 & 0 & \dots & 1 \end{pmatrix}.$$

If $\boldsymbol{X}$ is a square matrix then $\text{diag}(\boldsymbol{X})$ is the [column] vector of the diagonal elements of $\boldsymbol{X}$, i.e.,the vector (x_{ii}). If $\boldsymbol{u}$ is a vector, $\text{diag}(\boldsymbol{u})$ is the diagonal matrix with the elements of $\boldsymbol{u}$ along the diagonal and 0s elsewhere. So $\text{diag}(\text{diag}(\boldsymbol{X}))$ is a square matrix formed by setting all off-diagonal elements of $\boldsymbol{X}$ to 0. Some texts will call $\text{diag}(\boldsymbol{X})$ this matrix but the form $\text{diag}(\text{diag}(\boldsymbol{X}))$ here conforms with **R** syntax.

The ***trace*** of a square matrix is the sum of all the diagonal elements of the matrix, i.e.,trace($\boldsymbol{X}$) = $\text{tr}(\boldsymbol{X}) = \text{tr}(x_{ij}) = \sum_{i=1}^{n} x_{ii}$. Note that $\text{tr}(\mathbf{I}_n) = n$.

2.3 Matrix Arithmetic

Addition and subtraction of matrices of the same order are performed element by element (just as with vectors):

$$X+Y=(x_{ij})+(y_{ij})=(x_{ij}+y_{ij}),$$

Note that $X+Y=Y+X$ (**commutativity**) and $(X+Y)+Z=X+(Y+Z)$ (***associativity***), provided X, Y and Z are all of the same order. It is not possible to add or subtract matrices which do not have the same dimensions or order.
Scalar multiplication of a matrix is element by element:

$$\lambda X=\lambda(x_{ij})=(\lambda x_{ij})$$

$$\begin{pmatrix} 1 & 2 \\ 3 & 4 \end{pmatrix}+\begin{pmatrix} 5 & 6 \\ 7 & 8 \end{pmatrix}=\begin{pmatrix} 6 & 8 \\ 10 & 12 \end{pmatrix}=2\begin{pmatrix} 3 & 4 \\ 5 & 6 \end{pmatrix}.$$

2.3.1 Matrix multiplication

If A and B are matrices then we can multiply A by B (to get AB) **only if the number of columns of A equals the number of rows of B**. So if A is an $m\times n$ matrix and B is a $n\times p$ matrix, the product AB can be defined (but not the product BA). The result is a $m\times p$ matrix, i.e.,the number of rows of the first matrix and the number of columns of the second. The $(i,k)^{th}$ element of AB is obtained by summing the products of the elements of the ith row of A with the elements of the kth column of B, $AB=\left(\sum_{j=1}^{n}a_{ij}b_{jk}\right)$.

If C is $m\times n$ and D is $p\times q$ then the product CD can only be defined if $n=p$, in which case C and D are said to be ***conformable***. If C and D are such that CD is not defined then they are ***non-conformable***.

If x and y are vectors of order m and n respectively, i.e.,$m\times 1$ and $n\times 1$ matrices, then x and y' are conformable and so the product xy' is defined and is a $m\times n$ matrix with $(i,j)^{th}$ element x_iy_j, $i=1,\ldots m$, $j=1,\ldots n$. This is termed the ***outer product*** of x and y. The outer product of the sum vector ι_n with itself, i.e., $\iota_n\iota_n'$, is the $n\times n$ matrix J_n with all elements 1.

Example 2.2:
If $U=\begin{pmatrix} 1 & 2 \\ 3 & 4 \end{pmatrix}$, $V=\begin{pmatrix} 5 & 6 \\ 7 & 8 \end{pmatrix}$ then U is 2×2 and V is 2×2 so UV is $2\times 2\times 2\times 2\equiv 2\times 2$ and VU is $2\times 2\times 2\times 2\equiv 2\times 2$.

So,

$$\begin{aligned} \boldsymbol{UV} &= \begin{pmatrix} 1 & 2 \\ 3 & 4 \end{pmatrix}\begin{pmatrix} 5 & 6 \\ 7 & 8 \end{pmatrix} \\ &= \begin{pmatrix} 1\times5+2\times7 & 1\times6+2\times8 \\ 3\times5+4\times7 & 3\times6+4\times8 \end{pmatrix} \\ &= \begin{pmatrix} 5+14 & 6+16 \\ 15+28 & 18+32 \end{pmatrix} \\ &= \begin{pmatrix} 19 & 22 \\ 43 & 50 \end{pmatrix} \end{aligned}$$

and

$$\boldsymbol{VU} = \begin{pmatrix} 5 & 6 \\ 7 & 8 \end{pmatrix}\begin{pmatrix} 1 & 2 \\ 3 & 4 \end{pmatrix} = \begin{pmatrix} 5+18 & 10+24 \\ 7+24 & 14+32 \end{pmatrix} = \begin{pmatrix} 23 & 34 \\ 31 & 46 \end{pmatrix}.$$

. Notice that $\boldsymbol{UV} \neq \boldsymbol{VU}$ because $\begin{pmatrix} 19 & 22 \\ 43 & 50 \end{pmatrix} \neq \begin{pmatrix} 23 & 34 \\ 31 & 46 \end{pmatrix}$.

If $\boldsymbol{A} = \begin{pmatrix} 1 & 2 & 3 \\ 4 & 5 & 6 \end{pmatrix}$ and $\boldsymbol{B} = \begin{pmatrix} 1 & 2 \\ 3 & 4 \\ 5 & 6 \end{pmatrix}$, then $\boldsymbol{A}$ is a 2×3 matrix and $\boldsymbol{B}$ is a 3×2 matrix, so $\boldsymbol{AB}$ is $2 \times 3 \times 3 \times 2 \equiv 2 \times 2$ and $\boldsymbol{BA}$ is $3 \times 2 \times 2 \times 3 \equiv 3 \times 3$:

$$\begin{aligned} \boldsymbol{AB} &= \begin{pmatrix} 1 & 2 & 3 \\ 4 & 5 & 6 \end{pmatrix}\begin{pmatrix} 1 & 2 \\ 3 & 4 \\ 5 & 6 \end{pmatrix} \\ &= \begin{pmatrix} 1\times1+2\times3+3\times5 & 1\times2+2\times4+3\times6 \\ 4\times1+5\times3+6\times5 & 4\times2+5\times4+6\times6 \end{pmatrix} = \begin{pmatrix} 22 & 28 \\ 49 & 64 \end{pmatrix}. \end{aligned}$$

$$\begin{aligned} \boldsymbol{BA} &= \begin{pmatrix} 1 & 2 \\ 3 & 4 \\ 5 & 6 \end{pmatrix}\begin{pmatrix} 1 & 2 & 3 \\ 4 & 5 & 6 \end{pmatrix} \\ &= \begin{pmatrix} 1\times1+2\times4 & 1\times2+2\times5 & 1\times3+2\times6 \\ 3\times1+4\times4 & 3\times2+4\times5 & 3\times3+4\times6 \\ 5\times1+6\times4 & 5\times2+6\times5 & 5\times3+6\times6 \end{pmatrix} = \begin{pmatrix} 9 & 12 & 15 \\ 19 & 26 & 33 \\ 29 & 40 & 51 \end{pmatrix}. \end{aligned}$$

Notice that $\boldsymbol{AB} \neq \boldsymbol{BA}$ and indeed $\boldsymbol{AB}$ is a 2×2 matrix and $\boldsymbol{BA}$ is a 3×3 matrix.

$\boldsymbol{U}$ is 2×2, $\boldsymbol{A}$ is 2×3 and $\boldsymbol{B}$ is 3×2 so $\boldsymbol{U}$ and $\boldsymbol{A}$ are conformable and $\boldsymbol{UA}$ is defined (and is a $2 \times 2 \times 2 \times 3 \equiv 2 \times 3$ matrix) but $\boldsymbol{U}$ and $\boldsymbol{B}$ are non-conformable and you cannot calculate $\boldsymbol{UB}$ because it would be $2{\times}2{\times}3{\times}2 \not\cong$ anything. We have $\boldsymbol{UA} = \begin{pmatrix} 1 & 2 \\ 3 & 4 \end{pmatrix}\begin{pmatrix} 1 & 2 & 3 \\ 4 & 5 & 6 \end{pmatrix} = \begin{pmatrix} 9 & 12 & 15 \\ 19 & 26 & 33 \end{pmatrix}$. Note that the transpose of $\boldsymbol{B}$, $\boldsymbol{B}'$, is 2×3 so $\boldsymbol{U}$ and $\boldsymbol{B}'$ are conformable and we have $\boldsymbol{UB}' = \begin{pmatrix} 1 & 2 \\ 3 & 4 \end{pmatrix}\begin{pmatrix} 1 & 2 \\ 3 & 4 \\ 5 & 6 \end{pmatrix}' = \begin{pmatrix} 1 & 2 \\ 3 & 4 \end{pmatrix}\begin{pmatrix} 1 & 3 & 5 \\ 2 & 4 & 6 \end{pmatrix} = \begin{pmatrix} 5 & 11 & 17 \\ 11 & 25 & 39 \end{pmatrix}$.

In the product $\boldsymbol{AB}$ we say that $\boldsymbol{B}$ is ***premultiplied*** by $\boldsymbol{A}$ and $\boldsymbol{A}$ is ***postmultiplied*** by $\boldsymbol{B}$.

In general for two matrices $\boldsymbol{X}$ and $\boldsymbol{Y}$, we have $\boldsymbol{XY} \neq \boldsymbol{YX}$ even if $\boldsymbol{X}$ and $\boldsymbol{Y}$ are mutually conformable (i.e.,both of the products are defined). If we have matrices such that both $\boldsymbol{XY}$ and $\boldsymbol{YX}$ are defined and if we have $\boldsymbol{XY} = \boldsymbol{YX}$, we say that $\boldsymbol{X}$ and $\boldsymbol{Y}$ ***commute***. If $\boldsymbol{X}$ is $m \times n$ and $\boldsymbol{Y}$ is $p \times q$ and if both $\boldsymbol{XY}$ and $\boldsymbol{YX}$ are defined then we must have $p = n$ and $q = m$ so $\boldsymbol{XY}$ is $m \times n \times n \times m \equiv m \times m$ and $\boldsymbol{YX}$ is $n \times m \times m \times n \equiv n \times n$. Thus if $\boldsymbol{X}$ and $\boldsymbol{Y}$ commute then $\boldsymbol{XY} = \boldsymbol{YX}$ and in particular $\boldsymbol{XY}$ and $\boldsymbol{YX}$ must have the same *orders* so we must have $m = n$ and thus two matrices can only commute if they are ***square matrices of the same order***. Note that square matrices do not in general commute (e.g.,$\boldsymbol{U}$ and $\boldsymbol{V}$ above).

If $\boldsymbol{W} = \begin{pmatrix} 2 & 2 \\ 3 & 5 \end{pmatrix}$ then $\boldsymbol{UW} = \boldsymbol{WU}$ and $\boldsymbol{U}$ and $\boldsymbol{W}$ commute (check this as an exercise). The ***identity matrix*** $\mathbf{I_n}$ commutes with all $n \times n$ matrices (check this).

Note that $(\boldsymbol{A}+\boldsymbol{B})^2 = \boldsymbol{A}(\boldsymbol{A}+\boldsymbol{B}) + \boldsymbol{B}(\boldsymbol{A}+\boldsymbol{B}) = \boldsymbol{A}^2 + \boldsymbol{AB} + \boldsymbol{BA} + \boldsymbol{B}^2$ and so $(\boldsymbol{A}+\boldsymbol{B})^2 = \boldsymbol{A}^2 + 2\boldsymbol{AB} + \boldsymbol{B}^2$ ***only if*** $\boldsymbol{A}$ and $\boldsymbol{B}$ ***commute***.

2.3.2 Example 2.2 in R

Brief definitions of matrix operators in **R** and type illustrations of their use were given in the previous chapter. This section gives further illustrations using the specific matrices in the examples above, this time using **R**. These examples were given with sufficient detail to perform the calculations 'by hand'. It is of course important to understand the principles of matrix multiplication 'by hand' and there is no substitute for trying a few simple examples until the principle is absorbed. However, hand matrix calculations are undeniably tedious for matrices of anything other than orders two or three and use of **R** for them should become the norm.

The operators illustrated here are `+`, `-`, `%*%` (for matrix multiplication) and `*` (for scalar multiplication). Note that use of `*` for multiplication of matrices results in an element-by-element product which is rarely required (but see §6.3.0.1 and §8.4).

```
> U<-matrix(c(1,2,3,4),2,2,byrow=T)
> V<-matrix(c(5,6,7,8),2,2,byrow=T)
> W<-matrix(c(2,2,3,5),2,2,byrow=T)
> A<-matrix(c(1,2,3,4,5,6),2,3,byrow=T)
> B<-matrix(c(1,2,3,4,5,6),3,2,byrow=T)
```

```
> U ; V; W
     [,1] [,2]
[1,]    1    2
[2,]    3    4
     [,1] [,2]
[1,]    5    6
[2,]    7    8
     [,1] [,2]
[1,]    2    2
[2,]    3    5
```

```
> A;B
     [,1] [,2] [,3]
[1,]    1    2    3
[2,]    4    5    6
     [,1] [,2]
[1,]    1    2
[2,]    3    4
[3,]    5    6
```

Note the use of the repeat semicolon symbol ; in the examples above to enter several commands on one line.

```
> U+V ;2*U
     [,1] [,2]
[1,]    6    8
[2,]   10   12
     [,1] [,2]
[1,]    2    4
[2,]    6    8
> U%*%V ;V%*%U
     [,1] [,2]
[1,]   19   22
[2,]   43   50
     [,1] [,2]
[1,]   23   34
[2,]   31   46
```

```
> A%*%B ; B%*%A
     [,1] [,2]
[1,]   22   28
[2,]   49   64
     [,1] [,2] [,3]
[1,]    9   12   15
[2,]   19   26   33
[3,]   29   40   51
> U%*%A ; U%*%t(B)
     [,1] [,2] [,3]
[1,]    9   12   15
[2,]   19   26   33
     [,1] [,2] [,3]
[1,]    5   11   17
[2,]   11   25   39
```

```
> U%*%W ;W%*%U
     [,1] [,2]
[1,]    8   12
[2,]   18   26
     [,1] [,2]
[1,]    8   12
[2,]   18   26
```

```
> V%*%W ; W%*%V
     [,1] [,2]
[1,]   28   40
[2,]   38   54
     [,1] [,2]
[1,]   24   28
[2,]   50   58
```

2.4 Transpose and Trace of Sums and Products

If $\boldsymbol{A}+\boldsymbol{B}$ is defined (i.e., $\boldsymbol{A}$ and $\boldsymbol{B}$ have the same orders) then $(\boldsymbol{A}+\boldsymbol{B})' = \boldsymbol{A}'+\boldsymbol{B}'$ and tr($\boldsymbol{A}+\boldsymbol{B}$)=tr($\boldsymbol{A}$)+tr($\boldsymbol{B}$). If $\boldsymbol{A}$ is $m\times n$ and $\boldsymbol{B}$ is $n\times p$ then $\boldsymbol{AB}$ is $m\times n\times n\times p \equiv m\times p$ so $(\boldsymbol{AB})'$ is $p\times m$. It is easy to show that $(\boldsymbol{AB})' = \boldsymbol{B}'\boldsymbol{A}'$:

$$(\boldsymbol{AB})' = ((a_{ij})(b_{jk}))' = \left(\sum_{j=1}^{n} a_{ij}b_{jk}\right)' = \left(\sum_{j=1}^{n} a_{kj}b_{ji}\right) = (b_{jk})'(a_{ij})' = \boldsymbol{B}'\boldsymbol{A}'.$$

Note that $\boldsymbol{A}'$ is $n\times m$ and $\boldsymbol{B}'$ is $p\times n$ so the product $\boldsymbol{A}'\boldsymbol{B}'$ is not defined (unless $p=m$) but $\boldsymbol{B}'\boldsymbol{A}'$ is defined.

Clearly $\text{tr}(\boldsymbol{A}) = \text{tr}(\boldsymbol{A}')$ and if $\boldsymbol{A}$ and $\boldsymbol{B}$ are $m\times n$ and $n\times m$ matrices (so that both $\boldsymbol{AB}$ and $\boldsymbol{BA}$ are defined) then $\text{tr}(\boldsymbol{AB}) = \text{tr}(\boldsymbol{BA})$ because $\text{tr}(\boldsymbol{AB}) = \sum_{i=1}^{m}\sum_{j=1}^{n} a_{ij}b_{ij} = \sum_{j=1}^{n}\sum_{i=1}^{m} a_{ij}b_{ij} = \text{tr}(\boldsymbol{BA})$.

If λ is any scalar then $(\lambda\boldsymbol{A})' = \lambda\boldsymbol{A}'$.

2.5 Special Matrices

2.5.1 Symmetric and skew-symmetric matrices

A square $n\times n$ matrix $\boldsymbol{A} = (a_{ij})$ is ***symmetric*** if $\boldsymbol{A}' = \boldsymbol{A}$, i.e., $a_{ij} = a_{ji}$ for all i, j. A square matrix $\boldsymbol{B} = (b_{ij})$ is ***skew-symmetric*** if $\boldsymbol{B}' = -\boldsymbol{B}$, i.e.,$b_{ij} = -b_{ji}$ for all i, j. It is easy to see that all skew-symmetric matrices have zero elements on the diagonals.

Any square matrix $\boldsymbol{X}$ can be expressed as $\boldsymbol{X} = \frac{1}{2}(\boldsymbol{X}+\boldsymbol{X}') + \frac{1}{2}(\boldsymbol{X}-\boldsymbol{X}')$. Since $(\boldsymbol{X}+\boldsymbol{X}')' = (\boldsymbol{X}+\boldsymbol{X}')$ and $(\boldsymbol{X}-\boldsymbol{X}')' = -(\boldsymbol{X}-\boldsymbol{X}')$ we have that any square matrix can be expressed as the sum of a *symmetric part* and a *skew-symmetric part*.

Let $\boldsymbol{A} = \begin{pmatrix} 1 & 1 \\ 1 & 0 \end{pmatrix}$, $\boldsymbol{B} = \begin{pmatrix} 0 & 1 \\ 1 & 1 \end{pmatrix}$ then both $\boldsymbol{A}$ and $\boldsymbol{B}$ are symmetric but $\boldsymbol{AB} = \begin{pmatrix} 1 & 1 \\ 1 & 0 \end{pmatrix}\begin{pmatrix} 0 & 1 \\ 1 & 1 \end{pmatrix} = \begin{pmatrix} 1 & 2 \\ 0 & 1 \end{pmatrix}$ which is **not** symmetric.

Consider $\boldsymbol{Z} = \boldsymbol{A}'\boldsymbol{BA}$, if $\boldsymbol{B}$ is symmetric then $\boldsymbol{Z}$ is symmetric, because $(\boldsymbol{A}'\boldsymbol{BA})' = \boldsymbol{A}'\boldsymbol{B}'(\boldsymbol{A}')' = \boldsymbol{A}'\boldsymbol{BA}$ if $\boldsymbol{B}$ is symmetric.

However, the converse is not true: let $\boldsymbol{A} = \begin{pmatrix} 0 & 0 \\ 1 & 0 \end{pmatrix}$ then $\boldsymbol{A}' = \begin{pmatrix} 0 & 1 \\ 0 & 0 \end{pmatrix}$ and

$$\begin{aligned} \boldsymbol{A}'\boldsymbol{B}\boldsymbol{A} &= \begin{pmatrix} 0 & 0 \\ 1 & 0 \end{pmatrix}\begin{pmatrix} b_{11} & b_{12} \\ b_{21} & b_{22} \end{pmatrix}\begin{pmatrix} 0 & 1 \\ 1 & 0 \end{pmatrix} \\ &= \begin{pmatrix} 0 & 0 \\ b_{11} & b_{12} \end{pmatrix}\begin{pmatrix} 0 & 1 \\ 0 & 0 \end{pmatrix} = \begin{pmatrix} 0 & 0 \\ 0 & b_{11} \end{pmatrix} \end{aligned}$$

which is symmetric whether or not $\boldsymbol{B}$ is symmetric.

2.5.2 Products with transpose AA'

If $\boldsymbol{A}$ is $m \times n$ then the products $\boldsymbol{A}\boldsymbol{A}'$ and $\boldsymbol{A}'\boldsymbol{A}$ are both defined resulting in $m \times m$ and $n \times n$ square matrices respectively. Both are symmetric because, for example, $(\boldsymbol{A}\boldsymbol{A}')' = (\boldsymbol{A}')'\boldsymbol{A} = \boldsymbol{A}\boldsymbol{A}'$. If the columns of $\boldsymbol{A}$ are $a_{.j}$ then $\boldsymbol{A}\boldsymbol{A}' = \sum_j a_{.j}a'_{.j}$. (For a general product $\boldsymbol{A}\boldsymbol{B} = \sum_j a_{.j}b'_{j.}$ where the $b'_{j.}$ are the rows of $\boldsymbol{B}$.) $\boldsymbol{A}'\boldsymbol{A}$ is known as the ***cross-product*** of $\boldsymbol{A}$ and more generally $\boldsymbol{A}'\boldsymbol{B}$ is the cross-product of $\boldsymbol{A}$ and $\boldsymbol{B}$.

2.5.3 Orthogonal matrices

A $p \times p$ square matrix $\boldsymbol{A}$ is ***orthogonal*** if $\boldsymbol{A}'\boldsymbol{A} = \boldsymbol{A}\boldsymbol{A}' = \mathbf{I_p}$. Note that for square matrices if $\boldsymbol{A}'\boldsymbol{A} = \mathbf{I_p}$ and if $\boldsymbol{A}$ is non-singular (and so $\boldsymbol{A}'$ possesses an inverse $(\boldsymbol{A}')^{-1}$, see §5.2.1) then necessarily we have $\boldsymbol{A}\boldsymbol{A}' = \mathbf{I_p}$ since if $\boldsymbol{A}'\boldsymbol{A} = \mathbf{I_p}$ then $(\boldsymbol{A}')^{-1}\boldsymbol{A}'\boldsymbol{A}\boldsymbol{A}' = (\boldsymbol{A}')^{-1}\mathbf{I_p}\boldsymbol{A}' = \mathbf{I_p}$ (see Chapter 5 for definition of inverses). Also $\boldsymbol{A}^{-1} = \boldsymbol{A}'$ and if $\boldsymbol{A}$ is orthogonal then $\boldsymbol{A}'$ is also orthogonal.
If $\boldsymbol{A}$ and $\boldsymbol{B}$ are both orthogonal and both $p \times p$ then $\boldsymbol{A}\boldsymbol{B}$ is orthogonal because $(\boldsymbol{A}\boldsymbol{B})'\boldsymbol{A}\boldsymbol{B} = \boldsymbol{B}'\boldsymbol{A}'\boldsymbol{A}\boldsymbol{B} = \boldsymbol{B}'\mathbf{I_p}\boldsymbol{B} = \boldsymbol{B}'\boldsymbol{B} = \mathbf{I_p}$. It is possible to have a $m \times n$ matrix $\boldsymbol{B}$ such that $\boldsymbol{B}'\boldsymbol{B} = \mathbf{I_n}$ but $\boldsymbol{B}\boldsymbol{B}' \neq \mathbf{I_m}$, e.g.,the 2×1 matrix $\boldsymbol{B} = (1,0)'$ has $\boldsymbol{B}'\boldsymbol{B} = 1$ and $\boldsymbol{B}\boldsymbol{B}' \neq \mathbf{I}_2$.
Beware: two conformable matrices $\boldsymbol{U}$ and $\boldsymbol{V}$ are sometimes said to be *orthogonal* if $\boldsymbol{U}\boldsymbol{V} = \mathbf{0}$. This is using the term orthogonal in the same sense as two vectors being orthogonal (see §2.1.1, Page 22). More strictly it is better to say that $\boldsymbol{U}$ is orthogonal to $\boldsymbol{V}$; otherwise saying $\boldsymbol{U}$ and $\boldsymbol{V}$ are orthogonal could mean that both are orthogonal matrices.

Example 2.3

$$\boldsymbol{A} = \frac{1}{\sqrt{2}}\begin{pmatrix} 1 & -1 \\ 1 & 1 \end{pmatrix}, \quad \boldsymbol{B} = \begin{pmatrix} \frac{1}{2} & -\frac{\sqrt{3}}{2} \\ -\frac{\sqrt{3}}{2} & -\frac{1}{2} \end{pmatrix}$$

are both orthogonal.

$$\boldsymbol{A} = \begin{pmatrix} \cos(\theta) & -\sin(\theta) \\ \sin(\theta) & \cos(\theta) \end{pmatrix}, \quad \boldsymbol{B} = \begin{pmatrix} \cos(\theta) & -\sin(\theta) \\ -\sin(\theta) & -\cos(\theta) \end{pmatrix}$$

are both orthogonal for any value of θ and it can be shown that any 2×2 orthogonal matrix is of one of these two forms. Taking $\theta = \pi/4,\ \pi/3$ respectively gives the two orthogonal matrices above.

$$A = \frac{1}{\sqrt{2}}\begin{pmatrix} 1 & -1 & 1 \\ 0 & 1 & 2 \\ 1 & 1 & -1 \end{pmatrix} \text{ is orthogonal.}$$

If $\boldsymbol{A}$ is a $p \times p$ matrix with rows $\boldsymbol{a}_{1\cdot}, \boldsymbol{a}_{2\cdot}, \ldots, \boldsymbol{a}_{p\cdot}$ and columns $\boldsymbol{a}_{\cdot 1}, \boldsymbol{a}_{\cdot 2}, \ldots, \boldsymbol{a}_{\cdot p}$ then $\boldsymbol{a}'_{i\cdot}\boldsymbol{a}_{j\cdot} = 1$ if $i = j$ and 0 if $i \neq j$ and $\boldsymbol{a}_{i\cdot}\boldsymbol{a}'_{j\cdot} = 1$ if $i = j$ and 0 if $i \neq j$, i.e.,the rows and columns of $\boldsymbol{A}$ are orthogonal.

It will be seen in §4.4 that orthogonal matrices are either rotation or reflection matrices which means that in certain senses multiplying a data matrix by an orthogonal matrix does not change the intrinsic statistical properties of the data since it essentially means that the data are rotated onto a new set of orthogonal axes, possibly followed by a reflection.

An orthogonal matrix whose elements are all $+1$ or -1 is known as a ***Hadamard matrix***. These have a role in statistical experimental design.

2.5.4 Normal matrices

A $p \times p$ matrix is normal if $\boldsymbol{AA}' = \boldsymbol{A}'\boldsymbol{A}$, i.e.,if $\boldsymbol{A}$ commutes with $\boldsymbol{A}'$. Clearly all symmetric, all skew-symmetric and all orthogonal matrices are normal.

2.5.5 Permutation matrices

The $n \times n$ matrix $\boldsymbol{A}$ is a ***permutation matrix*** if each row and each column has exactly one 1 and the rest of the entries are zero. All permutation matrices are orthogonal.

Example 2.4:

$$\boldsymbol{A}_1 = \begin{pmatrix} 1 & 0 \\ 0 & 1 \end{pmatrix}, \boldsymbol{A}_2 = \begin{pmatrix} 0 & 1 \\ 1 & 0 \end{pmatrix}, \boldsymbol{B}_1 = \begin{pmatrix} 0 & 0 & 1 \\ 0 & 1 & 0 \\ 1 & 0 & 0 \end{pmatrix}, \boldsymbol{B}_2 = \begin{pmatrix} 0 & 1 & 0 \\ 1 & 0 & 0 \\ 0 & 0 & 1 \end{pmatrix}$$

and all identity matrices of any order $\mathbf{I_2}, \mathbf{I_3}, \mathbf{I_4}, \ldots, \mathbf{I_p}$. The columns of a permutation matrix are the complete set of unit vectors $\boldsymbol{e}_i$ taken in some order, i.e.,not necessarily $\boldsymbol{e}_1, \boldsymbol{e}_2, \ldots, \boldsymbol{e}_p$. For example $\boldsymbol{B}_1 = (\boldsymbol{e}_3, \boldsymbol{e}_2, \boldsymbol{e}_1), \boldsymbol{B}_2 = (\boldsymbol{e}_2, \boldsymbol{e}_1, \boldsymbol{e}_3)$. The effect of premultiplication of a matrix $\boldsymbol{X}$ by a permutation matrix is to permute the rows of $\boldsymbol{X}$; postmultiplication permutes the columns.

2.5.6 Idempotent matrices

A $p \times p$ matrix $\boldsymbol{A}$ is ***idempotent*** if $\boldsymbol{A}^2 = \boldsymbol{A}$. Clearly $\mathbf{I_p}$ and $\mathbf{0_{p \times p}}$ are idempotent and so is $\boldsymbol{xx}'$ with $\boldsymbol{x}'\boldsymbol{x} = 1$. Idempotent matrices play a key role in statistics because they can be regarded as projections and indeed are termed projection matrices. It can be shown (e.g. Banerjee and Roy, 2014, Theorem 8.6) that a matrix $\boldsymbol{P}$ is an orthogonal projection if and only if it is idempotent and symmetric.

2.5.6.1 The centering matrix $\mathbf{H_n}$

Let $\mathbf{H_n} = \mathbf{I_n} - \frac{1}{n}\iota_\mathbf{n}\iota_\mathbf{n}'$ then $\mathbf{H_n^2} = \left(\mathbf{I_n} - \frac{1}{n}\iota_\mathbf{n}\iota_\mathbf{n}'\right)^2 = \mathbf{I_n} - \frac{1}{n}\iota_\mathbf{n}\iota_\mathbf{n}' - \frac{1}{n}\iota_\mathbf{n}\iota_\mathbf{n}'(\mathbf{I_n} - \frac{1}{n}\iota_\mathbf{n}\iota_\mathbf{n}')$ $= \mathbf{I_n} - \frac{1}{n}\iota_\mathbf{n}\iota_\mathbf{n}' - \frac{1}{n}\iota_\mathbf{n}\iota_\mathbf{n}'\mathbf{I_n} + \frac{1}{n^2}\iota_\mathbf{n}(\iota_\mathbf{n}'\iota_\mathbf{n})\iota_\mathbf{n}'$ and $\iota_\mathbf{n}'\iota_\mathbf{n}$ is $1\times n\times n\times 1 \equiv 1\times 1$, i.e.,a scalar and $= n$ noting that $\iota_\mathbf{n}$ is the *sum vector*, which sums the elements of a vector when postmultiplying it (or premultiplying it by its transpose). So $\mathbf{H_n^2} = \mathbf{H_n} - \frac{1}{n}\iota_\mathbf{n}\iota_\mathbf{n}' + \frac{1}{n}\iota_\mathbf{n}\iota_\mathbf{n}' = \mathbf{H_n}$ and $\mathbf{H_n}$ is idempotent. $\mathbf{H_n}$ is called the ***centering matrix***. Premultiplying a $n\times 1$ vector by $\mathbf{H_n}$ subtracts the overall mean from each element of the vector. Postmultiplying a $n\times n$ matrix by $\mathbf{H_n}$ subtracts its column mean from each element. Note that $\mathbf{H_n}$ can be written as $\mathbf{H_n} = \mathbf{I_n} - \frac{1}{n}\mathbf{J_n}$.

2.5.7 Nilpotent matrices

A matrix $\boldsymbol{A} \neq \mathbf{0}$ is said to be ***nilpotent*** if $\boldsymbol{A}^2 = \mathbf{0}$, or more strictly *nilpotent of index 2*. A matrix such that $\boldsymbol{A}^r = \mathbf{0}$ but $\boldsymbol{A}^{r-1} \neq \mathbf{0}$ is nilpotent of index r.

2.5.8 Unipotent matrices

A $n\times n$ matrix $\boldsymbol{A}$ such that $\boldsymbol{A}^2 = \mathbf{I_n}$ is said to be ***unipotent***. Simple examples of unipotent matrices are all identity matrices and $\boldsymbol{A} = \begin{pmatrix} 1 & x \\ 0 & -1 \end{pmatrix}$.

2.5.9 Similar matrices

Two matrices $\boldsymbol{A}$ and $\boldsymbol{B}$ are sad to be ***similar*** if there is a non-singular matrix $\boldsymbol{C}$ such that $\boldsymbol{C}^{-1}\boldsymbol{A}\boldsymbol{C} = \boldsymbol{B}$ (see §5 for definition of $\boldsymbol{C}^{-1}$). Similar matrices share many properties, notably they have identical eigenvalues; see §6.4.3.

2.6 Partitioned Matrices

2.6.1 Sub-matrices

A ***sub-matrix*** of a $m\times n$ matrix $\boldsymbol{A}$ is a rectangular $m_1\times n_1$ section of it forming a $m_1\times n_1$ matrix $\boldsymbol{A}_{11}$. Note that some texts will regard any matrix formed by a subset of rows and columns (not necessarily contiguous) as a sub-matrix of $\boldsymbol{A}$. A sub matrix $\boldsymbol{A}_{ij}$ of $\boldsymbol{A}$ can be expressed as a product $\boldsymbol{EAF}$ where $\boldsymbol{E}$ and $\boldsymbol{F}$ are matrices with all entries either 0 or 1. For example, if suppose $\boldsymbol{A}_{11}$ is the top left hand corner of $\boldsymbol{A}$ consisting of m_1 rows and n_1 columns. Let $\boldsymbol{E}$ be the $m_1\times n$ matrix with a 1 in cells $(1,1),(2,2),\ldots,(m_1,m_1)$ and 0 elsewhere and let $\boldsymbol{F}$ be the $m\times n_1$ matrix with 1 in cells $(1,1),(2,2),\ldots,(n_1,n_1)$ and 0 elsewhere. Then $\boldsymbol{EAF} = \boldsymbol{A}_{11}$.

Example 2.5

```
>  A<- matrix(c(1,2,3,4,5,6,7,
+ 8,9),3,3,byrow=T)
> A
     [,1] [,2] [,3]
[1,]    1    2    3
[2,]    4    5    6
[3,]    7    8    9
> E<-matrix(c(1,0,0,0,1,0),
+ 2,3,byrow=T)
> E
     [,1] [,2] [,3]
[1,]    1    0    0
[2,]    0    1    0
```

```
> E%*%A
     [,1] [,2] [,3]
[1,]    1    2    3
[2,]    4    5    6
> F<- matrix(c(1,0,0,1,0,0),
+ 3,2,byrow=T)
> F
     [,1] [,2]
[1,]    1    0
[2,]    0    1
[3,]    0    0
```

```
 > A%*%F
     [,1] [,2]
[1,]    1    2
[2,]    4    5
[3,]    7    8
```

```
> E%*%A%*%F
     [,1] [,2]
[1,]    1    2
[2,]    4    5
```

2.6.2 Manipulation of partitioned matrices

A matrix $\boldsymbol{A}$ could be partitioned in four sub-matrices $\boldsymbol{A}_{11}$,$\boldsymbol{A}_{12}$,$\boldsymbol{A}_{21}$ and $\boldsymbol{A}_{22}$ with orders $m_1 \times n_1, m_1 \times n_2, m_2 \times n_1$ and $m_2 \times n_2$, with $m_1 + m_2 = m$ and $n_1 + n_2 = n$, so we would have

$$\boldsymbol{A} = \begin{pmatrix} \boldsymbol{A}_{11} & \boldsymbol{A}_{12} \\ \boldsymbol{A}_{21} & \boldsymbol{A}_{22} \end{pmatrix}.$$

This can be useful if some of the $\boldsymbol{A}_{ij}$ have a special form, e.g.,are $\boldsymbol{0}$ or diagonal or the identity matrix. Adding and multiplying matrices expressed in their partitioned form is performed by rules analogous to those when adding and multiplying element by element.

For example, if $\boldsymbol{B}$ is partitioned in four sub-matrices $\boldsymbol{B}_{11}, \boldsymbol{B}_{12}, \boldsymbol{B}_{21}$ and $\boldsymbol{B}_{22}$ with orders $m_1 \times n_1, m_1 \times n_2, m_2 \times n_1$ and $m_2 \times n_2$, $\boldsymbol{C}$ is partitioned in four sub-matrices $\boldsymbol{C}_{11}, \boldsymbol{C}_{12}, \boldsymbol{C}_{21}$ and $\boldsymbol{C}_{22}$ with orders $n_1 \times p_1, n_1 \times p_2, n_2 \times p_1$ and $n_2 \times p_2$, (so $\boldsymbol{A}_{ij}$ and $\boldsymbol{B}_{ij}$ are *addition conformable* and $\boldsymbol{A}_{ij}$ and $\boldsymbol{C}_{ij}$ are *multiplication conformable*), we have

$$\begin{aligned} \boldsymbol{A}+\boldsymbol{B} &= \begin{pmatrix} \boldsymbol{A}_{11}+\boldsymbol{B}_{11} & \boldsymbol{A}_{12}+\boldsymbol{B}_{12} \\ \boldsymbol{A}_{21}+\boldsymbol{B}_{21} & \boldsymbol{A}_{22}+\boldsymbol{B}_{22} \end{pmatrix}, \\ \boldsymbol{AC} &= \begin{pmatrix} \boldsymbol{A}_{11}\boldsymbol{C}_{11}+\boldsymbol{A}_{12}\boldsymbol{C}_{21} & \boldsymbol{A}_{12}\boldsymbol{C}_{12}+\boldsymbol{A}_{12}\boldsymbol{C}_{22} \\ \boldsymbol{A}_{21}\boldsymbol{C}_{11}+\boldsymbol{A}_{22}\boldsymbol{C}_{21} & \boldsymbol{A}_{22}\boldsymbol{C}_{12}+\boldsymbol{A}_{21}\boldsymbol{C}_{22} \end{pmatrix} \end{aligned}$$

which symbolically are identical to the forms we would have if $m_1 = m_2 = n_1 = n_2 = p_1 = p_2 = 1$. Further we have $\boldsymbol{A}' = \begin{pmatrix} \boldsymbol{A}_{11} & \boldsymbol{A}_{12} \\ \boldsymbol{A}_{21} & \boldsymbol{A}_{22} \end{pmatrix}' = \begin{pmatrix} \boldsymbol{A}'_{11} & \boldsymbol{A}'_{21} \\ \boldsymbol{A}'_{12} & \boldsymbol{A}'_{22} \end{pmatrix}$ (*note the interchange of the two off-diagonal blocks*).

Example 2.6

A matrix can be partitioned into its individual columns:

Let $\boldsymbol{X} = (\boldsymbol{x}_1, \boldsymbol{x}_2, \ldots, \boldsymbol{x}_n)$ where the $\boldsymbol{x}_i$ are p-vectors, i.e.,$\boldsymbol{X}$ is $p \times n$ matrix. Then

$$\boldsymbol{x}_i = \begin{pmatrix} x_{i1} \\ x_{i2} \\ \vdots \\ x_{ip} \end{pmatrix} \text{ and } \boldsymbol{X}' = \begin{pmatrix} x'_1 \\ x'_2 \\ \vdots \\ x'_n \end{pmatrix}$$

is a $n \times p$ matrix partitioned into n rows. Further, we have $\boldsymbol{XX}' = \sum_{k=1}^{n} \boldsymbol{x}_k \boldsymbol{x}'_k$ and $\boldsymbol{X}'\boldsymbol{X} = (\boldsymbol{x}'_i \boldsymbol{x}_j)$ (note that $\boldsymbol{XX}'$ is a $p \times p$ matrix whose $(i,j)^{th}$ element is $\sum_{k=1}^{n} x_{ki} x_{kj}$ and $\boldsymbol{X}'\boldsymbol{X}$ is a $n \times n$ matrix whose $(i,j)^{th}$ element is $\boldsymbol{x}_i \boldsymbol{x}_j$ i.e.,$\sum_{k=1}^{p} x_{ik} x_{jk}$).

A partitioned matrix of the form $\begin{pmatrix} \boldsymbol{Z}_{11} & 0 & 0 & 0 \\ 0 & \boldsymbol{Z}_{22} & 0 & 0 \\ 0 & 0 & \ddots & 0 \\ 0 & 0 & 0 & \boldsymbol{Z}_{rr} \end{pmatrix}$ where $\boldsymbol{Z}_{ii}$ is a $m_i \times n_i$ matrix and the sero sub-matrices are of conformable orders is called a ***block diagonal matrix***.

2.6.3 Implementation of partitioned matrices in R

Matrices of conformable dimensions can be joined together horizontally and vertically by command `cbind(.,.)` and `rbind(.,.)`. A sub-matrix of a $m \times n$ matrix $\boldsymbol{A}$ of dimensions $m_1 \times n_1$ [in the top left corner] can be specified by `A[1:m_1,1:n_1]`.

```
>A<-matrix(c(1:6),2,3); A
      [,1] [,2] [,3]
[1,]     1    3    5
[2,]     2    4    6

> U<-cbind(A,B)
> U
   [,1] [,2] [,3] [,4] [,5] [,6]
[1,] 1    3    5    1    2    3
[2,] 2    4    6    4    5    6
```

```
> B<-matrix(c(1:6),2,3,
+ byrow=T);B
      [,1] [,2] [,3]
[1,]     1    2    3
[2,]     4    5    6

> U[1:2,3:6]
     [,1] [,2] [,3] [,4]
[1,]    5    1    2    3
[2,]    6    4    5    6
```

```
> U[,3:6]
     [,1] [,2] [,3] [,4]
[1,]    5    1    2    3
[2,]    6    4    5    6

> V<- rbind(A,B)
> V
     [,1] [,2] [,3]
[1,]    1    3    5
[2,]    2    4    6
[3,]    1    2    3
[4,]    4    5    6
```

```
> V[1:2,1:3]
     [,1] [,2] [,3]
[1,]    1    3    5
[2,]    2    4    6

> V[2:4,]
     [,1] [,2] [,3]
[1,]    2    4    6
[2,]    1    2    3
[3,]    4    5    6
```

Note that because $\boldsymbol{U}$ has two rows, `U[1:2,3:6]` and `U[,3:6]` refer to the same sub-matrices (as do `V[2:4,]` and `V[2:4,1:3]` since $\boldsymbol{V}$ has three columns).

2.7 Algebraic Manipulation of Matrices

2.7.1 General expansion of products

An expression such as $(\boldsymbol{A}+\boldsymbol{B})(\boldsymbol{X}+\boldsymbol{Y})$ can be multiplied out term by term but remember to preserve the order of the multiplication of the matrices.
So $(\boldsymbol{A}+\boldsymbol{B})(\boldsymbol{X}+\boldsymbol{Y}) = \boldsymbol{A}(\boldsymbol{X}+\boldsymbol{Y})+\boldsymbol{B}(\boldsymbol{X}+\boldsymbol{Y}) = \boldsymbol{AX}+\boldsymbol{AY}+\boldsymbol{BX}+\boldsymbol{BY}$ and this cannot in general be simplified further. In particular, $(\boldsymbol{A}+\boldsymbol{B})^2 = (\boldsymbol{A}+\boldsymbol{B})(\boldsymbol{A}+\boldsymbol{B}) = \boldsymbol{A}(\boldsymbol{A}+\boldsymbol{B})+\boldsymbol{B}(\boldsymbol{A}+\boldsymbol{B}) = \boldsymbol{A}^2+\boldsymbol{AB}+\boldsymbol{BA}+\boldsymbol{B}^2$ and unless $\boldsymbol{A}$ and $\boldsymbol{B}$ commute we can go no further in collating terms.

2.8 Useful Tricks

2.8.1 Tracking dimensions and 1×1 matrices

It can happen that in an expression involving the product of several matrices and vectors, some element or sub-product is a 1×1 matrix (i.e.,a scalar). This permits two operations of just this part of the product:

(1) It can commute with other terms.

(2) It can be replaced by its transpose.

For example, suppose $\boldsymbol{S}$ is $p\times p$ and $\boldsymbol{x}$ and $\boldsymbol{y}$ are both p-vectors (i.e.,p). Let $\boldsymbol{A} = \boldsymbol{Sxx}'$; then $\boldsymbol{A}$ is $p\times p\times p\times 1\times 1\times p = p\times p$. Let $\boldsymbol{B} = \boldsymbol{Sx}$ then $\boldsymbol{B}$ is $p\times p\times p\times 1 = p\times 1$.

$\boldsymbol{AB} = (\boldsymbol{Sxx}')\boldsymbol{Sx} = \boldsymbol{Sx}(\boldsymbol{x}'\boldsymbol{Sx})$ which is $(p\times p\times p\times 1)\times(1\times p\times p\times p\times p\times 1)$. The second factor $\boldsymbol{x}'\boldsymbol{Sx}$ is $(1\times p\times p\times p\times p\times 1) = 1\times 1$, i.e.,a scalar, and so commutes with the first factor. So $(\boldsymbol{Sxx}')\boldsymbol{Sx} = (\boldsymbol{x}'\boldsymbol{Sx})\boldsymbol{Sx}$. This is of the form of matrix $\times$ vector = scalar $\times$ vector (the same vector) and is referred to as an *eigenequation* (see §6.2).

Another example is used in the next section: consider $\boldsymbol{x}'\boldsymbol{Ax}$ which is 1×1, i.e.,a scalar, and so is symmetric, since $(\boldsymbol{x}'\boldsymbol{Ax})' = \boldsymbol{x}'\boldsymbol{Ax}$, we have $\boldsymbol{x}'\boldsymbol{Ax} = \frac{1}{2}(\boldsymbol{x}'\boldsymbol{Ax}+\boldsymbol{x}'\boldsymbol{A}'\boldsymbol{x}) = \frac{1}{2}\boldsymbol{x}'(\boldsymbol{A}+\boldsymbol{A}')\boldsymbol{x}$ and the matrix $\frac{1}{2}(\boldsymbol{A}+\boldsymbol{A}')$ is symmetric.

2.8.2 Trace of products

Recall that $\mathrm{tr}(\boldsymbol{AB}) = \mathrm{tr}(\boldsymbol{BA})$ (if both products are defined) (see §2.4). This is useful either if one of $\boldsymbol{AB}$ and $\boldsymbol{BA}$ is simple (e.g.,a scalar) or some other simplification is possible. For example if $\boldsymbol{x}$ is a p-vector then $\mathrm{tr}(\boldsymbol{xx}') = \mathrm{tr}(\boldsymbol{x}'\boldsymbol{x})$ and $\boldsymbol{xx}'$ is $p\times p$ but $\boldsymbol{x}'\boldsymbol{x}$ is 1×1, i.e.,a scalar, and so possibly easier to evaluate.

An example in the other direction is evaluating $\boldsymbol{y}'\boldsymbol{Sy}$ by noting that $\boldsymbol{y}'\boldsymbol{Sy} = \mathrm{tr}(\boldsymbol{y}'\boldsymbol{Sy}) = \mathrm{tr}(\boldsymbol{yy}'\boldsymbol{S}) = \mathrm{tr}(\boldsymbol{Syy}')$, for symmetric $\boldsymbol{S}$. A trick like this is used in working with the maximized likelihood of the multivariate normal distribution.

2.9 Linear and Quadratic Forms

If $\boldsymbol{a}$ and $\boldsymbol{x}$ are p-vectors then the inner product $\boldsymbol{a}'\boldsymbol{x} = a_1x_1 + a_2x_2 + +a_px_p$ is termed a ***linear form*** in $\boldsymbol{x}$; it is presumed that $\boldsymbol{a}$ is a known vector of numbers and $\boldsymbol{x}$ is a variable. If $\boldsymbol{A}$ is a $p\times p$ matrix then $\boldsymbol{x}'\boldsymbol{Ax}$ is a ***quadratic form*** in $\boldsymbol{x}$. Again it is presumed that $\boldsymbol{A}$ is a known matrix of numbers and $\boldsymbol{x}$ is a variable. Note that $\boldsymbol{x}'\boldsymbol{Ax}$ is 1×1, i.e.,a scalar, and so is symmetric, so $\boldsymbol{x}'\boldsymbol{Ax} = (\boldsymbol{x}'\boldsymbol{Ax})' = \boldsymbol{x}'\boldsymbol{A}'\boldsymbol{x}$, and we have $\boldsymbol{x}'\boldsymbol{Ax} = \frac{1}{2}(\boldsymbol{x}'\boldsymbol{Ax}+\boldsymbol{x}'\boldsymbol{A}'\boldsymbol{x}) = \frac{1}{2}\boldsymbol{x}'(\boldsymbol{A}+\boldsymbol{A}')\boldsymbol{x}$ and the matrix $\frac{1}{2}(\boldsymbol{A}+\boldsymbol{A}')$ is symmetric. We need only consider the properties of quadratic forms which involve a symmetric matrix.

If $\boldsymbol{x}'\boldsymbol{Ax} > 0$ whatever the value of $\boldsymbol{x}$ then $\boldsymbol{A}$ is said to be *positive definite* and if $\boldsymbol{x}'\boldsymbol{Ax} \geq 0$ for all $\boldsymbol{x}$ then $\boldsymbol{A}$ is said to be *positive semi-definite*, similarly *negative definite* and *negative semi-definite* if $\boldsymbol{x}'\boldsymbol{Ax} < 0$ or ≤ 0. It is always assumed that if these terms are used, then $\boldsymbol{A}$ is symmetric.

Example 2.7 Suppose $p = 3$, then

$$\begin{aligned}
\boldsymbol{x}'\boldsymbol{Ax} &= a_{11}x_1^2 + a_{22}x_2^2 + a_{33}x_3^2 \\
&\quad + a_{12}x_1x_2 + a_{21}x_2x_1 + a_{13}x_1x_3 + a_{31}x_3x_1 + a_{23}x_2x_3 + a_{32}x_3x_2 \\
&= a_{11}x_1^2 + a_{22}x_2^2 + a_{33}x_3^2 \\
&\quad + (a_{12}+a_{21})x_1x_2 + (a_{13}+a_{31})x_1x_3 + (a_{23}+a_{32})x_2x_3 \\
&= \frac{1}{2}\boldsymbol{x}'(\boldsymbol{A}+\boldsymbol{A}')\boldsymbol{x}.
\end{aligned}$$

If $A = \begin{pmatrix} 1 & 4 & 7 \\ 2 & 5 & 8 \\ 3 & 5 & 9 \end{pmatrix}$ then

$$\begin{aligned} \boldsymbol{x}'\boldsymbol{A}\boldsymbol{x} &= x_1^2 + 5x_2^2 + 9x_3^2 + 6x_1x_2 + 10x_1x_3 + 13x_2x_3 \\ &= x_1^2 + 5x_2^2 + 9x_3^2 + 2\times 3x_1x_2 + 2\times 5x_1x_3 + 2\times 6.5x_2x_3 \end{aligned}$$

so we have $\boldsymbol{x}' \begin{pmatrix} 1 & 4 & 7 \\ 2 & 5 & 8 \\ 3 & 5 & 9 \end{pmatrix} \boldsymbol{x} = \boldsymbol{x}' \begin{pmatrix} 1 & 3 & 5 \\ 3 & 5 & 6.5 \\ 5 & 6.5 & 9 \end{pmatrix} \boldsymbol{x}$, replacing the matrix $\boldsymbol{A}$ with a symmetric one.

2.10 Creating Matrices in R

Small matrices can be entered directly using the function `matrix(.)`. Large matrices may best be entered by reading the data from a file into a `dataframe` (e.g.,with `read.table(.)` or `read.csv(.)` or `scan(.)`) and then converting to a matrix.
$\mathtt{A} < -\mathtt{matrix}(\mathtt{c}(x_{11},\ldots,x_{m1},x_{12},\ldots,x_{m2},\ldots,x_{1n},\ldots,x_{mn}),$
`+ nrow=m, ncol=n, byrow=F)`
creates an $m \times n$ matrix taking the values column by column.
$\mathtt{B} < -\mathtt{matrix}(\mathtt{c}(x_{11},\ldots,x_{m1},x_{12},\ldots,x_{m2},\ldots,x_{1n},\ldots,x_{mn}),$
`+ nrow=m, ncol=n, byrow=T)`
creates an $m \times n$ matrix taking the values row by row. If the `byrow` is omitted, then `byrow=F` is assumed and the values are taken column by column. If the terms `nrow=` and `ncol=` are omitted (and just the values `m` and `n` given), then it is assumed they are in the order row and column).
Thus $\mathtt{A} < -\mathtt{matrix}(\mathtt{c}(x_{11},\ldots,x_{m1},x_{12},\ldots,x_{m2},\ldots,x_{1n},\ldots,x_{mn}),\mathtt{m},\mathtt{n})$ creates an $m \times n$ matrix taking the values column by column. See the next section for more examples.
If data are read into a ***dataframe*** then this can be converted to a matrix by `data.matrix(.)` or `as.matrix(.)`:

```
X<-read.table(filename)
X<-data.matrix(X)
```

Initially $\boldsymbol{X}$ is of class `"dataframe"` and then is converted to class `"matrix"`; `as.matrix(X)` will have the same effect but can also be used on objects which are not entirely numeric. The importance of the class of an object is that some commands will accept arguments only of certain classes.
There are many other ways of creating matrices; some other functions give a matrix as a result, e.g.,`eigen(.)` produces a matrix of eigenvectors, `cbind(.,.,.)` will join vectors together into a matrix. Details of these are not given here and the `help(.)` system is generally most informative.

Example 2.8

```
> A<-matrix(c(1,2,3,4,5,6),nrow=2,ncol=3,byrow=F)
> B<-matrix(c(1,2,3,4,5,6),nrow=2,ncol=3,byrow=T)
```

```
> A
     [,1] [,2] [,3]
[1,]    1    3    5
[2,]    2    4    6
```

```
> B
     [,1] [,2] [,3]
[1,]    1    2    3
[2,]    4    5    6
```

Note that the columns of ***A*** were filled successively (because `byrow=F`) and the rows of ***B*** were filled successively (because `byrow=T`).

```
> C<-matrix(c(1,2,3,4,5,6),
+ 2,3)
> C
     [,1] [,2] [,3]
[1,]    1    3    5
[2,]    2    4    6
```

Note that the columns of ***C*** were filled successively (because `byrow=F` was assumed by default).

```
> D<-matrix(c(1,2,3,4,5,6),
+ 2,3,byrow=T)
> D
     [,1] [,2] [,3]
[1,]    1    2    3
[2,]    4    5    6
```

The rows of ***D*** were filled successively (because `byrow=T` was specified). In both cases two rows and three columns were taken since the order rows, columns was assumed.

The order can be overridden by specifying the parameters:

```
> E<-matrix(c(1,2,3,4,5,6),
+ncol=2,nrow=3,byrow=T)
> E
     [,1] [,2]
[1,]    1    2
[2,]    3    4
[3,]    5    6
```

If the parameters are omitted entirely then `ncol=1` is assumed and a column vector is created (which has class `matrix` because of the `matrix(.)` function).

```
> F<-matrix(c(1,2,3,4,5,6))
> F
     [,1]
[1,]    1
[2,]    2
[3,]    3
[4,]    4
[5,]    5
[6,]    6
```

2.10.1 Ambiguity of vectors

Note that entering vectors without specifying whether they are column or row vectors can lead to ambiguities: if we create the vector $\boldsymbol{a}$ by

```
> a<-c(1,2,3)
```

then it may be interpreted as a column vector (i.e.,a 3×1 matrix) or as a row vector (i.e.,a 1×3 matrix) according to what operation is being attempted. **R** will do its very best to produce some answer and avoid giving an error message. For example, if $\boldsymbol{X}$ is a 3×3 matrix, premultiplying $\boldsymbol{X}$ by a will cause $\boldsymbol{a}$ to be assumed to be a row vector but postmultiplying $\boldsymbol{X}$ by a will cause **R** to regard $\boldsymbol{a}$ as a column vector. See §2.11.2.7. This ambiguity can also arise in vectors produced as a result of a function, for example, the vector of eigenvalues held in `eigen(X)$values` (see §6.3) may be treated as a row or column vector according to context.

2.11 Matrix Arithmetic in R

2.11.1 Addition, subtraction and transpose

```
> A+C
     [,1] [,2] [,3]
[1,]    2    6   10
[2,]    4    8   12
```

```
> A-D
     [,1] [,2] [,3]
[1,]    0    1    2
[2,]   -2   -1    0
```

Beware:

```
> A+E
Error in A + E : non-conformable arrays
```

```
> E
     [,1] [,2]
[1,]    1    2
[2,]    3    4
[3,]    5    6
```

```
>t(E)
     [,1] [,2] [,3]
[1,]    1    3    5
[2,]    2    4    6
```

So A and E' are conformable:

```
> A+t(E)
     [,1] [,2] [,3]
[1,]    2    6   10
[2,]    4    8   12
```

2.11.2 Multiplication

2.11.2.1 Standard multiplication

```
> A<-matrix(c(1,2,3,4,5,6),nrow=2,ncol=3,byrow=T)
> B<-matrix(c(1,2,3,4,5,6),nrow=3,ncol=2,byrow=T)
```

```
> A; A%*%B; B; B%*%A

     [,1] [,2] [,3]
[1,]    1    2    3
[2,]    4    5    6
     [,1] [,2]
[1,]   22   28
[2,]   49   64
```

```
     [,1] [,2]
[1,]    1    2
[2,]    3    4
[3,]    5    6
     [,1] [,2] [,3]
[1,]    9   12   15
[2,]   19   26   33
[3,]   29   40   51
```

2.11.2.2 Element-by-element (Hadamard) multiplication

BEWARE: If A and B have the same numbers of rows and columns then $A*B$ gives element-by-element multiplication which is rarely required (but see §6.3.0.1).

```
> t(B)
     [,1] [,2] [,3]
[1,]    1    3    5
[2,]    2    4    6
```

```
> A*t(B)
     [,1] [,2] [,3]
[1,]    1    6   15
[2,]    8   20   36
```

2.11.2.3 Non-commuting matrices

```
> U<-matrix(c(1,2,3,4),2,2,byrow=T)
> V<-matrix(c(5,6,7,8),2,2,byrow=T)
```

```
> U
     [,1] [,2]
[1,]    1    2
[2,]    3    4
> U%*%V
     [,1] [,2]
[1,]   19   22
[2,]   43   50
```

```
> V
     [,1] [,2]
[1,]    5    6
[2,]    7    8
> V%*%U
     [,1] [,2]
[1,]   23   34
[2,]   31   46
```

2.11.2.4 Commuting matrices

```
> W<-matrix(c(2,2,3,5),2,2,byrow=T)
```

```
> W
     [,1] [,2]
[1,]    2    2
[2,]    3    5
```

```
> U%*%W
     [,1] [,2]
[1,]    8   12
[2,]   18   26
```

```
>  W%*%U
     [,1] [,2]
[1,]    8   12
[2,]   18   26
```

Beware:

```
> U%*%A
     [,1] [,2] [,3]
[1,]    9   12   15
[2,]   19   26   33
```

But:

```
> B
     [,1] [,2]
[1,]    1    2
[2,]    3    4
[3,]    5    6
```

```
> U%*%B
Error in U %*% B : non-conformable arguments
```

```
> t(B)
     [,1] [,2] [,3]
[1,]    1    3    5
[2,]    2    4    6
```

```
> U%*%t(B)
     [,1] [,2] [,3]
[1,]    5   11   17
[2,]   11   25   39
```

2.11.2.5 Transpose of products

```
> t(U%*%V)
     [,1] [,2]
[1,]   19   43
[2,]   22   50
```

```
> t(V)%*%t(U)
     [,1] [,2]
[1,]   19   43
[2,]   22   50
```

```
> t(U)%*%t(V)
     [,1] [,2]
[1,]   23   31
[2,]   34   46
```

So $(\boldsymbol{UV})' = \boldsymbol{V}'\boldsymbol{U}' \neq \boldsymbol{U}'\boldsymbol{V}'$

```
> t(U%*%W)
     [,1] [,2]
[1,]    8   18
[2,]   12   26
```

```
> t(W)%*%t(U)
     [,1] [,2]
[1,]    8   18
[2,]   12   26
```

```
> t(U)%*%t(W)
     [,1] [,2]
[1,]    8   18
[2,]   12   26
```

Note that $\boldsymbol{U}$ and $\boldsymbol{W}$ commute so it follows that $\boldsymbol{U}'$ and $\boldsymbol{W}'$ also commute.

```
> U%*%t(W)
     [,1] [,2]
[1,]    6   13
[2,]   14   29
```

```
> t(W)%*%U
     [,1] [,2]
[1,]   11   16
[2,]   17   24
```

But it does not follow that because $\boldsymbol{U}$ and $\boldsymbol{W}$ commute then $\boldsymbol{W}'$ also commutes with $\boldsymbol{U}$ as the above example demonstrates.

2.11.2.6 Cross-products

```
> crossprod(A); t(A)%*%A
     [,1] [,2] [,3]
[1,]   17   22   27
[2,]   22   29   36
[3,]   27   36   45
     [,1] [,2] [,3]
[1,]   17   22   27
[2,]   22   29   36
[3,]   27   36   45
```

```
> tcrossprod(A) ; A%*%t(A)
     [,1] [,2]
[1,]   14   32
[2,]   32   77
     [,1] [,2]
[1,]   14   32
[2,]   32   77
```

```
> crossprod(A,D); t(A)%*%D
     [,1] [,2] [,3]
[1,]    9   19   29
[2,]   12   26   40
[3,]   15   33   51
     [,1] [,2] [,3]
[1,]    9   19   29
[2,]   12   26   40
[3,]   15   33   51
```

```
> tcrossprod(A,D); A%*%t(D)
     [,1] [,2]
[1,]   22   28
[2,]   49   64
     [,1] [,2]
[1,]   22   28
[2,]   49   64
```

2.11.2.7 Ambiguity of vectors

Consider the following:

```
> X<-matrix(c(1,2,3,4,5,6,7,8,9),3,3,byrow=T)
> X
     [,1] [,2] [,3]
[1,]    1    2    3
```

```
[2,]    4    5    6
[3,]    7    8    9
```

```
> a<-c(1,2,3)
> a%*%X
     [,1] [,2] [,3]
[1,]   30   36   42
```

```
> b<-c(4,5,6)
> X%*%b
     [,1]
[1,]   32
[2,]   77
[3,]  122
```

So ***a*** is interpreted as a row vector but ***b*** as column vector above.

```
> a%*%b
     [,1]
[1,]   32
```

```
> b%*%a
     [,1]
[1,]   32
```

and here the inner product of ***a*** and ***b*** is returned, whatever the order of the product. To force ***a*** and ***b*** to column and row vectors see

```
> a<-matrix(a,3,1)
> a
     [,1]
[1,]    1
[2,]    2
[3,]    3
> b%*%a
     [,1]
[1,]   32
```

```
> b<-matrix(b,1,3)
> b
     [,1] [,2] [,3]
[1,]    4    5    6

> a%*%b
     [,1] [,2] [,3]
[1,]    4    5    6
[2,]    8   10   12
[3,]   12   15   18
```

2.11.3 Diagonal matrices

2.11.3.1 Creating a diagonal matrix from a list

```
> F<-diag(c(1,2,3,4,5))
> F
     [,1] [,2] [,3] [,4] [,5]
[1,]    1    0    0    0    0
[2,]    0    2    0    0    0
[3,]    0    0    3    0    0
[4,]    0    0    0    4    0
[5,]    0    0    0    0    5
```

2.11.3.2 Extracting the diagonal

```
> E%*%A
     [,1] [,2] [,3]
[1,]    9   12   15
[2,]   19   26   33
[3,]   29   40   51
> diag(E%*%A)
[1]  9 26 51
```

2.11.3.3 Converting a matrix to diagonal

```
> diag(diag(E%*%A))
     [,1] [,2] [,3]
[1,]    9    0    0
[2,]    0   26    0
[3,]    0    0   51
```

2.11.4 Trace

2.11.4.1 Trace of square matrix

```
> sum(diag(E%*%A))
[1] 86
> sum(diag(U))
[1] 5
> V
     [,1] [,2]
[1,]    5    6
[2,]    7    8
```

```
> sum(diag(V))
[1] 13
```

But be careful because sum(V) gives the sum of all elements in the matrix, not just the diagonals.

```
> sum(V)
[1] 26
```

2.11.4.2 Trace of transpose and products

```
> sum(diag(U)); sum(diag(t(U)))
[1] 5
[1] 5
```

```
> U%*%V
     [,1] [,2]
[1,]   19   22
[2,]   43   50
```

```
> V%*%U
     [,1] [,2]
[1,]   23   34
[2,]   31   46
```

But

```
> sum(diag(U%*%V)); sum(diag(V%*%U))
[1] 69
[1] 69
```

2.11.4.3 Creating a function for trace of a matrix

A useful facility in **R** is the creation of functions to execute a sequence of commands on arguments supplied to it. For example, to create a function `tr(.)` to calculate the trace of a matrix, first store the function in the object `tr` by

```
> tr<-function(X) { tr<-sum(diag(X))
return(tr)
}
```

Here, the arguments of the function are indicated by the dummy objects in the first pair of braces (`{}`) following `function` and the sequence of commands to be used is contained between the second pair of braces; in this case they are `sum()` and `diag()`. Then the trace can be calculated by using `tr(.)`. For example:

```
> tr(U)
[1] 5
> tr(t(U))
[1] 5
```

```
> tr(U%*%V);tr(V%*%U)
[1] 69
[1] 69
```

2.12 Initial Statistical Applications

2.12.1 Introduction

Here we introduce some illustrations of uses of vectors and matrices in statistics. This will only be an initial presentation since further analysis depends on material in later chapters, especially vector and matrix calculus in Chapter 7 which is used for obtaining maximum likelihood estimates. Also, determinants, inverses and eigenanalyses are all required for the full treatment of the topics and so the full discussion will be given after these topics have been covered in subsequent chapters. We begin with the basic formulation of linear models and multivariate analysis.

Unfortunately, major texts on these subjects do not conform to a single standard system of notation. This is especially true of multivariate analysis where, for example, Gnanadesikan (1997) and Mardia et al. (1979) define the data matrix consisting of n observations on a p-dimensional random variable differently. Another difference between these two texts is that the former uses a divisor $(n-1)$ in the sample variance matrix but the latter uses n. This forces a choice to be made and here the system used broadly conforms with that of Gnanadesikan (1997). Consequently, readers more familiar with Mardia et al. (1979) (or who are following courses based on this or a similar text) will find that some results will appear to be unfamiliar with $\boldsymbol{X}'$ instead of $\boldsymbol{X}$ or factors of n instead of $(n-1)$. This choice has particular consequence when implementing the results in **R** since **R** more naturally conforms with the Mardia et al. (1979) convention for the data matrix but with the Gnanadesikan (1997) convention for the sample variance. This warning first arises in §2.12.3.2 below.

2.12.2 Linear models

The simple linear model expressing the dependence of a variable y on independent or regressor variables $x_1, x_2, \ldots x_p$ can be written as

$$y = \beta_1 x_1 + \beta_2 x_2 + \cdots + \beta_p x_p + \varepsilon$$

where ε is a random error term with $\mathrm{E}[\varepsilon] = 0$ and $\mathrm{var}(\varepsilon) = \sigma^2$. If n observations of y are available corresponding to n observations on each of the x_i then we can write

$$y_i = \beta_1 x_{i1} + \beta_2 x_{i2} + \cdots + \beta_p x_{ip} + \varepsilon_i.$$

If we let the $n \times 1$ vector $\boldsymbol{y} = (y_1, y_2, \ldots, y_n)'$, the $p \times 1$ vector $\beta = (\beta_1, \ldots, \beta_p)'$, $\boldsymbol{X}$ be the $n \times p$ matrix with $(i, j)^{th}$ element $x_{i,j}$; $i = 1, \ldots, n$; $j = 1, \ldots, p$ and the $n \times 1$ vector $\varepsilon = (\varepsilon_1, \ldots, \varepsilon_n)'$, we can write this succinctly as $\boldsymbol{y} = \boldsymbol{X}\beta + \varepsilon$.

We can then use this expression for investigating the statistical properties of the model and obtaining estimates of the unknown parameters β. For example, since $\boldsymbol{X}\beta$ is a constant, we have that $\mathrm{E}[\boldsymbol{y}] = \mathrm{E}[\boldsymbol{X}\beta + \varepsilon] = \boldsymbol{X}\beta + \mathrm{E}[\varepsilon] = \boldsymbol{X}\beta$ and more generally if $\boldsymbol{G}$ is a $p \times n$ matrix then $\mathrm{E}[\boldsymbol{G}\boldsymbol{y}] = \mathrm{E}[\boldsymbol{G}\boldsymbol{X}\beta + \boldsymbol{G}\varepsilon] = \boldsymbol{G}\boldsymbol{X}\beta + \boldsymbol{G}\mathrm{E}[\varepsilon] = \boldsymbol{G}\boldsymbol{X}\beta$ so we might be interested in matrices $\boldsymbol{G}$ such that $\boldsymbol{GX}\beta = \beta$ because then we would have a linear function of the observations $\boldsymbol{Gy}$ which is an unbiased estimator of β.

We will return to this topic in a later chapter.

2.12.3 Multivariate analysis

In the previous section we introduced the idea of writing n observations on p variables $x_1, x_2, \ldots, x_p$ in the form of a matrix. In the context of a simple linear model, these variables are regarded as non-random or *design variables* and the matrix of observations as the *design matrix* in the sense that specific values of them can be chosen to take observations of the variable y which depends upon them, the form of the dependence being a linear function plus some random error.

Here we will regard the variables $x_1, x_2, \ldots, x_p$ as random variables with means μ_i, variances σ_{ii} and covariances σ_{ij}. We will write this as the p-dimensional random variable $\boldsymbol{x} = (x_1, x_2, \ldots, x_p)'$ with mean $\mathrm{E}[\boldsymbol{x}] = \mu$ and variance $\mathrm{var}(\boldsymbol{x}) = \Sigma$ where Σ is the symmetric matrix with $(i,j)^{th}$ element σ_{ij}. Note that we have $\Sigma = \mathrm{var}(\boldsymbol{x}) = \mathrm{E}[(\boldsymbol{x} - \mathrm{E}[\boldsymbol{x}])(\boldsymbol{x} - \mathrm{E}[\boldsymbol{x}])'] = \mathrm{E}[(\boldsymbol{x} - \mu)(\boldsymbol{x} - \mu)']$. Generally we will assume that Σ is a non-singular positive definite symmetric matrix. The definition can be extended to handle cases where the variance matrix is singular (i.e.,positive semi-definite) but we do not do so here.

If $\boldsymbol{A}$ is a $n \times p$ matrix, and $\boldsymbol{b}$ a $p \times 1$ vector, then $\mathrm{E}[\boldsymbol{Ax} + \boldsymbol{b}] = \boldsymbol{A}\mathrm{E}[\boldsymbol{x}] + \boldsymbol{b}$ and $\mathrm{var}(\boldsymbol{Ax} + \boldsymbol{b}) = \boldsymbol{A}\mathrm{var}(\boldsymbol{x})\boldsymbol{A}'$. Note that $\mathrm{var}(\boldsymbol{Ax}) = \boldsymbol{A}\mathrm{var}(\boldsymbol{x})\boldsymbol{A}'$ because $\mathrm{var}(\boldsymbol{Ax}) = \mathrm{E}[(\boldsymbol{Ax} - \mathrm{E}[\boldsymbol{Ax}])(\boldsymbol{Ax} - \mathrm{E}[\boldsymbol{Ax}])'] = \mathrm{E}[\boldsymbol{A}(\boldsymbol{x} - \mu)(\boldsymbol{x} - \mu)'\boldsymbol{A}'] = \boldsymbol{A}\mathrm{E}[(\boldsymbol{x} - \mu)(\boldsymbol{x} - \mu)']\boldsymbol{A}' = \boldsymbol{A}\mathrm{var}(\boldsymbol{x})\boldsymbol{A}'$.

2.12.3.1 Random samples

Suppose $\boldsymbol{x}_1, \boldsymbol{x}_2, \ldots, \boldsymbol{x}_n$ are independent observations of $\boldsymbol{x}$, and $\boldsymbol{x}_i = (x_{i1}, x_{i2}, \ldots, x_{ip})'$. Define the **data matrix** $\boldsymbol{X}'$ as the matrix with i^{th} row given by $\boldsymbol{x}_i'$. So $\boldsymbol{X}'$ is a $n \times p$ matrix and $\boldsymbol{X}$ is the $p \times n$ matrix with columns given by $\boldsymbol{x}_1, \boldsymbol{x}_2, \ldots, \boldsymbol{x}_n$:

$$\boldsymbol{X} = \begin{pmatrix} x_{11} & x_{21} & \cdots & x_{n1} \\ x_{12} & x_{22} & \cdots & x_{n2} \\ \vdots & \vdots & \ddots & \vdots \\ x_{1p} & x_{2p} & \cdots & x_{np} \end{pmatrix}.$$

Define the sample mean vector $\bar{\boldsymbol{x}} = (\bar{x}_1, \bar{x}_2, \ldots, \bar{x}_p)' = \frac{1}{n}\boldsymbol{X}\iota_{\mathbf{n}}$. Let $\overline{\boldsymbol{X}} = \bar{\boldsymbol{x}}\iota_{\mathbf{n}}'$ (so $\overline{\boldsymbol{X}}$ is a $p \times n$ matrix with all columns equal to $\bar{\boldsymbol{x}}$). Define the sample variance

$$\boldsymbol{S} = \tfrac{1}{(n-1)}(\boldsymbol{X} - \overline{\boldsymbol{X}})(\boldsymbol{X} - \overline{\boldsymbol{X}})' = \tfrac{1}{(n-1)}\left\{\sum_1^n \boldsymbol{x}_i\boldsymbol{x}_i' - n\bar{\boldsymbol{x}}\bar{\boldsymbol{x}}'\right\}.$$

This last equality follows because the $(\boldsymbol{X} - \overline{\boldsymbol{X}})$ has columns $\boldsymbol{x}_i - \bar{\boldsymbol{x}}$ and so $(\boldsymbol{X} - \overline{\boldsymbol{X}})(\boldsymbol{X} - \overline{\boldsymbol{X}})' = \sum_i^n(\boldsymbol{x}_i - \bar{\boldsymbol{x}})(\boldsymbol{x}_i - \bar{\boldsymbol{x}})' = \sum_i^n \boldsymbol{x}_i\boldsymbol{x}_i' - 2\bar{\boldsymbol{x}}\sum_1^n \boldsymbol{x}_i' + n\bar{\boldsymbol{x}}\bar{\boldsymbol{x}}' = \sum_1^n \boldsymbol{x}_i\boldsymbol{x}_i' - n\bar{\boldsymbol{x}}\bar{\boldsymbol{x}}$.

Clearly $\mathrm{E}[\bar{\boldsymbol{x}}] = \mu$ and $\mathrm{var}(\bar{\boldsymbol{x}}) = \frac{1}{n^2}\sum_1^n \mathrm{var}(\boldsymbol{x}_i) = \frac{1}{n}\mathrm{var}(\boldsymbol{x}) = \frac{1}{n}\Sigma$. By showing that

$$\boldsymbol{S} = \tfrac{1}{(n-1)}\left\{(1-\tfrac{1}{n})\sum_{i=1}^{n}(\boldsymbol{x}_i-\mu)(\boldsymbol{x}_i-\mu)' - \tfrac{1}{n}\sum\sum_{i\neq j}(\boldsymbol{x}_i-\mu)(\boldsymbol{x}_j-\mu)'\right\}$$

and noting that $\mathrm{E}[(\boldsymbol{x}_i-\mu)(\boldsymbol{x}-\mu)'] = \Sigma$ if $i = j$ and 0 if $i \neq j$ we have that $\mathrm{E}[\boldsymbol{S}] = \Sigma$, and thus $\bar{\boldsymbol{x}}$ and $\boldsymbol{S}$ are unbiased estimators of μ and Σ.

2.12.3.2 Sample statistics in R

Suppose a dataframe A consists of n observations on p variables and so is arranged with rows corresponding to observations and columns to variables. The convention in this text is to refer to this as the data matrix $\boldsymbol{X}'$ following Gnanadesikan (1997); note in particular the transpose.

```
X<-t(as.matrix(A))      ###set up data matrix
xbar<-X%*%matrix(rep(1,n),n)/n  ### calculate mean vector
Xbar<-xbar%*%t(matrix(rep(1,n),n))  ### Calculate X-bar matrix
S<-(X-Xbar)%*%t(X-Xbar)/(n-1) ### Calculate the sample variance
```

The **R** code above conforms with the formulation given in the previous section but **R** actually has built-in functions to perform these although calculation of the mean requires use of the function `apply(.,.,.)` (type `help(apply)` to find out the details). Using these allows

```
xbar<-matrix(apply(X,1,mean),p)
S<-var(t(X))
```

Note: This may seem unfamiliar to some more experienced readers who would expect these commands to read `xbar<-matrix(apply(A,2,mean),p);` `S<-var(A)` but it is a consequence of defining the data matrix X as the transpose of the dataframe A.

The sample correlation matrix $\boldsymbol{R}$ is given by $\boldsymbol{D}^{-1/2}\boldsymbol{S}\boldsymbol{D}^{-1/2}$ where $\boldsymbol{D} = \mathrm{diag}(\mathrm{diag}(\boldsymbol{S}))$ which can be calculated in **R** by

```
Dh<-diag(1/sqrt(diag(S)))
R<Dh%*%S%*%Dh
```

or more easily by `cor(t(X))`.

2.13 Exercises

(1) Let $\boldsymbol{a} = \begin{pmatrix}1\\2\\3\end{pmatrix}$, $\boldsymbol{b} = \begin{pmatrix}4\\5\\6\end{pmatrix}$, $\boldsymbol{u} = \begin{pmatrix}3\\2\\1\end{pmatrix}$, $\boldsymbol{v} = \begin{pmatrix}6\\5\\4\end{pmatrix}$, $\boldsymbol{w} = (7,8,9)$.

(a) Calculate $\boldsymbol{a}+\boldsymbol{b}$, $\boldsymbol{v}-\boldsymbol{a}$, $\boldsymbol{w}'+\boldsymbol{b}$, $3\boldsymbol{u}$, $\boldsymbol{w}'-\boldsymbol{a}$, $\boldsymbol{v}/3$, $\boldsymbol{ab}'$ and $\boldsymbol{ba}'$.

(b) Repeat the calculations in (a) using **R**.

(2) Let $\boldsymbol{x} = \begin{pmatrix} 2 \\ 2 \\ -3 \end{pmatrix}$ and $\boldsymbol{y} = \begin{pmatrix} 1 \\ -2 \\ 1 \end{pmatrix}$.

(a) Which of $\boldsymbol{a}, \boldsymbol{b}, \boldsymbol{u}, \boldsymbol{v}$ in Exercise (1) are orthogonal to $\boldsymbol{x}$?

(b) Which of $\boldsymbol{a}, \boldsymbol{b}, \boldsymbol{u}, \boldsymbol{v}$ in Exercise (1) are orthogonal to $\boldsymbol{y}$?

(c) Check the answers to (a) and (b) using **R**.

(3) Let $\boldsymbol{A} = \begin{pmatrix} 1 & 2 & 3 \\ 4 & 5 & 6 \end{pmatrix}$, $\boldsymbol{B} = \begin{pmatrix} 1 & 2 \\ 3 & 4 \\ 5 & 6 \end{pmatrix}$, $\boldsymbol{U} = \begin{pmatrix} 1 & 2 \\ 3 & 4 \end{pmatrix}$, $\boldsymbol{V} = \begin{pmatrix} 5 & 6 \\ 7 & 8 \end{pmatrix}$, $\boldsymbol{W} = \begin{pmatrix} 2 & 2 \\ 3 & 5 \end{pmatrix}$ and $\boldsymbol{Z} = \begin{pmatrix} 3 & 2 \\ 3 & 6 \end{pmatrix}$ (and use the vectors from Exercises (1) and (2)).

(a) Find $\boldsymbol{AB}$, $\boldsymbol{B}'\boldsymbol{A}'$, $\boldsymbol{BA}$, $\boldsymbol{a}'\boldsymbol{A}$, $\boldsymbol{a}'\boldsymbol{Aa}$, $\boldsymbol{V}\text{diag}(\boldsymbol{U})$, $\text{diag}(\boldsymbol{B}'\boldsymbol{A}')$, $\boldsymbol{UVWZ}$, $\text{diag}(\text{diag}(\boldsymbol{UV}))$, $\text{diag}((\text{diag}(\boldsymbol{U})))\text{diag}(\text{diag}(\boldsymbol{V}))$.

(b) Verify that $\boldsymbol{U}$ and $\boldsymbol{V}$ do not commute but $\boldsymbol{U}$ and $\boldsymbol{W}$ commute and $\boldsymbol{U}$ and $\boldsymbol{Z}$ commute. Do $\boldsymbol{W}$ and $\boldsymbol{Z}$ commute? (Guess and verify.)

(4) Use the matrices from Exercise (3), and let $\boldsymbol{z} = (2,5)'$.

(a) Calculate $\boldsymbol{z}'\boldsymbol{Uz}$, $\boldsymbol{z}$'$\boldsymbol{Vz}$,$\boldsymbol{x}'\boldsymbol{BAx}$ and $\boldsymbol{x}'\boldsymbol{A}'\boldsymbol{B}'\boldsymbol{x}$.

(b) Write the four results in the form $\boldsymbol{x}'\boldsymbol{Sx}$ where $\boldsymbol{S}$ is ***symmetric***.

(5) Let

$\boldsymbol{A} = \begin{pmatrix} 0 & 1 \\ -1 & 0 \end{pmatrix}$, $\boldsymbol{B} = \begin{pmatrix} 0 & 1 \\ 0 & 0 \end{pmatrix}$, $\boldsymbol{C} = \begin{pmatrix} 1 & 1 \\ 1 & -1 \end{pmatrix}$, $\boldsymbol{D} = \begin{pmatrix} 1 & -1 \\ -1 & 1 \end{pmatrix}$, $\boldsymbol{E} = \begin{pmatrix} 1 & 1 \\ 1 & 1 \end{pmatrix}$ and $\boldsymbol{F} = \begin{pmatrix} 1 & 1 \\ -1 & -1 \end{pmatrix}$ then show

(a) $\boldsymbol{A}^2 = -\mathbf{I_2}$ (so $\boldsymbol{A}$ is 'like' the square root of -1).

(b) $\boldsymbol{B}^2 = 0$ (but $\boldsymbol{B} \neq 0$), i.e.,$\boldsymbol{B}$ is nilpotent; see §2.5.7.

(c) $\boldsymbol{CD} = -\boldsymbol{DC}$ (but $\boldsymbol{CD} \neq 0$).

(d) $\boldsymbol{EF} = 0$ (but $\boldsymbol{E} \neq 0$ and $\boldsymbol{F} \neq 0$).

(6) Show that $\text{tr}(\boldsymbol{xy}') = \boldsymbol{x}'\boldsymbol{y}$.

(7) Use the **R** help system to find out what the **R** functions `rep(.)` and `seq(.)` do by typing `help(rep)` and `help(seq)`.

(8) (a) Construct the sum vector ι_4 in **R**.

(b) Construct the identity matrix $\mathbf{I_5}$.

(c) Construct the vector $\boldsymbol{e}_4$ of length 23 (i.e.,the vector of length 23 with a 1 in the fourth place and zeros elsewhere).

(d) Construct the unit matrix $\mathbf{J_6}$.

(e) Construct the centering matrix $\mathbf{H_3}$.

(f) Construct the vector containing all even numbers in order from 2 to 28.

(9) Suppose $\boldsymbol{A}$ is a non-singular $n \times n$ idempotent matrix. Show that $\mathbf{I_n} - \boldsymbol{A}$ is idempotent.

(10) Suppose $\boldsymbol{A}$ is a non-singular $n \times n$ idempotent matrix. Show that $\boldsymbol{A} = \mathbf{I_n}$.

(11) Suppose $\boldsymbol{A}$ and $\boldsymbol{B}$ are idempotent matrices. Show that $(\boldsymbol{A}+\boldsymbol{B})$ is idempotent if and only if $\boldsymbol{AB} = \boldsymbol{BA} = \mathbf{0}$.

(12) If $\boldsymbol{A}$ is either symmetric or skew-symmetric show that $\boldsymbol{A}^2$ is symmetric.

(13) Suppose $\boldsymbol{x}_1, \boldsymbol{x}_2, \ldots, \boldsymbol{x}_n$ are p-dimensional observations with sample mean and variance $\bar{\boldsymbol{x}}_n$ and $\boldsymbol{S}_n$ and nfx_n is a further observation. Show that the sample mean and variance of the augmented sample $\boldsymbol{x}_1, \boldsymbol{x}_2, \ldots, \boldsymbol{x}_n, \boldsymbol{x}_{n+1}$ are given by

$$\bar{\boldsymbol{x}}_{n+1} = \frac{n\bar{\boldsymbol{x}}_n + \boldsymbol{x}_{n+1}}{n+1} \quad \text{and} \quad \boldsymbol{S}_{n+1} = \frac{1}{n}\left\{(n-1)\boldsymbol{S}_n + \frac{n}{n+1}(\boldsymbol{x}_{n+1} - \bar{\boldsymbol{x}}_n)(\boldsymbol{x}_{n+1} - \bar{\boldsymbol{x}}_n)'\right\}.$$

(These are known as the ***updating formulæ*** for mean and variance. They are appreciably more numerically stable when calculating sample variances for large quantities of data since the formulæ avoid the subtraction of two similarly sized large numbers.)

2.13.1 Notes

Exercise (5) illustrates that some rules of scalar multiplication do not carry over to matrix multiplication. However there are some analogies:

(i) If a real square matrix $\boldsymbol{A}$ is such that $\boldsymbol{A}'\boldsymbol{A} = 0$, then we must have $\boldsymbol{A} = 0$ and the $(i,j)^{th}$ element of $\boldsymbol{A}'\boldsymbol{A}$ is $\sum_{k=1}^{n} a_{kj}^2$ so if $\boldsymbol{A}'\boldsymbol{A} = 0$, then in particular the diagonal elements of $\boldsymbol{A}'\boldsymbol{A}$ are all zero so we must have $\sum_{k=1}^{n} a_{kj}^2 = 0$ and so $a_{kj} = 0$ for all k and j and so $\boldsymbol{A} = 0$.

(ii) $\boldsymbol{AB} = 0$ if and only if $\boldsymbol{A}'\boldsymbol{AB} = 0$ since if $\boldsymbol{A}'\boldsymbol{AB} = 0$ then $\boldsymbol{B}'\boldsymbol{A}'\boldsymbol{AB} = 0$ so $(\boldsymbol{AB})'(\boldsymbol{AB}) = 0$ and the results follow from note (i).

(iii) $\boldsymbol{AB} = \boldsymbol{AC}$ if and only if $\boldsymbol{A}'\boldsymbol{AB} = \boldsymbol{A}'\boldsymbol{AC}$ which follows by replacing $\boldsymbol{B}$ by $\boldsymbol{B} - \boldsymbol{C}$ in note (ii).

3

Rank of Matrices

3.1 Introduction and Definitions

An $m \times n$ matrix $\boldsymbol{X}$ has n columns, $\boldsymbol{x}_1, \boldsymbol{x}_2, \ldots, \boldsymbol{x}_n$, each of which are [column] vectors of length m (or more technically they are elements of $\Re^m$) and it has m rows, all of which are [row] vectors of length n. Two vectors $\boldsymbol{x}_1$ and $\boldsymbol{x}_2$ are ***linearly independent*** if $a_1\boldsymbol{x}_1 + a_2\boldsymbol{x}_2 = 0$ (where a_1 and a_2 are real numbers) implies $a_1 = a_2 = 0$. A set of vectors $\boldsymbol{x}_1, \boldsymbol{x}_2, \ldots, \boldsymbol{x}_r$ is linearly independent if $\sum a_i \boldsymbol{x}_i = 0$ implies all $a_i = 0$ or, in words, they are linearly independent *if there are no non-trivial linear combinations of them which equal zero.*

The **column rank** of $\boldsymbol{X}$ is the maximum number of linearly independent columns of $\boldsymbol{X}$. The **row rank** of $\boldsymbol{X}$ is the maximum number of linearly independent rows of $\boldsymbol{X}$. The row rank of $\boldsymbol{X}$ is clearly the same as the column rank of $\boldsymbol{X}'$ (the transpose of $\boldsymbol{X}$).

A key theorem, which is non-trivial to prove, is that the row rank and the column rank of a matrix are equal. Thus we can talk unambiguously about the **rank** of a matrix $\boldsymbol{X}$ (written $\rho(\boldsymbol{X})$) without specifying whether we mean row rank or column rank. (The most straightforward proof relies on the notions of a dimension of a vector space and is beyond the immediate needs of this introductory algebraic material for statistics.) Given this result, clearly $\rho(\boldsymbol{A}) = \rho(\boldsymbol{A}')$ and further it can be shown that $\rho(\boldsymbol{A}\boldsymbol{A}') = \rho(\boldsymbol{A}'\boldsymbol{A}) = \rho(\boldsymbol{A}) = \rho(\boldsymbol{A}')$ (e.g. Banerjee and Roy, 2014, p. 132).

Clearly the [column-] rank of $\boldsymbol{X} \leq n$ and also the [row-] rank of $\boldsymbol{X} \leq m$ so we have $\rho(\boldsymbol{X}) \leq \min(m, n)$. The rank of the zero matrix $\boldsymbol{0}$ is zero and $\rho(\boldsymbol{X}) = 0$ only if $\boldsymbol{X} = \boldsymbol{0}$.

Example 3.1

(1) Let $\boldsymbol{X} = \begin{pmatrix} 1 & 3 & 5 \\ 2 & 4 & 6 \end{pmatrix}$ so $\boldsymbol{X}$ is a 2×3 matrix so $\rho(\boldsymbol{X}) \leq \min(2, 3) = 2$, so $\rho(\boldsymbol{X}) =$ either 1 or 2. If $\rho(\boldsymbol{X}) = 1$ then the rows of $\boldsymbol{X}$ are *linearly dependent*, i.e., there are constants a_1 and a_2 such that $a_1(1, 3, 5) + a_2(2, 4, 6) = 0$. Thus we need $a_1 + 2a_2 = 0, 3a_1 + 4a_2 = 0$ and $5a_1 + 6a_2 = 0$. Subtracting three times the first equation from the second yields $2a_2 = 0$ so we have $a_1 = a_2 = 0$ and so the rows of $\boldsymbol{X}$ are linearly independent and $\rho(\boldsymbol{X}) \geq 2$, thus $\rho(\boldsymbol{X}) = 2$.

(2) Let $X = \begin{pmatrix} 4 & 6 \\ 6 & 9 \end{pmatrix}$. So X is 2×2 so $\rho(X) \leq 2$. If $a_1(4,6) + a_2(6,9) = 0$ (i.e., $2a_1 + 3a_2 = 0$) then we have $4a_1 + 6a_2 = 0$ and $6a_1 + 9a_2 = 0$, i.e., $2a_1 + 3a_2 = 0$ (again) so we can take $a_1 = 3$ and $a_2 = -2$ and so the columns of X are linearly dependent and thus $\rho(X) < 2$, but $\rho(X) \geq 1$ and so we conclude $\rho(X) = 1$.

(3) Let $X = \begin{pmatrix} 1 & 2 & 3 & 2 \\ 4 & 5 & 6 & -1 \\ 5 & 7 & 9 & 1 \end{pmatrix}$. Then X is 3×4 so $\rho(X) \leq \min(3,4) = 3$. Looking at X it is easy to see that the first row plus the second row is equal to the third row so the rows are not linearly independent, thus $\rho(X) < 3$. If $\rho(X) = 1$ then each row must be a multiple (possibly fractional) of every other row and again it is easy to see that this is not so and thus $\rho(X) \geq 2$ and we conclude that $\rho(X) = 2$.

(4) Let $X = \begin{pmatrix} 1 & 2 & 3 \\ 5 & 1 & 5 \\ 6 & 4 & 5 \\ 3 & 1 & 4 \end{pmatrix}$. Then X is 4×3 so $\rho(X) \leq \min(4,3) = 3$. It is easy to see that the first two columns of X are linearly independent (otherwise one would be a multiple of the other) (so certainly $\rho(X) \geq 2$) but not so easy to tell whether all three columns are linearly independent. Suppose $X(a_1, a_2, a_3)' = 0$ then we have $a_1 + 2a_2 + 3a3 = 0$, $5a_1 + a_2 + 5a_3 = 0$, $6a_1 + 4a_2 + 5a_3 = 0$ and $3a_1 + a_2 + 4a_3 = 0$. Subtracting multiples of the first from the second and third gives $-9a_2 - 10a_3 = 0$ and $-8a_2 - 13a_3 = 0$. Eliminating a_2 from these shows $a_3 = 0$ and hence $a_2 = a_3 = 0$ and so the columns of X are linearly independent and thus $\rho(X) = 3$.

(5) Let $X = \begin{pmatrix} 1 & 5 & 6 \\ 2 & 6 & 8 \\ 7 & 1 & 8 \end{pmatrix}$. Then X is 3×3 so $\rho(X) \leq 3$ necessarily and it is easy to see that the first two columns of X are linearly independent and so $\rho(X) \geq 2$. Suppose $X(a_1, a_2, a_3)' = 0$, then we have $a_1 + 5a_2 + 6a_3 = 0$, $2a_1 + 6a_2 + 8a_3 = 0$ and $7a_1 + a_2 + 8a_3 = 0$. Using the first equation to eliminate a_1 from the second and third gives $-4a_2 - 4a_3 = 0$ and $-34a_2 - 34a_3 = 0$ and so we can take $(a_1, a_2, a_3)' = (1, 1, -1)$ to satisfy $X(a_1, a_2, a_3)' = 0$ non-trivially showing $\rho(X) < 3$, thus $\rho(X) = 2$.

3.1.1 Notes

An $m \times n$ matrix X with $\rho(X) = \min(m,n)$ is said to be of ***full rank*** (sometimes full row rank or full column rank as appropriate). It is clear that it is not always easy to determine the rank of a matrix, nor even whether it is of full rank, using elementary definitions as above. In practice, the easiest method is to use results which come later in these notes. In particular, to determine whether a square $n \times n$ matrix is of full rank, one can evaluate its determinant (using `det(X)` in **R**) and the result that the determinant is non-zero if and only if X is of full rank; see §4.3.2 on Page 62. To find the exact rank of any symmetric square matrix the result that the rank is equal to the number of non-zero eigenvalues (see §6.4.7 on Page 91) is useful and this can be easily checked in **R** with the function `eigen(X)`. For general $m \times n$ matrices a lower bound for the rank is provided by the number of non-zero singular values of $\boldsymbol{X}$ given by `svd(X)`, see §6.7.2, or as the number of non-zero eigenvalues either $\boldsymbol{XX}'$ or $\boldsymbol{X}'\boldsymbol{X}$; see §6.4.7. Some downloadable **R** packages contain routines for calculating the rank of any matrix (e.g., `rk(X)` in package `fBasics`). In fact the function `qr(X)` will return the rank of $\boldsymbol{X}$ as a byproduct of obtaining the $\boldsymbol{QR}$ decomposition of $\boldsymbol{X}$; see §8.2.1.

Example 3.2

(1) If $\mathbf{I_n}$ is the $n \times n$ identity matrix (i.e.,$n \times n$, diagonal elements all 1 and all off-diagonal elements 0, then $\rho(\mathbf{I_n}) = n$ since $\mathbf{I_n}\boldsymbol{a} = \boldsymbol{a}$ for all vectors $\boldsymbol{a}$ so $\mathbf{I_n}\boldsymbol{a} = 0$ implies $\boldsymbol{a} = \boldsymbol{0}$ and the rows of $\mathbf{I_n}$ are linearly independent and it is thus of full rank n.

(2) If $\boldsymbol{D}$ is a diagonal matrix, then $\rho(\boldsymbol{D})$ = number of non-zero diagonal elements in $\boldsymbol{D}$. This follows by an argument similar to that in (1).

(3) $\rho(\boldsymbol{X}) = 0$ if and only if $\boldsymbol{X} = \mathbf{0}$. If the rank is zero, then there are no linearly independent columns and so $\boldsymbol{X} = \mathbf{0}$. If $\boldsymbol{X} = \mathbf{0}$, then any column $\boldsymbol{x}_i$ of $\boldsymbol{X}$ is $\boldsymbol{0}$ and so we have $\boldsymbol{ax}_i = 0$ for any $\boldsymbol{a}$ (including at least one $\boldsymbol{a} \neq \boldsymbol{0}$) so $\boldsymbol{X}$ has no linearly independent columns and thus $\rho(\boldsymbol{X}) = 0$.

(4) $\rho(\lambda\boldsymbol{X}) = \lambda\rho(\boldsymbol{X})$ if $\lambda \neq 0$, [obviously].

3.2 Rank Factorization

3.2.1 Matrices of rank 1

Suppose $\boldsymbol{X}$ is $m \times n$ and $\rho(\boldsymbol{X}) = 1$; then let the columns of $\boldsymbol{X}$ be $\boldsymbol{x}_1, \boldsymbol{x}_2, \ldots, \boldsymbol{x}_n$ and suppose (with no loss in generality) that $\boldsymbol{x}_1 \neq 0$. Since $\boldsymbol{X}$ has rank 1, every column $\boldsymbol{x}_j$ $(2 \leq j \leq n)$ of $\boldsymbol{X}$ must be linearly dependent on $\boldsymbol{x}_1$. So for each j, $2 \leq j \leq n$ we have $a_1\boldsymbol{x}_1 + a_j\boldsymbol{x}_j = 0$, or $\boldsymbol{x}_j = -(a_1/a_j)\boldsymbol{x}_1$, noting that $a_j \neq 0$ because if $a_j = 0$ we have $a_1\boldsymbol{x}_1 = 0$ which implies $a_1 = 0$ because we know $\boldsymbol{x}_1 \neq 0$ but we cannot have both a_1 and a_j zero. Thus each column of $\boldsymbol{X}$ is a multiple of the first and we can write $\boldsymbol{X} = (a_1\boldsymbol{x}_1, a_2\boldsymbol{x}_1, \ldots, a_n\boldsymbol{x}_1) = \boldsymbol{x}_1\boldsymbol{a}'$ which is of the form $\boldsymbol{xy}'$ where $\boldsymbol{x}$ is an m-vector and $\boldsymbol{y}$ an n-vector, i.e., the outer product of $\boldsymbol{x}$ and $\boldsymbol{y}'$. Conversely if $\boldsymbol{X} = \boldsymbol{xy}'$, then $\boldsymbol{X} = (y_1\boldsymbol{x}, y_2\boldsymbol{x}, \ldots, y_n\boldsymbol{x})$ and so all columns are linearly dependent upon on an $m \times 1$ vector and so $\rho(\boldsymbol{X}) = 1$.

Thus if a matrix is of rank 1, it can be written as $\boldsymbol{xy}'$ for some vectors $\boldsymbol{x}$ and $\boldsymbol{y}$, (i.e., the outer product of $\boldsymbol{x}$ and $\boldsymbol{y}$; see §2.3.1 on Page 25). Clearly the converse is also true.

3.2.2 $m \times n$ matrices of rank r

The above is a special case of the result that any $m \times n$ matrix of rank r can be written as $\boldsymbol{UV}'$ where $\boldsymbol{U}$ is $m \times r$ and $\boldsymbol{V}$ is $n \times r$ and each has rank r. Since $\boldsymbol{X}$ is of rank r it has r linearly independent columns, say $\boldsymbol{u}_1, \boldsymbol{u}_2, \ldots, \boldsymbol{u}_r$ and each column $\boldsymbol{x}_i$ of $\boldsymbol{X}$ is a linear combination of these, so $\boldsymbol{x}_i = \sum_{j=1}^{r} v_{ij}\boldsymbol{u}_j$ for some constants v_{ij}. Letting $\boldsymbol{U} = (\boldsymbol{u}_1, \boldsymbol{u}_2, \ldots, \boldsymbol{u}_r)$ and $\boldsymbol{V} = (v_{ij})$, the result follows. In passing, note that if $\boldsymbol{V} = (\boldsymbol{v}_1, \boldsymbol{v}_2, \ldots, \boldsymbol{v}_r)$ we have $\boldsymbol{X} = \sum_{j=1}^{r} \boldsymbol{u}_j\boldsymbol{v}'_j$, a sum of r $m \times n$ matrices each of rank 1.

3.3 Rank Inequalities

3.3.1 Sum and difference

$\rho(\boldsymbol{X}+\boldsymbol{Y}) \leq \rho(\boldsymbol{X}) + \rho(\boldsymbol{Y})$ because if $\rho(\boldsymbol{X}) = r$ and $\rho(\boldsymbol{Y}) = s$ then let $\boldsymbol{x}_1, \boldsymbol{x}_2, \ldots, \boldsymbol{x}_r$ be r linearly independent columns of $\boldsymbol{X}$ and $\boldsymbol{y}_1, \boldsymbol{y}_2, \ldots, \boldsymbol{y}_s$ be s linearly independent columns of $\boldsymbol{Y}$. Then, since every column of $\boldsymbol{X}$ can be expressed as a linear combination of the $\boldsymbol{x}_i$ and likewise every column of $\boldsymbol{Y}$ in terms of the $\boldsymbol{y}_j$, every column of $\boldsymbol{X}$+$\boldsymbol{Y}$ can be expressed in terms of a linear combination of the $r+s$ vectors $\boldsymbol{x}_1, \boldsymbol{x}_2, \ldots, \boldsymbol{x}_r, \boldsymbol{y}_1, \boldsymbol{y}_2, \ldots, \boldsymbol{y}_s$. So $\rho(\boldsymbol{X}+\boldsymbol{Y}) \leq \rho(\boldsymbol{X}) + \rho(\boldsymbol{Y})$.

$\rho(\boldsymbol{X}-\boldsymbol{Y}) \geq |\rho(\boldsymbol{X}) - \rho(\boldsymbol{Y})|$ follows from above by replacing $\boldsymbol{X}$ by $\boldsymbol{X}-\boldsymbol{Y}$ and noting $\rho(\boldsymbol{X}-\boldsymbol{Y}) = \rho(\boldsymbol{Y}-\boldsymbol{X})$.

Note also that $\rho(\boldsymbol{X}-\boldsymbol{Y}) = \rho(\boldsymbol{X}+(-\boldsymbol{Y})) \leq \rho(\boldsymbol{X}) + \rho(\boldsymbol{Y}) = \rho(\boldsymbol{X}) + \rho(\boldsymbol{Y})$, i.e., $\rho(\boldsymbol{X}-\boldsymbol{Y}) \leq \rho(\boldsymbol{X}) + \rho(\boldsymbol{Y})$

3.3.2 Products

$\rho(\boldsymbol{XY}) \leq \min(\rho(\boldsymbol{X}), \rho(\boldsymbol{Y}))$ because if $\boldsymbol{y}_1, \boldsymbol{y}_2, \ldots, \boldsymbol{y}_r$ are a set of r linearly independent columns of $\boldsymbol{Y}$ (where $\rho(\boldsymbol{Y}) = r$ and presuming that these are the first r columns of Y, without losing generality), any column $\boldsymbol{y}_j$ of $\boldsymbol{Y}$ can be expressed as a linear combination of these r columns. If the columns of $\boldsymbol{Z} = \boldsymbol{XY}$ are $z_1, z_2, \ldots$ and noting $z_j = \boldsymbol{X}\boldsymbol{y}_j$, any column z_j of $\boldsymbol{Z}$ can be expressed as a linear combination of $z_1, z_2, \ldots, z_r$ and so $\rho(\boldsymbol{Z}) \leq r = \rho(\boldsymbol{Y})$. Similarly $\rho(\boldsymbol{X}) \leq \rho(\boldsymbol{Z})$ and we have $\rho(\boldsymbol{XY}) \leq \min(\rho(\boldsymbol{X}), \rho(\boldsymbol{Y}))$.

3.3.2.1 Product with orthogonal matrix

If $\boldsymbol{C}$ is an orthogonal matrix then $\rho(\boldsymbol{AC}) = \rho(\boldsymbol{A})$ because $\rho(\boldsymbol{A}) = \rho(\boldsymbol{ACC}') \leq \rho(\boldsymbol{AC}) \leq \rho(\boldsymbol{A})$.

3.3.3 Sub-matrices

If $\boldsymbol{A}_{ij}$ is a sub-matrix of $\boldsymbol{A}$ then $\rho(\boldsymbol{A}_{ij}) \leq \rho(\boldsymbol{A})$ because if we express $\boldsymbol{A}_{ij} = \boldsymbol{EAF}$ (see §2.6.1 on Page 32) $\rho(\boldsymbol{A}_{ij}) = \rho(\boldsymbol{EAF}) \leq \rho(\boldsymbol{EA}) \leq \rho(\boldsymbol{A})$.

3.4 Rank in Statistics

In a variety of statistical applications, the rank of a matrix plays a crucial role in the form of the analysis. For example, in the simple linear model $\boldsymbol{y} = \boldsymbol{X}\beta + \varepsilon$ considered in §2.12.2, the matrix $\boldsymbol{X}$ has dimensions $n \times p$ where [usually] $n > p$. If this matrix has rank p, then the $p \times p$ matrix $\boldsymbol{X}'\boldsymbol{X}$ has rank p which we will see in the next chapter means that it possesses an inverse and this considerably simplifies obtaining useful estimates of the unknown parameters β. If $\boldsymbol{X}$ is not of full column rank, estimates may still be obtainable but may require the use of generalized inverses, a topic we return to in Chapter 8. This is closely related to considering the system of linear equations $\boldsymbol{y} = \boldsymbol{Ax}$ which may have a unique solution for $\boldsymbol{x}$ depending on the dimensions and rank of $\boldsymbol{A}$ or it may be possible to find a *least squares* solution. This also is considered in Chapter 8.

In multivariate analysis where the $n \times p$ data matrix $\boldsymbol{X}'$ of n observations of each of p random variables $\boldsymbol{x}_1, \boldsymbol{x}_2, \ldots, \boldsymbol{x}_p$, the rank of $\boldsymbol{X}'$ is important because certain statistical techniques rest on $\boldsymbol{X}'$ having full column rank, i.e., whether the columns of $\boldsymbol{X}'$ are linearly independent. Note the distinction here between [mathematical] linear independence and statistical independence. It is quite possible that the p random variables $\boldsymbol{x}_i$ are statistically independent but the columns of $\boldsymbol{X}'$ are not linearly independent (e.g., if $n < p$) and conversely.

Further, if the p random variables $\boldsymbol{x}_i$ themselves are linearly dependent (i.e., there are constants a_i such that $\sum_i a_i \boldsymbol{x}_i = 0$), then inevitably the columns of observations of $\boldsymbol{X}'$ will be linearly dependent. For example if $\boldsymbol{x}_3 = \boldsymbol{x}_1 - \boldsymbol{x}_2$, the complete set of the $\boldsymbol{x}_i$ will be linearly dependent. Usually in multivariate analysis it is presumed that there are no such structural linear dependencies (since the variable $\boldsymbol{x}_3$ gives no additional information, being totally determined by the variables $\boldsymbol{x}_1$ and $\boldsymbol{x}_2$, it can be removed from the set without really changing the interpretation of and conclusions drawn from the data). Further, it is usually presumed that if $n > p$ then the columns of $\boldsymbol{X}'$ are linearly independent though, of course, since they are random observations, it is possible that by 'bad luck' they might not be, especially with random variables taking a small set of discrete values (not a situation often considered in multivariate analysis).

3.5 Exercises

(1) Let $X_1 = \begin{pmatrix} 1.3 & 9.1 \\ 1.2 & 8.4 \end{pmatrix}$, $X_2 = \begin{pmatrix} 1.2 & 9.1 \\ 1.3 & 8.4 \end{pmatrix}$, $X_3 = \begin{pmatrix} 1 & 2 & 3 \\ 2 & 1 & 9 \end{pmatrix}$

$X_4 = \begin{pmatrix} 1 & 2 \\ 3 & 9 \\ 2 & 1 \end{pmatrix}$, $X_5 = \begin{pmatrix} 1 & 2 & 9 \\ 2 & 1 & 3 \\ 9 & 3 & 0 \end{pmatrix}$ and $X_6 = \begin{pmatrix} 6 & 2 & 8 \\ 5 & 1 & 6 \\ 1 & 7 & 8 \end{pmatrix}$.

(a) What is the rank of each of $X_1, \ldots, X_6$?

(b) Find constants a_1, a_2, a_3 such that $a_1\boldsymbol{c}_{31} + a_2\boldsymbol{c}_{32} + a_3\boldsymbol{c}_{33} = 0$ where $\boldsymbol{c}_{3j}$, $j = 1,2,3$ are the three columns of X_3.

(c) Find constants a_1, a_2, a_3 such that $a_1\boldsymbol{r}_{41} + a_2\boldsymbol{r}_{42} + a_3\boldsymbol{r}_{43} = 0$ where $\boldsymbol{r}_{4j}$, $j = 1,2,3$ are the three rows of X_4.

(2) Let $X_7 = \begin{pmatrix} 4 & 5 & 6 \\ 8 & 10 & 12 \\ 12 & 15 & 18 \end{pmatrix}$ and $X_8 = \begin{pmatrix} 4 & 12 & 8 \\ 6 & 18 & 12 \\ 5 & 15 & 10 \end{pmatrix}$.

(a) Show that X_7 and X_8 are both of rank 1.

(b) Find vectors $\boldsymbol{a}$ and $\boldsymbol{b}$ such that $X_7 = \boldsymbol{ab}'$.

(c) Find vectors $\boldsymbol{u}$ and $\boldsymbol{v}$ such that $X_8 = \boldsymbol{uv}'$.

(3) Let $X_9 = X_3X_4$ and $X_{10} = X_4X_3$.

(a) Evaluate X_9 and X_{10} in **R** .

(b) What is the rank of X_9?

(c) What is the rank of X_{10}?

(4) If $\boldsymbol{x}$ is a $n \times 1$ vector show that $\rho(\boldsymbol{xx}' - \boldsymbol{x}'\boldsymbol{x}\mathbf{I_n}) < n$.

(5) If $\rho(\boldsymbol{X}) < n$ and $\boldsymbol{Xx} = \lambda\boldsymbol{x}$ show that (a) $\rho(\boldsymbol{X} + \lambda\boldsymbol{xx}') < n$ and (b) $\rho(\boldsymbol{X} + \boldsymbol{xy}') < n$ for any $n \times 1$ vector $\boldsymbol{y}$.

(6) and $\rho(\boldsymbol{AB}) = m$ show that $\rho(\boldsymbol{A}) = \rho(\boldsymbol{B}) = m$ and thus $\boldsymbol{BA}$ is singular unless both $\boldsymbol{A}$ and $\boldsymbol{B}$ are square.

(7) If $\boldsymbol{A}$ is $m \times n$ with $m \geq n$ and $\rho(\boldsymbol{A}) = n$ show that $\rho(\boldsymbol{AB}) = \rho(\boldsymbol{B})$ for any conformable matrix $\boldsymbol{B}$.

(8) If $\boldsymbol{X} = \boldsymbol{AB}$ where $\boldsymbol{A}$ is $m \times n$, $\boldsymbol{B}$ is $n \times m$ with $\rho(\boldsymbol{A}) = \rho(\boldsymbol{B}) = n$ show that $\rho(\boldsymbol{X}) = \rho(\boldsymbol{X}^2)$ if and only if $\rho(\boldsymbol{BA}) = n$.

(9) Suppose $\boldsymbol{A}$ is $m \times n$ and let $\boldsymbol{B}$ and $\boldsymbol{C}$ be $m \times m$ and $n \times n$ non-singular matrices.

(i) Show that $\rho(\boldsymbol{BAC}) = \rho(\boldsymbol{A})$.

(ii) Deduce $\rho(\boldsymbol{BA}) = \rho(\boldsymbol{AC}) = \rho(\boldsymbol{A})$.

(10) Show that $\rho\begin{pmatrix} \boldsymbol{A} & \boldsymbol{0} \\ \boldsymbol{0} & \boldsymbol{D} \end{pmatrix} = \rho(\boldsymbol{A}) + \rho(\boldsymbol{D})$.

NB: Some parts of the following exercises require use of properties (§4.3.2).

(11) Suppose $\boldsymbol{A}$ is $m \times m$ and non-singular, $\boldsymbol{D}$ is $n \times n$.
Let $\boldsymbol{Z} = \begin{pmatrix} \boldsymbol{A} & \boldsymbol{B} \\ \boldsymbol{0} & \boldsymbol{D} \end{pmatrix}$ and $\boldsymbol{W} = \begin{pmatrix} \mathbf{I_m} & -\boldsymbol{A}^{-1}\boldsymbol{B} \\ \boldsymbol{0} & \mathbf{I_n} \end{pmatrix}$.

(i) Show that $\boldsymbol{W}$ is non-singular.

(ii) Show that $\boldsymbol{ZW} = \begin{pmatrix} \boldsymbol{A} & \boldsymbol{0} \\ \boldsymbol{0} & \boldsymbol{D} \end{pmatrix}$.

(iii) Show that $\rho(\boldsymbol{Z}) = \rho(\boldsymbol{A}) + \rho(\boldsymbol{D})$.

(12) Suppose $\boldsymbol{Z} = \begin{pmatrix} \boldsymbol{A} & \boldsymbol{B} \\ \boldsymbol{0} & \boldsymbol{D} \end{pmatrix}$ with $\boldsymbol{B}$ non-singular.

(i) Show that
$$\begin{pmatrix} \mathbf{I_m} & \boldsymbol{0} \\ -\boldsymbol{DB}^{-1} & \mathbf{I_n} \end{pmatrix}\begin{pmatrix} \boldsymbol{A} & \boldsymbol{B} \\ \boldsymbol{0} & \boldsymbol{D} \end{pmatrix}\begin{pmatrix} \boldsymbol{0} & \mathbf{I_p} \\ \mathbf{I_m} & -\boldsymbol{B}^{-1}\boldsymbol{A} \end{pmatrix} = \begin{pmatrix} \boldsymbol{B} & \boldsymbol{0} \\ \boldsymbol{0} & -\boldsymbol{DB}^{-1}\boldsymbol{A} \end{pmatrix}.$$

(ii) Show that the first and third matrices in part (i) are non-singular.

(iii) Show that $\rho(Z) = \rho(\boldsymbol{B}) + \rho(\boldsymbol{DB}^{-1}\boldsymbol{A})$.

(13) Suppose $\boldsymbol{Z} = \begin{pmatrix} \boldsymbol{A} & \boldsymbol{B} \\ \boldsymbol{0} & \boldsymbol{D} \end{pmatrix}$ with $\boldsymbol{A}$ $m \times n$, $\boldsymbol{B}$ $m \times q$ and $\boldsymbol{D}$ $p \times q$ with neither $\boldsymbol{A}$, $\boldsymbol{B}$ nor $\boldsymbol{D}$ necessarily non-singular. Show that $\rho(\boldsymbol{Z}) \geq \rho(\boldsymbol{A}) + \rho(\boldsymbol{D})$.

(14) Suppose $\boldsymbol{A}$, $\boldsymbol{B}$ and $\boldsymbol{C}$ are matrices such that the product $\boldsymbol{ABC}$ is defined.

(i) Show that
$$\begin{pmatrix} \mathbf{I_m} & -\boldsymbol{A} \\ \boldsymbol{0} & \mathbf{I_n} \end{pmatrix}\begin{pmatrix} \boldsymbol{0} & \boldsymbol{AB} \\ \boldsymbol{BC} & \boldsymbol{B} \end{pmatrix}\begin{pmatrix} \mathbf{I_q} & \boldsymbol{0} \\ -\boldsymbol{C} & \mathbf{I_p} \end{pmatrix} = \begin{pmatrix} -\boldsymbol{ABC} & \boldsymbol{0} \\ \boldsymbol{0} & \boldsymbol{B} \end{pmatrix}.$$

(ii) Show that the first and third matrices in part (i) are non-singular.

(iii) Show that $\rho\begin{pmatrix} \boldsymbol{0} & \boldsymbol{AB} \\ \boldsymbol{BC} & \boldsymbol{B} \end{pmatrix} = \rho(\boldsymbol{ABC}) + \rho(\boldsymbol{B})$.

(15) Suppose $\boldsymbol{A}$, $\boldsymbol{B}$ and $\boldsymbol{C}$ are matrices such that the product $\boldsymbol{ABC}$ is defined. Show that $\rho(\boldsymbol{ABC}) \geq \rho(\boldsymbol{AB}) + \rho(\boldsymbol{BC}) - \rho(\boldsymbol{B})$ (the ***Frobenius inequality***).

(16) Suppose $\boldsymbol{A}$ is $m \times n$ and $\boldsymbol{B}$ is $n \times p$ show $\rho(\boldsymbol{AB}) \geq \rho(\boldsymbol{A}) + \rho(\boldsymbol{B}) - n$.
(This is known as ***Sylvester's inequality***.)

(17) If $\boldsymbol{A}$ is $m \times n$ and $\boldsymbol{B}$ is $n \times p$ and $\boldsymbol{AB} = \boldsymbol{0}$ show that $\rho(\boldsymbol{A}) \leq n - \rho(\boldsymbol{B})$.

(18) If $\rho(\boldsymbol{A}^k) = \rho(\boldsymbol{A}^{k+1})$ show that $\rho(\boldsymbol{A}^{k+1}) = \rho(\boldsymbol{A}^{k+2})$.

(19) If $\boldsymbol{A}$ is $n \times n$ show that there is a k, $0 < k \leq n$, such that
$\rho(\boldsymbol{A}) > \rho(\boldsymbol{A}^2) > \ldots > \rho(\boldsymbol{A}^k) = \rho(\boldsymbol{A}^{k+1}) = \ldots$.

(20) If $\boldsymbol{A}$ is $n \times n$ show that $\rho(\boldsymbol{A}^{k+1}) - 2\rho(\boldsymbol{A}^k) + \rho(\boldsymbol{A}^{k-1}) \geq 0$.

4

Determinants

4.1 Introduction and Definitions

With every square $n \times n$ matrix $\boldsymbol{A} = (a_{ij})$, there is a value $|\boldsymbol{A}|$ or det($\boldsymbol{A}$), the determinant of $\boldsymbol{A}$, calculated from its elements (a_{ij}). Although at first sight determinants may seem to be mathematical curiosities, they have many crucial roles in both mathematics and statistics, notably in calculating inverses of non-singular square matrices (Chapter 5) and in performing eigenanlyses of matrices. (Chapter 6). In some ways they provide a measure of 'size' of a matrix though this analogy should not be taken too literally.

In statistics, most interest focuses on the determinant of symmetric positive definite (or positive semi-definite) matrices, especially a variance matrix introduced in §2.12.3. Sometimes the determinant of the [population] variance matrix Σ of a p-dimensional random variable $\boldsymbol{x}$ is termed the generalized variance of $\boldsymbol{x}$. This occurs in the multivariate joint probability density function of the multivariate normal distribution (see §9.2) and when changing variables in multiple integrals, a particular determinant, the Jacobean (see §7.4), is required to complete the transformation. This arises in finding the [population] mean and variance of the multivariate nrmal distribution, see §9.2.1. In statistical experimental design (a topic beyond the scope of this text) the determinant of the symmetric matrix $\boldsymbol{X}'\boldsymbol{X}$ (where $\boldsymbol{X}$ is the design matrix) plays a role in finding certain types of optimal designs.

After some examples in **R**, we illustrate some of the basic properties, followed by considering partitioned matrices allowing some key properties to be proved. The final section §4.6.2 contains a surprising result which greatly simplifies many statistical calculations and is used in several places in Chapter 9.

Example 4.1

(i) 2×2 matrices: $\begin{vmatrix} a_{11} & a_{12} \\ a_{21} & a_{22} \end{vmatrix} = a_{11}a_{22} - a_{12}a_{21}.$

(ii) 3×3 matrices:

$$\begin{vmatrix} a_{11} & a_{12} & a_{13} \\ a_{21} & a_{22} & a_{23} \\ a_{31} & a_{32} & a_{33} \end{vmatrix}$$

$$= a_{11}a_{22}a_{33} - a_{11}a_{23}a_{32} - a_{12}a_{21}a_{33} + a_{12}a_{23}a_{31} + a_{13}a_{21}a_{32} - a_{13}a_{22}a_{31}$$

$$= a_{11}\begin{vmatrix} a_{22} & a_{23} \\ a_{32} & a_{33} \end{vmatrix} - a_{12}\begin{vmatrix} a_{21} & a_{23} \\ a_{31} & a_{33} \end{vmatrix} + a_{13}\begin{vmatrix} a_{21} & a_{22} \\ a_{31} & a_{32} \end{vmatrix}.$$

In (ii) above, the 3×3 determinant has been expanded along the first row with each element in the row multiplying the determinant of the 2×2 sub-matrix obtained by deleting the row and column containing that element. Note further that the signs alternate in the expansion. In fact, the 3×3 determinant could have been expanded along the first column:

$$= a_{11}a_{22}a_{33} - a_{11}a_{23}a_{32} - a_{12}a_{21}a_{33} + a_{13}a_{21}a_{32} + a_{12}a_{23}a_{31} - a_{13}a_{22}a_{31}$$

$$= a_{11}\begin{vmatrix} a_{22} & a_{23} \\ a_{32} & a_{33} \end{vmatrix} - a_{21}\begin{vmatrix} a_{12} & a_{13} \\ a_{32} & a_{33} \end{vmatrix} + a_{31}\begin{vmatrix} a_{12} & a_{13} \\ a_{22} & a_{23} \end{vmatrix}.$$

Note that expanding along the first column gives just the same terms as expanding along the first row but in a different order.

Again each element of the column multiplies the determinant of the 2×2 sub matrix obtained by deleting the row and column containing that element, with alternating signs.
In fact, the 3×3 determinant could be expanded using any row or column in the same way with signs alternating starting with a $+$ or a $-$ according as the row or column number is odd or even, respectively. For example, expanding along the second row gives

$$|A| = -a_{21}\left|\boldsymbol{A}_{(21)}\right| + a_{22}\left|\boldsymbol{A}_{(22)}\right| - a_{23}\left|\boldsymbol{A}_{(23)}\right|,$$

where $\boldsymbol{A}_{(ij)}$ is the matrix obtained by deleting row i and column j. The quantity $c_{ij} = (1)^{i+j}\left|\boldsymbol{A}_{(ij)}\right|$ is termed the ***cofactor*** of a_{ij}. The ***cofactor matrix*** of $\boldsymbol{A}$ is the matrix $\boldsymbol{C} = (c_{ij})$. $\boldsymbol{C}'$, the transpose of $\boldsymbol{C}$, is the ***adjoint*** of $\boldsymbol{A}$ and is denoted by $\boldsymbol{A}^{\#}$.

(iii) A 4×4 matrix can be evaluated by expanding it using any row or column with each element multiplying a 3×3 matrix.

(iv) If $\boldsymbol{A}$ is any $n \times n$ matrix then the determinant of $\boldsymbol{A}$ is $|\boldsymbol{A}| = \sum_{k=1}^{n} a_{ik}c_{ik} = \sum_{k=1}^{n} a_{kj}c_{kj}$ for any choice of i or j (remember c_{ik} is the determinant of a $(n-1) \times (n-1)$ matrix). Each c_{ik} can in turn be expressed as a sum of terms in any row and their cofactors. The final expression will contain all of the possible products of n elements of $\boldsymbol{A}$ taken from distinct rows and columns (i.e., no pair of terms in any product occurs in the same row or same column of $\boldsymbol{A}$).

4.1.1 Notes

(i) If (a_{11}, a_{21}) and (a_{12}, a_{22}) are the coordinates of points in a plane, then $|\boldsymbol{A}|$ is the area of the parallelogram formed by the vectors $(a_{11}, a_{21})'$ and $(a_{12}, a_{22})'$. Similarly if $\boldsymbol{A}$ is a 3×3 matrix, then $|\boldsymbol{A}|$ is the volume of the parallelepiped formed by the three columns of $\boldsymbol{A}$.

(ii) More generally for an $n \times n$ matrix, $|\boldsymbol{A}|$ represents the volume of the parallelotope formed by the columns of $\boldsymbol{A}$. In this sense $|\boldsymbol{A}|$ reflects the ***size of the matrix A***.

(iii) Note that each term in the expansion of the determinant is a product of n elements, no two of which are in the same row or column. The sign of the term depends on whether the sequence of columns [when expanding by rows] is an even or odd permutation of the integers $1, 2, \ldots, n$.

Example 4.2

(i) Let $\boldsymbol{X} = \begin{pmatrix} 4 & 6 \\ 6 & 9 \end{pmatrix}$ then $|\boldsymbol{X}| = 4 \times 9 - 6 \times 6 = 0$.

(ii) Let $\boldsymbol{X} = \begin{pmatrix} 3 & -1 \\ 2 & 2 \end{pmatrix}$ then $|\boldsymbol{X}| = 3 \times 2 - (-1) \times 2 = 8$.

(iii) Let $\boldsymbol{X} = \begin{pmatrix} 1 & 5 & 6 \\ 2 & 6 & 8 \\ 7 & 1 & 8 \end{pmatrix}$,
then $|\boldsymbol{X}| = 1 \times (6 \times 8 - 1 \times 8) - 2 \times (5 \times 8 - 1 \times 6) + 7 \times (5 \times 8 - 6 \times 6) = 40 - 68 + 28 = 0$.

(iv) Let $\boldsymbol{X} = \begin{pmatrix} 1 & -2 & 2 \\ 2 & 0 & 1 \\ 1 & 1 & -2 \end{pmatrix}$, then (expanding by the middle row)
$|\boldsymbol{X}| = -2 \times ((-2) \times (-2) - 2) + 0 - 1 \times (1 \times 1 - (-2) \times 1) = -4 - 3 = -7$.

(v) If $\boldsymbol{A} = \mathbf{I_n}$ then $|\boldsymbol{A}| = 1$ (expand successively by any row or any column).

4.2 Implementation in R

The determinant of a square matrix in **R** is provided by the function `det(.)`.

Example 4.3

```
> det(matrix(c(1,0,0,1),2,2,byrow=T))
[1] 1
> det(matrix(c(1,0,0,0,1,0,0,0,1),3,3,byrow=T)
+ )
[1] 1
> det(matrix(c(4,6,6,9),2,2,byrow=T))
[1] -2e-15
> det(matrix(c(3,-1,2,2),2,2,byrow=T))
[1] 8
> det(matrix(c(1,5,6,2,6,8,7,1,8),3,3,byrow=T))
[1] -4.64e-14
> det(matrix(c(1,-2,2,2,0,1,1,1,-2),3,3,byrow=T))
[1] -7
```

(Note that these are the matrices considered in the section above. Values such as `-2E-15` and `-4.64e-14` should be taken as zero and may appear to be slightly different on different platforms.)

4.3 Properties of Determinants

4.3.1 Elementary row and column operations

(i) If we multiply a single row(or column) of $\boldsymbol{A}$ by a scalar λ, then the determinant is multiplied by λ (follows from definition).

(ii) If we interchange two rows (columns) of a matrix A, then the determinant but not absolute value changes sign (proof not given, see examples below).

(iii) If we add a scalar multiple of one row (column) to another row (column) the determinant does not change (proof not given, see examples below). This is useful in evaluation of matrices.

4.3.2 Other properties of determinants

(i) $|\boldsymbol{A}'| = |\boldsymbol{A}|$ (this follows from the fact that a determinant can be expanded either by rows or by columns).

(ii) If $\boldsymbol{A}$ is $n \times n$ and λ is a scalar then $|\lambda \boldsymbol{A}| = \lambda^n |\boldsymbol{A}|$ (this follows directly from the definition).

(iii) If a complete row [or column] of $\boldsymbol{A}$ consists of zeroes (i.e., if $a_{ij}=0$ for all j [or for all i]), then $|\boldsymbol{A}|=0$ (consider expanding the determinant by that row [column]).

(iv) If $\boldsymbol{A}$ has two identical rows [columns] then $|\boldsymbol{A}|=0$ (replace one of the rows [columns] by the difference between the two identical rows [columns]).

(v) If $\boldsymbol{A}$ is $n\times n$ and $\rho(\boldsymbol{A})<n$ then $|\boldsymbol{A}|=0$ (if $\boldsymbol{A}$ is not of full rank then there is a linear combination of rows [columns] that is zero, so replace any row [column] by this linear combination). The converse is also true, i.e., if $\rho(\boldsymbol{A})=n$, then $|\boldsymbol{A}|\neq 0$ (see next chapter).

(vi) If $\boldsymbol{D}=\mathrm{diag}(d_1,d_2,\ldots,d_n)$, i.e., diagonal matrix with elements $d_1,d_2,\ldots,d_n$ down the diagonal, then $|\boldsymbol{D}|=d_1d_2\ldots d_n$ (expand $|\boldsymbol{D}|$ successively by leftmost columns).

(vii) If $\boldsymbol{T}$ is a triangular matrix with elements $t_1,t_2,\ldots,t_n$ down the diagonal (i.e.,if $\boldsymbol{T}$ is upper [lower] triangular then all elements below [above] the diagonal are zero) then $|\boldsymbol{T}|=t_1t_2\ldots t_n$ (expand $|\boldsymbol{T}|$ successively by leftmost [rightmost] columns).

(viii) $|\boldsymbol{AB}|=|\boldsymbol{A}|\,|\boldsymbol{B}|$ for $n\times n$ matrices $\boldsymbol{A}$ and $\boldsymbol{B}$ (proof given in §4.5.1 (vii) below).

4.3.3 Illustrations of properties of determinants

(i) If $\boldsymbol{X}=\begin{pmatrix}3 & -1\\ 2 & 2\end{pmatrix}$ then $\boldsymbol{X}'=\begin{pmatrix}3 & 2\\ -1 & 2\end{pmatrix}$ so $|\boldsymbol{X}'|=3\times 2-2\times(-1)=6+2=8=|\boldsymbol{X}|$.

(ii) If $\boldsymbol{X}=\begin{pmatrix}3 & -1\\ 2 & 2\end{pmatrix}$ then $\begin{vmatrix}3\times 3 & -1\\ 3\times 2 & 2\end{vmatrix}=3\times 3\times 2-3\times 2\times(-1)=18+6=24=3\times 8=3|\boldsymbol{X}|$.

(iii) If $\boldsymbol{X}=\begin{pmatrix}3 & -1\\ 2 & 2\end{pmatrix}$ then $\begin{vmatrix}-1 & 3\\ 2 & 2\end{vmatrix}=-1\times 2-3\times 2=-8=-|\boldsymbol{X}|$.

(iv) If $\boldsymbol{X}=\begin{pmatrix}1 & -2 & 2\\ 2 & 0 & 1\\ 1 & 1 & -2\end{pmatrix}$ and $\boldsymbol{Y}=\begin{pmatrix}2 & 0 & 1\\ 1 & -2 & 2\\ 1 & 1 & -2\end{pmatrix}$ then

(expanding by the top row)

$|\boldsymbol{Y}|=2\times(-2)\times(-2)-2)-0+1\times(1\times 1-(-2)\times 1)=4\times 3=|\boldsymbol{X}|$.

(v) If $\boldsymbol{X}=\begin{pmatrix}1 & -2 & 2\\ 2 & 0 & 1\\ 1 & 1 & -2\end{pmatrix}$ and $\boldsymbol{Y}=\begin{pmatrix}1-2\times 2 & -2 & 2\\ 2-2\times 1 & 0 & 1\\ 1-2\times(-2) & 1 & -2\end{pmatrix}$

(subtracting twice the third column from the first column) then

$$\boldsymbol{Y} = \begin{pmatrix} -3 & -2 & 2 \\ 0 & 0 & 1 \\ 5 & 1 & -2 \end{pmatrix} \text{ so } |\boldsymbol{Y}| = \begin{vmatrix} -3 & 2 \\ 5 & 1 \end{vmatrix} = -7 = |\boldsymbol{X}|.$$

(vi) $\boldsymbol{X} = \begin{pmatrix} a_{11} & a_{12} \\ a_{21} & a_{22} \end{pmatrix}$ and $\boldsymbol{Y} = \begin{pmatrix} b_{11} & b_{12} \\ b_{21} & b_{22} \end{pmatrix}$ then

$$\boldsymbol{XY} = \begin{pmatrix} a_{11}b_{11} + a_{12}b_{21} & a_{11}b_{12} + a_{12}b_{22} \\ a_{21}b_{11} + a_{22}b_{21} & a_{21}b_{12} + a_{22}b_{22} \end{pmatrix} \text{ so}$$

$$\begin{aligned} |\boldsymbol{XY}| &= a_{11}a_{21}b_{11}b_{12} + a_{11}a_{22}b_{11}b_{22} + a_{12}a_{21}b_{21}b_{12} + a_{12}a_{22}b_{21}b_{22} \\ &\quad - a_{11}a_{21}b_{11}b_{12} - a_{21}a_{12}b_{11}b_{22} - a_{11}a_{22}b_{12}b_{21} - a_{12}a_{22}b_{21}b_{22} \\ &= a_{11}a_{22}b_{11}b_{22} - a_{11}a_{22}b_{12}b_{21} - a_{21}a_{12}b_{11}b_{22} + a_{12}a_{21}b_{21}b_{12} \\ &= (a_{11}a_{22} - a_{12}a_{21})(b_{11}b_{22} - b_{12}b_{21}) = |\boldsymbol{X}|\,|\boldsymbol{Y}|. \end{aligned}$$

(vii) $\boldsymbol{X} = \begin{pmatrix} 1 & \rho & \rho & \cdots & \rho \\ \rho & 1 & \rho & \cdots & \rho \\ \vdots & \vdots & \ddots & \vdots & \vdots \\ \rho & \cdots & \cdots & \ddots & \rho \\ \rho & \rho & \cdots & \rho & 1 \end{pmatrix}$ then

$$|\boldsymbol{X}| = \begin{vmatrix} 1 & \rho & \rho & \cdots & \rho \\ \rho & 1 & \rho & \cdots & \rho \\ \vdots & \vdots & \ddots & \vdots & \vdots \\ \rho & \cdots & \cdots & \ddots & \rho \\ \rho & \rho & \cdots & \rho & 1 \end{vmatrix} = \begin{vmatrix} 1 & \rho & \rho & \cdots & \rho \\ -1+\rho & 1-\rho & 0 & \cdots & 0 \\ \vdots & \vdots & \ddots & \vdots & \vdots \\ -1+\rho & \cdots & \cdots & \ddots & 0 \\ -1+\rho & 0 & \cdots & 0 & 1-\rho \end{vmatrix}$$

(subtracting the first row from each of the subsequent rows)

$$= \begin{vmatrix} 1+(n-1)\rho & \rho & \rho & \cdots & \rho \\ 0 & 1-\rho & 0 & \cdots & 0 \\ \vdots & \vdots & \ddots & \vdots & \vdots \\ 0 & \cdots & \cdots & \ddots & 0 \\ 0 & 0 & \cdots & 0 & 1-\rho \end{vmatrix} = [1+(n-1)\rho](1-\rho)^{(n-1)}$$

(replacing the first column by the sum of all the columns and then noting the matrix is upper triangular so the determinant is the product of the diagonal elements). $\boldsymbol{X}$ is known as the ***equicorrelation matrix***.

4.4 Orthogonal Matrices

A is orthogonal if $AA' = A'A = \mathbf{I_n}$, then since $|A| = |A'|$ and $|\mathbf{I_n}| = 1$ we have that if A is orthogonal then $|A| = \pm 1$. If $|A| = +1$ then A is termed a ***rotation matrix*** and if $|A| = -1$ then A is a ***reflection matrix***.

Example 4.4 The matrices $A_1 = \begin{pmatrix} 1 & 0 \\ 0 & 1 \end{pmatrix}$, $A_2 = \begin{pmatrix} 0 & 1 \\ -1 & 0 \end{pmatrix}$ and

$A_3 = \begin{pmatrix} \cos\theta & \sin\theta \\ -\sin\theta & \cos\theta \end{pmatrix}$ are all orthogonal rotation matrices and

$B_1 = \begin{pmatrix} 1 & 0 \\ 0 & -1 \end{pmatrix}$, $B_2 = \begin{pmatrix} 0 & 1 \\ 1 & 0 \end{pmatrix}$ and $B_3 = \begin{pmatrix} \cos\theta & \sin\theta \\ \sin\theta & -\cos\theta \end{pmatrix}$

are all orthogonal reflection matrices.

4.5 Determinants of Partitioned Matrices

4.5.1 Some basic results

Consider the partitioned matrix $\begin{pmatrix} A & B \\ C & D \end{pmatrix}$ where the dimensions of the sub-matrices match suitably. Consider first some special cases where some of $A,\ldots,D$ are either $\mathbf{0}$ or identity matrices.

(i) If A and B are square matrices then [clearly] $\begin{vmatrix} A & B \\ \mathbf{0} & D \end{vmatrix} = |A|\,|D|$.

(ii) $\begin{vmatrix} \mathbf{0} & \mathbf{I_m} \\ \mathbf{I_n} & \mathbf{0} \end{vmatrix} = (-1)^{mn}$ (obtained by column interchanges to convert this matrix to the identity matrix, each interchange changes the sign of the determinant).

(iii) If B and C are both square matrices then $\begin{vmatrix} \mathbf{0} & B \\ C & \mathbf{0} \end{vmatrix} = (-1)^{mn}\,|B|\,|C|$

noting $\begin{pmatrix} \mathbf{0} & B \\ C & \mathbf{0} \end{pmatrix} = \begin{pmatrix} B & \mathbf{0} \\ \mathbf{0} & C \end{pmatrix} \begin{pmatrix} \mathbf{0} & \mathbf{I_m} \\ \mathbf{I_n} & \mathbf{0} \end{pmatrix}$.

It can be shown that if B and C are not square, then the matrix must be singular and so has zero determinant.

(iv) $\begin{vmatrix} \mathbf{I_m} & B \\ \mathbf{0} & \mathbf{I_n} \end{vmatrix} = 1$ (since the matrix is [upper] triangular).

(v) Noting (if $\boldsymbol{C} = 0$ and $\boldsymbol{B} \neq 0$)

$$\begin{pmatrix} \mathbf{I_m} & \mathbf{0} \\ \mathbf{0} & \boldsymbol{D} \end{pmatrix} \begin{pmatrix} \mathbf{I_m} & \boldsymbol{B} \\ \mathbf{0} & \mathbf{I_n} \end{pmatrix} \begin{pmatrix} \boldsymbol{A} & \mathbf{0} \\ \mathbf{0} & \mathbf{I_n} \end{pmatrix} = \begin{pmatrix} \boldsymbol{A} & \boldsymbol{B} \\ \mathbf{0} & \boldsymbol{D} \end{pmatrix}$$

(and similarly if $\boldsymbol{B} = 0$ and $\boldsymbol{C} \neq 0$) gives $\begin{vmatrix} \boldsymbol{A} & \boldsymbol{B} \\ \mathbf{0} & \boldsymbol{D} \end{vmatrix} = \begin{vmatrix} \boldsymbol{A} & \mathbf{0} \\ \boldsymbol{C} & \boldsymbol{D} \end{vmatrix} = |\boldsymbol{A}|\,|\boldsymbol{D}|$.

(vi) $\begin{vmatrix} \boldsymbol{A} & \boldsymbol{B} \\ \mathbf{0} & \boldsymbol{D} \end{vmatrix} = (-1)^m\,|\boldsymbol{D}|\,|\boldsymbol{B}\boldsymbol{D}^{-1}\boldsymbol{C}|$ (where $\boldsymbol{B}$ is $m \times n$)

because $\begin{vmatrix} \boldsymbol{A} & \boldsymbol{B} \\ \mathbf{0} & \boldsymbol{D} \end{vmatrix} \begin{vmatrix} \mathbf{I_m} & -\boldsymbol{B}\boldsymbol{D}^{-1} \\ \mathbf{0} & \mathbf{I_n} \end{vmatrix} = \begin{vmatrix} -\boldsymbol{B}\boldsymbol{D}^{-1}\boldsymbol{C} & \mathbf{0} \\ \boldsymbol{C} & \boldsymbol{D} \end{vmatrix}$

and $|-\boldsymbol{B}\boldsymbol{D}^{-1}\boldsymbol{C}| = (-1)^m\,|\boldsymbol{B}\boldsymbol{D}^{-1}\boldsymbol{C}|$.

(vii) (Proof that $|\boldsymbol{AB}| = |\boldsymbol{A}|\,|\boldsymbol{B}|$).

Let $\boldsymbol{P} = \begin{pmatrix} \mathbf{I_n} & \boldsymbol{A} \\ \mathbf{0} & \mathbf{I_n} \end{pmatrix}$, $\boldsymbol{Q} = \begin{pmatrix} \boldsymbol{A} & \mathbf{0} \\ -\mathbf{I_n} & \boldsymbol{B} \end{pmatrix}$, $\boldsymbol{R} = \begin{pmatrix} \mathbf{0} & \boldsymbol{AB} \\ -\mathbf{I_n} & \boldsymbol{B} \end{pmatrix}$

then $\boldsymbol{PQ} = \boldsymbol{R}$, $|\boldsymbol{P}| = 1$ and $|\boldsymbol{Q}| = |\boldsymbol{A}|\,|\boldsymbol{B}|$. Now, premultiplying $\boldsymbol{Q}$ by $\boldsymbol{P}$ only multiplies the last n rows of $\boldsymbol{Q}$ by $\boldsymbol{A}$ and adds them to the first n rows, i.e., it is essentially a combination of elementary row and column operations and so this leaves the determinant unchanged, i.e., $|\boldsymbol{PQ}| = |\boldsymbol{Q}| = |\boldsymbol{A}|\,|\boldsymbol{B}|$.

Further, $|\boldsymbol{R}| = (-1)^n\,|\boldsymbol{B}|\,|\boldsymbol{ABB}^{-1}\mathbf{I_n}| = (-1)^{2n}\,|\boldsymbol{AB}| = |\boldsymbol{AB}|$.

Thus $|\boldsymbol{AB}| = |\boldsymbol{R}| = |\boldsymbol{PQ}| = |\boldsymbol{Q}| = |\boldsymbol{A}|\,|\boldsymbol{B}|$.

4.6 A Key Property of Determinants

The results in this section are of key importance in simplifying evaluation of the determinant of a sum of a single matrix with a product of two matrices. The proofs rely on non-obvious manipulation of partitioned matrices that is essentially a slick trick which needs to be memorized.

4.6.1 General result

$$\begin{vmatrix} \boldsymbol{A} & \boldsymbol{B} \\ \boldsymbol{C} & \boldsymbol{D} \end{vmatrix} = |\boldsymbol{A}|\,|\boldsymbol{D} - \boldsymbol{C}\boldsymbol{A}^{-1}\boldsymbol{B}| = |\boldsymbol{D}|\,|\boldsymbol{A} - \boldsymbol{B}\boldsymbol{D}^{-1}\boldsymbol{C}|$$

(provided $\boldsymbol{A}$ [$\boldsymbol{D}$] is non-singular because

$$\begin{vmatrix} A & B \\ C & D \end{vmatrix} = \left| \begin{pmatrix} \mathbf{I_m} & \mathbf{0} \\ -CA^{-1} & \mathbf{I_n} \end{pmatrix} \begin{pmatrix} A & B \\ C & D \end{pmatrix} \right| = \begin{vmatrix} A & B \\ \mathbf{0} & D - CA^{-1}B \end{vmatrix} \text{ and}$$

$$\begin{vmatrix} A & B \\ C & D \end{vmatrix} = \left| \begin{pmatrix} \mathbf{I_m} & -BD^{-1} \\ \mathbf{0} & \mathbf{I_n} \end{pmatrix} \begin{pmatrix} A & B \\ C & D \end{pmatrix} \right| = \begin{vmatrix} A - BD^{-1}C & \mathbf{0} \\ C & D \end{vmatrix}.$$

4.6.1.1 Note

Replacing C by $-C$ gives $|A|\,|D + CA^{-1}B| = |D|\,|A + BD^{-1}C|$.

4.6.2 Important special cases

(i) Putting $A = \mathbf{I_m}$ and $D = \mathbf{I_n}$ in the note above gives $|\mathbf{I_m} - BC| = |\mathbf{I_n} - CB|$ where B is $m \times n$ and C is $n \times m$.

(ii) Similarly we have $|\mathbf{I_m} + BC| = |\mathbf{I_n} + CB|$.

(iii) Putting $C = x$ and $B = y'$ where x and y are n-vectors gives $|\mathbf{I_n} + xy'| = |\mathbf{I_1} + y'x| = (1 + \sum xy_i)$.

(iv) Putting $y = x$ gives $|\mathbf{I_n} + xx'| = |\mathbf{I_1} + x'x| = (1 + \sum x_i^2)$.

(v) Putting $x = \mathbf{I_n}$ gives $|\mathbf{I_n} + \iota_n \iota_n{}'| = (n + 1)$.

4.7 Exercises

The following should be evaluated 'by hand' and then checked using **R** where possible.

(1) Find the determinants of $A = \begin{pmatrix} 2 & 3 \\ 2 & 4 \end{pmatrix}$ and $B = \begin{pmatrix} -4 & 3 \\ 2 & -2 \end{pmatrix}$.

(2) Find the determinant of $X = \begin{pmatrix} 4 & 5 & 6 \\ 8 & 10 & 12 \\ 12 & 15 & 18 \end{pmatrix}$.

(3) Find the determinant of $X = \begin{pmatrix} 1 & 2 & 9 \\ 2 & 1 & 3 \\ 9 & 3 & 0 \end{pmatrix}$.

(4) Find the determinants of $\boldsymbol{X} = \begin{pmatrix} 2 & 3 & 3 \\ 3 & 3 & 3 \\ 3 & 3 & 2 \end{pmatrix}$ and $\boldsymbol{Y} = \begin{pmatrix} 3 & 2 & 3 \\ 2 & 6 & 6 \\ 3 & 6 & 11 \end{pmatrix}$.

(5) Find the determinants of $\boldsymbol{S} = \begin{pmatrix} 1+\alpha & 1 & \beta \\ 1 & 1+\alpha & \beta \\ \beta & \beta & \alpha+\beta^2 \end{pmatrix}$.

(6) Find the determinants of $\boldsymbol{S} = \begin{pmatrix} 2 & 1 & 3 \\ 1 & 2 & 3 \\ 3 & 3 & 10 \end{pmatrix}$.

(7) (Equicorrelation matrix) If $\boldsymbol{X} = \sigma^2 \begin{pmatrix} 1 & \rho & \rho & \cdots & \rho \\ \rho & 1 & \rho & \cdots & \rho \\ \vdots & \vdots & \ddots & \vdots & \vdots \\ \rho & \cdots & \cdots & \ddots & \rho \\ \rho & \rho & \cdots & \rho & 1 \end{pmatrix}$ show that $\boldsymbol{X} = \sigma^2[(1-\rho)\mathbf{I_n} + \rho\iota_\mathbf{n}\iota_\mathbf{n}']$ and hence that $|\boldsymbol{X}| = \sigma^{2n}[1+(n-1)\rho](1-\rho)^{(n-1)}$.

(8) If $\boldsymbol{X} = \begin{pmatrix} 1+\alpha & 1 & \beta \\ 1 & 1+\alpha & \beta \\ \beta & \beta & \alpha+\beta^2 \end{pmatrix}$ show that $|\boldsymbol{X}| = \alpha^2(2+\alpha+\beta^2)$ by showing that $\boldsymbol{X} = \alpha\mathbf{I_3} + \boldsymbol{x}\boldsymbol{x}'$ for a suitable choice of $\boldsymbol{x}$.

(9) Use the results of the previous exercise to evaluate the determinant of $\boldsymbol{S} = \begin{pmatrix} 2 & 1 & 3 \\ 1 & 2 & 3 \\ 3 & 3 & 10 \end{pmatrix}$.

(10) To show $\begin{vmatrix} \boldsymbol{A} & \boldsymbol{B} \\ \boldsymbol{B} & \boldsymbol{A} \end{vmatrix} = |\boldsymbol{A}+\boldsymbol{B}|\,|\boldsymbol{A}-\boldsymbol{B}|$ for $n \times n$ matrices $\boldsymbol{A}$ and $\boldsymbol{B}$:

(a) Show $\begin{pmatrix} \mathbf{I_n} & \mathbf{I_n} \\ \mathbf{0} & \mathbf{I_n} \end{pmatrix} \begin{pmatrix} \boldsymbol{A} & \boldsymbol{B} \\ \boldsymbol{B} & \boldsymbol{A} \end{pmatrix} = \begin{pmatrix} \boldsymbol{A}+\boldsymbol{B} & \boldsymbol{B}+\boldsymbol{A} \\ \boldsymbol{B} & \boldsymbol{A} \end{pmatrix}$.

(b) Show $\begin{pmatrix} \boldsymbol{A}+\boldsymbol{B} & \mathbf{0} \\ \mathbf{0} & \mathbf{I_n} \end{pmatrix} \begin{pmatrix} \mathbf{I_n} & \mathbf{I_n} \\ \boldsymbol{B} & \boldsymbol{A} \end{pmatrix} = \begin{pmatrix} \boldsymbol{A}+\boldsymbol{B} & \boldsymbol{B}+\boldsymbol{A} \\ \boldsymbol{B} & \boldsymbol{A} \end{pmatrix}$.

(c) Show $\begin{pmatrix} \mathbf{I_n} & \mathbf{I_n} \\ \boldsymbol{B} & \boldsymbol{A} \end{pmatrix} = \begin{pmatrix} \mathbf{I_n} & \mathbf{0} \\ \boldsymbol{B} & \mathbf{I_n} \end{pmatrix} \begin{pmatrix} \mathbf{I_n} & \mathbf{I_n} \\ \mathbf{0} & \boldsymbol{A}-\boldsymbol{B} \end{pmatrix}$.

(d) Show $\begin{vmatrix} \boldsymbol{A} & \boldsymbol{B} \\ \boldsymbol{B} & \boldsymbol{A} \end{vmatrix} = |\boldsymbol{A}+\boldsymbol{B}|\,|\boldsymbol{A}-\boldsymbol{B}|$.

(11) Let $\boldsymbol{A} = \begin{pmatrix} 2 & 3 \\ 3 & 3 \\ 3 & 3 \end{pmatrix}$ and $\boldsymbol{B} = \begin{pmatrix} 3 \\ 3 \\ 2 \end{pmatrix}$ and $\boldsymbol{X} = \begin{pmatrix} \boldsymbol{A} & \boldsymbol{B} \end{pmatrix}$.

(a) Find $\begin{vmatrix} \mathbf{A}'\mathbf{A} & \mathbf{A}'\mathbf{B} \\ \mathbf{B}'\mathbf{A} & \mathbf{B}'\mathbf{B} \end{vmatrix}$.

(b) Find $|\mathbf{A}\mathbf{A}' + \mathbf{B}\mathbf{B}'|$.

(c) Find $|\mathbf{X}|$.

(12) If $\mathbf{A}$ and $\mathbf{B}$ are $n \times m$ and $n \times (n-m)$ matrices and $\mathbf{X} = \begin{pmatrix} \mathbf{A} & \mathbf{B} \end{pmatrix}$ prove that $|\mathbf{X}|^2 = |\mathbf{A}\mathbf{A}' + \mathbf{B}\mathbf{B}'| = \begin{vmatrix} \mathbf{A}'\mathbf{A} & \mathbf{A}'\mathbf{B} \\ \mathbf{B}'\mathbf{A} & \mathbf{B}'\mathbf{B} \end{vmatrix}$.

(13) Show that

$$\begin{vmatrix} 1+\lambda_1 & \lambda_2 & \cdots & \lambda_n \\ \lambda_1 & 1+\lambda_2 & \cdots & \lambda_n \\ \vdots & \vdots & \ddots & \vdots \\ \lambda_1 & \lambda_2 & \cdots & 1+\lambda_n \end{vmatrix} = 1 + \lambda_1 + \lambda_2 + \cdots + \lambda_n.$$

(14) By considering the matrix $\begin{pmatrix} \mathbf{E} & \mathbf{0} \\ \mathbf{0} & \mathbf{I_n} \end{pmatrix}$ show that

$$\begin{vmatrix} \mathbf{E}\mathbf{A} & \mathbf{E}\mathbf{B} \\ \mathbf{C} & \mathbf{D} \end{vmatrix} = |\mathbf{E}| \begin{vmatrix} \mathbf{A} & \mathbf{B} \\ \mathbf{C} & \mathbf{D} \end{vmatrix}.$$

(15) By considering the matrix $\begin{pmatrix} \mathbf{I_m} & \mathbf{0} \\ \mathbf{E} & \mathbf{I_n} \end{pmatrix}$ show that

$$\begin{vmatrix} \mathbf{A} & \mathbf{B} \\ \mathbf{C}+\mathbf{E}\mathbf{A} & \mathbf{D}+\mathbf{E}\mathbf{B} \end{vmatrix} = \begin{vmatrix} \mathbf{A} & \mathbf{B} \\ \mathbf{C} & \mathbf{D} \end{vmatrix}.$$

(16) Suppose $\mathbf{Z} = \begin{pmatrix} \mathbf{A} & \mathbf{B} \\ \mathbf{C} & \mathbf{D} \end{pmatrix}$ and $\mathbf{Z}^{-1} = \begin{pmatrix} \mathbf{P} & \mathbf{Q} \\ \mathbf{R} & \mathbf{S} \end{pmatrix}$.

(i) Express $\mathbf{Z}\mathbf{Z}^{-1}$ in terms of the eight sub-blocks of $\mathbf{Z}$ and $\mathbf{Z}^{-1}$.

(ii) Show that

$$\begin{pmatrix} \mathbf{A} & \mathbf{B} \\ \mathbf{C} & \mathbf{D} \end{pmatrix} \begin{pmatrix} \mathbf{I_m} & \mathbf{Q} \\ \mathbf{0} & \mathbf{S} \end{pmatrix} = \begin{pmatrix} \mathbf{A} & \mathbf{0} \\ \mathbf{C} & \mathbf{I_n} \end{pmatrix}.$$

(iii) Show that

$$\begin{pmatrix} \mathbf{A} & \mathbf{B} \\ \mathbf{C} & \mathbf{D} \end{pmatrix} \begin{pmatrix} \mathbf{P} & \mathbf{0} \\ \mathbf{R} & \mathbf{I_n} \end{pmatrix} = \begin{pmatrix} \mathbf{A} & \mathbf{0} \\ \mathbf{C} & \mathbf{I_n} \end{pmatrix}.$$

(iv) Show that

$$|\mathbf{Z}| = \frac{|\mathbf{A}|}{|\mathbf{S}|} = \frac{|\mathbf{D}|}{|\mathbf{P}|}.$$

(17) If $|\mathbf{A}| \neq 0$ and $\mathbf{A}$ and $\mathbf{C}$ commute show that $\begin{vmatrix} \mathbf{A} & \mathbf{B} \\ \mathbf{C} & \mathbf{D} \end{vmatrix} = |\mathbf{A}\mathbf{D} - \mathbf{C}\mathbf{B}|$.

5

Inverses

5.1 Introduction and Definitions

In this chapter we consider inverses of matrices which play a similar role to reciprocals of scalars. We begin with the inverse of a square matrix which has full rank (see §3.1.1) since this is the easiest and most commonly occurring case in routine statistical applications. Later (§8.3) we consider generalizations to matrices which are not square and also to matrices which are not of full rank.

We begin with definitions, examples (including in **R**) and basic properties before looking at some tricks for handling matrices which have various patterns. In the fifth section we consider partitioned matrices. This section is particularly useful in both linear models and in multivariate analysis when variables might divide into distinct groups and so it is convenient to partition the design matrix or the data matrix to reflect these groups. This may lead to interpreting the variance matrix as composed of blocks down the diagonal, giving the variance matrices of the subgroups of variables and the off-diagonal blocks as covariances between variables in different groups. This is illustrated in §9.5.1. Finally, we pick up the various statistical applications introduced in §2.12 and move a step further in their development.

A square $n \times n$ matrix $\boldsymbol{A}$ is ***non-singular*** if $\rho(\boldsymbol{A}) = n$; if $\rho(\boldsymbol{A}) < n$, $\boldsymbol{A}$ is ***singular***. If $\boldsymbol{A}$ is an $n \times n$ matrix and $\boldsymbol{B}$ is also an $n \times n$ matrix such that $\boldsymbol{AB} = \boldsymbol{BA} = \mathbf{I_n}$, the $n \times n$ identity matrix, then $\boldsymbol{B}$ is the inverse of $\boldsymbol{A}$ and is denoted by $\boldsymbol{A}^{-1}$.

Example 5.1

(i) If $\boldsymbol{A} = \begin{pmatrix} 2 & 3 \\ 3 & 4 \end{pmatrix}$ and $\boldsymbol{B} = \begin{pmatrix} -4 & 3 \\ 3 & -2 \end{pmatrix}$, then $\boldsymbol{AB} = \boldsymbol{BA} = \begin{pmatrix} 1 & 0 \\ 0 & 1 \end{pmatrix} = \mathbf{I_2}$,

so $\boldsymbol{A}^{-1} = \begin{pmatrix} -4 & 3 \\ 3 & -2 \end{pmatrix}$.

(ii) If $\boldsymbol{A} = \begin{pmatrix} 3 & 4 \\ 2 & 3 \end{pmatrix}$ and $\boldsymbol{B} = \begin{pmatrix} 3 & -2 \\ -4 & 3 \end{pmatrix}$, then $\boldsymbol{AB} = \boldsymbol{BA} = \begin{pmatrix} 1 & 0 \\ 0 & 1 \end{pmatrix} = \mathbf{I_2}$,

so $\boldsymbol{A}^{-1} = \begin{pmatrix} 3 & -2 \\ -4 & 3 \end{pmatrix}$.

(iii) If $A = \begin{pmatrix} 1 & 0 & 0 \\ 0 & 2 & 3 \\ 0 & 3 & 4 \end{pmatrix}$ and $B = \begin{pmatrix} 1 & 0 & 0 \\ 0 & -4 & 3 \\ 0 & 3 & -2 \end{pmatrix}$, then

$$AB = BA = \begin{pmatrix} 1 & 0 & 0 \\ 0 & 1 & 0 \\ 0 & 0 & 1 \end{pmatrix} = \mathbf{I_3} \text{ so } A^{-1} = \begin{pmatrix} 1 & 0 & 0 \\ 0 & -4 & 3 \\ 0 & 3 & -2 \end{pmatrix}.$$

(iv) If $A = \begin{pmatrix} 3 & 0 & 4 \\ 0 & 1 & 0 \\ 2 & 0 & 3 \end{pmatrix}$ and $B = \begin{pmatrix} 3 & 0 & -4 \\ 0 & 1 & 0 \\ -2 & 0 & 3 \end{pmatrix}$, then

$$AB = BA = \begin{pmatrix} 1 & 0 & 0 \\ 0 & 1 & 0 \\ 0 & 0 & 1 \end{pmatrix} = \mathbf{I_3} \text{ so } A^{-1} = \begin{pmatrix} 3 & 0 & -4 \\ 0 & 1 & 0 \\ -2 & 0 & 3 \end{pmatrix}.$$

(v) If $X = \begin{pmatrix} a & b \\ c & d \end{pmatrix}$ and $Y = \frac{1}{ad-bc}\begin{pmatrix} d & -c \\ -b & a \end{pmatrix}$, (with $ad - bc \neq 0$)

$$\text{then } XY = YX = \begin{pmatrix} 1 & 0 \\ 0 & 1 \end{pmatrix} = \mathbf{I_3} \text{ so } X^{-1} = \frac{1}{ad-bc}\begin{pmatrix} d & -c \\ -b & a \end{pmatrix}.$$

5.1.1 Notes

(a) Note the similarities between (i) and (iii) and also between (ii) and (iv).

(b) Note that (i) and (ii) follow from the general formula in (v) for 2×2 matrices.

(c) Note that the condition $ad - bc \neq 0$ in (v) is equivalent to $|X| \neq 0$.

5.2 Properties

5.2.1 Singular and non-singular matrices

An $n \times n$ matrix A only possesses an inverse (i.e., is invertible) if $\rho(A) = n$. If B is the inverse of A, $\mathbf{I_n} = AB$, so $n = \rho(\mathbf{I_n}) = \rho(AB) \leq \min(\rho(A), \rho(B)) \leq \rho(A) \leq n$; i.e., $n \leq \rho(A) \leq n$ so $\rho(A) = n$ and A is non-singular. The converse can also be proved [using ideas of vector spaces], i.e., if A is non-singular A must possess an inverse.

5.2.2 Uniqueness

Suppose A has an inverse B and also an inverse C, $AB = \mathbf{I_n}$ and also $CA = \mathbf{I_n}$, so $C = C(\mathbf{I_n}) = C(AB) = (CA)B = (\mathbf{I_n})B = B$ and the inverse of A, A^{-1}, is unique.

5.2.3 Inverse of inverse

$(A^{-1})^{-1} = A$ because $(A)A^{-1} = A^{-1}(A) = \mathbf{I_n}$ (because A^{-1} is the inverse of A), but this also means that $A(A^{-1}) = (A^{-1})A = \mathbf{I_n}$, showing A is the inverse of A^{-1}.

5.2.4 Determinant of inverse

$1 = |\mathbf{I_n}| = |AA^{-1}| = |A|\,|A^{-1}|$, so $|A^{-1}| = |A|^{-1}$. So, if A is non-singular (i.e., $\rho(A) = n$ if A is $n \times n$) $|A| \neq 0$ (since A must possess an inverse if it is non-singular, §5.2.1).

5.2.5 Inverse of transpose

$(A')^{-1} = (A^{-1})'$ because $A'(A^{-1})' = (A^{-1}A)' = \mathbf{I_n}' = \mathbf{I_n}$ [noting that the product of transposes is the transpose of the reverse product, i.e., $X'Y' = (YX)'$, see §2.11.2.5].

5.2.6 Inverse of product

If A and B are both $n \times n$ non-singular matrices, $(AB)^{-1} = B^{-1}A^{-1}$ because $(AB)B^{-1}A^{-1} = A\mathbf{I_n}A^{-1} = AA^{-1} = \mathbf{I_n}$ and similarly $B^{-1}A^{-1}(AB) = \mathbf{I_n}$.

5.2.6.1 Rank of product with non-singular matrix

(Generalization of §3.3.2.1). If C is a non-singular matrix, $\rho(AC) = \rho(A)$ since $\rho(A) = \rho(ACC^{-1}) \leq \rho(AC) \leq \rho(A)$.

5.2.7 Orthogonal matrices

(See also §2.5.3 and §4.4). An $n \times n$ matrix A is orthogonal if $A^{-1} = A'$, i.e., if $AA' = A'A = \mathbf{I_n}$. Clearly, if A is orthogonal, so is A'.

5.2.8 Scalar products of matrices

If A is an $n \times n$ matrix and λ is any scalar (i.e., real number or constant), $(\lambda A)^{-1} = (1/\lambda)A^{-1}$, because $\lambda A(1/\lambda)A^{-1} = \lambda(1/\lambda)AA^{-1} = 1\mathbf{I_n} = \mathbf{I_n}$.

5.2.9 Left and right inverses

If A is a non-square $m \times n$ matrix (i.e., with $m \neq n$), a $n \times m$ matrix B such that $BA = \mathbf{I_n}$ is termed a ***left inverse*** of A and a $n \times m$ matrix C such that $AC = \mathbf{I_m}$ is termed a ***right inverse*** of A. It can be shown (e.g., Banerjee and Roy, 2014, Theorem 5.5) that if A is of full column rank (i.e., $\rho(A) = n$), it must possess a left inverse and if it is of full row rank (i.e., $\rho(A) = m$), it must possess a right inverse. It will be shown in §8.3.1 that in these cases the left and right inverses are given by the Moore–Penrose inverse of A which is given by $(A'A)^{-1}A'$ and $A'(AA')^{-1}$ in the two cases respectively.

5.3 Implementation in R

The inverse of a non-singular matrix is provided in **R** by the function `solve(.)`. Two other functions in libraries will also produce inverses of non-singular matrices but they are designed to produce generalized inverses (see §8.3 below) of singular and non-rectangular matrices. These functions are `ginv(A)` (in the `MASS` library) and `MPinv(A)` (in the gnm library). The function `solve(.)` will generate a warning message if the matrix is singular.

Example 5.2

```
> A<-matrix(c(2,3,3,4),
+ 2,2,byrow=T)
> A; solve(A)
     [,1] [,2]
[1,]    2    3
[2,]    3    4
     [,1] [,2]
[1,]   -4    3
[2,]    3   -2
> A%*%solve(A)
     [,1] [,2]
[1,]    1    0
[2,]    0    1
```

```
> B<-matrix(c(3,4,2,3),
+ 2,2,byrow=T)
> B; solve(B)
     [,1] [,2]
[1,]    3    4
[2,]    2    3
     [,1] [,2]
[1,]    3   -4
[2,]   -2    3
> B%*%solve(B)
     [,1] [,2]
[1,]    1    0
[2,]    0    1
```

```
> C<-matrix(c(1,0,0,0,2,3,
+ 0,3,4),3,3,byrow=T)
> C; solve(C)
     [,1] [,2] [,3]
[1,]    1    0    0
[2,]    0    2    3
[3,]    0    3    4
     [,1] [,2] [,3]
[1,]    1    0    0
[2,]    0   -4    3
[3,]    0    3   -2
> C%*%solve(C)
     [,1] [,2] [,3]
[1,]    1    0    0
[2,]    0    1    0
[3,]    0    0    1
```

```
> D<-matrix(c(3,0,4,0,1,0,
+ 2,0,3),3,3,byrow=T)
> D; solve(D)
     [,1] [,2] [,3]
[1,]    3    0    4
[2,]    0    1    0
[3,]    2    0    3
     [,1] [,2] [,3]
[1,]    3    0   -4
[2,]    0    1    0
[3,]   -2    0    3
> D%*%solve(D)
     [,1] [,2] [,3]
[1,]    1    0    0
[2,]    0    1    0
[3,]    0    0    1
```

```
> E<-matrix(c(1.3,9.1,1.2,
+8.4),2,2,byrow=T)
> E; solve(E)
     [,1] [,2]
[1,]  1.3  9.1
[2,]  1.2  8.4
Error in solve.default(E) :
system is computationally
singular: reciprocal
condition number=1.598e-17
> F<-matrix(c(1.2,9.1,1.3,
+ 8.4),2,2,byrow=T)
```

```
> F; solve(F)
     [,1] [,2]
[1,]  1.2  9.1
[2,]  1.3  8.4
       [,1]   [,2]
[1,] -4.800  5.200
[2,]  0.743 -0.686
>> F%*%solve(F)
          [,1]       [,2]
[1,] 1.000e+00 1.0096e-15
[2,] 8.630e-17 1.0000e+00
```

Note that in first example of the two immediately above the matrix is singular and the result in the second is the identity matrix $\mathbf{I}_2$ to within rounding error of order 10^{-16}. To control the number of digits printed, use the function `options(.)` with digits specified, e.g.,`options(digits=3)`.

```
> U<-matrix(c(1,2,9,2,1,3,9,
+3,0),3,3,byrow=T)
> U; solve(U)
     [,1] [,2] [,3]
[1,]    1    2    9
[2,]    2    1    3
[3,]    9    3    0
       [,1]   [,2]   [,3]
[1,] -0.500  1.500 -0.167
[2,]  1.500 -4.500  0.833
[3,] -0.167  0.833 -0.167
> U%*%solve(U)
      [,1] [,2] [,3]
[1,]  1.00 0.00 0.00
[2,]  0.00 1.00 0.00
[3,]  0.00 0.00 1.00
```

```
> V<-matrix(c(6,2,8,5,1,6,
+1,7,8),3,3,byrow=T)
> V
     [,1] [,2] [,3]
[1,]    6    2    8
[2,]    5    1    6
[3,]    1    7    8
> solve(V)
Error in solve.default(V) :
system is computationally
singular: reciprocal
condition number= .177e-18
>
```

5.4 Inverses of Patterned Matrices

If a matrix has a particular pattern, it can be the case that the inverse has a similar pattern. So, in some cases it is possible to determine the inverse by guessing the form of the inverse up to a small number of unknown constants and then determining the constants so that the product of the matrix and the inverse is the identity matrix. For example, consider the 3×3 matrix

$$X = \begin{pmatrix} 2 & 3 & 3 \\ 3 & 2 & 3 \\ 3 & 3 & 2 \end{pmatrix};$$

this has identical elements down the diagonal and all off-diagonal elements are identical (but distinct from the diagonal). It is a sensible guess to look for an inverse with the same structure:

$$Y = \begin{pmatrix} a & b & b \\ b & a & b \\ b & b & a \end{pmatrix}.$$

If $XY = \mathbf{I_3}$, we have

$$\begin{pmatrix} 2a+6b & 3a+5b & 3a+5b \\ 3a+5b & 2a+6b & 3a+5b \\ 3a+5b & 3a+5b & 2a+6b \end{pmatrix} = \mathbf{I_3} = \begin{pmatrix} 1 & 0 & 0 \\ 0 & 1 & 0 \\ 0 & 0 & 1 \end{pmatrix}$$

so we require that $2a + 6b = 1$ and $3a + 5b = 0$, so $a = -5b/3$ and so $b = 1/(6 - 10/3) = 3/8$ and $a = 5/8$. Check (in **R**):

```
> a<- -5/8 ; b<- 3/8
> X<- matrix(c(2,3,3,3,2,3,3,3,2),3,3,byrow=T)
> Y<- matrix(c(a,b,b,b,a,b,b,b,a),3,3,byrow=T)
> X%*%Y          [,1] [,2] [,3]
[1,]    1    0    0
[2,]    0    1    0
[3,]    0    0    1
```

5.4.1 Matrices of form $\alpha\mathbf{I_n} + \beta\iota_n\iota_n'$

The example above is a special case of matrices of the form

$$\alpha\mathbf{I_n} + \beta\iota_n\iota_n' = \begin{pmatrix} \alpha+\beta & \beta & \cdots & \beta \\ \beta & \alpha+\beta & \vdots & \vdots \\ \vdots & \cdots & \ddots & \beta \\ \beta & \cdots & \beta & \alpha+\beta \end{pmatrix},$$

(in above $\alpha = -1$ and $\beta = 3$). Numerical matrices of this form are easy to recognise. Recall that $\iota_n\iota_n'$ is the $n \times n$ matrix $\boldsymbol{J}$ with all elements equal to 1. If the inverse has a similar form $a\mathbf{I_n} + b\iota_n\iota_n'$, then we need to find constants a and b such that $(\alpha\mathbf{I_n} + \beta\iota_n\iota_n')(a\mathbf{I_n} + b\iota_n\iota_n') = \mathbf{I_n}$ so we need $\mathbf{I_n} = \alpha a\mathbf{I_n}\mathbf{I_n} + \alpha b\mathbf{I_n}\iota_n\iota_n' + a\beta\iota_n\iota_n'\mathbf{I_n} + \beta b\iota_n\iota_n'\iota_n\iota_n' = \alpha a\mathbf{I_n} + (\alpha b + a\beta + nb\beta)\iota_n\iota_n'$ (noting $\iota_n'\iota_n = n$, see §2.8.1). Thus we need $\alpha a + \beta b + a\beta + nb\beta = 1$ and $\alpha b + a\beta + nb\beta = 0$, so $a = \alpha^{-1}$ and $b = -\beta/\alpha(\alpha + n\beta)$.

In the numerical example above we have $a = -1$ and $b = 3/(-1+9) = 3/8$ and then the inverse is $-\mathbf{I_3} + 3\iota_3\iota_3'/8$, i.e., with diagonal elements $1 - 3/8 = 5/8$ and off-diagonal elements $3/8$.

5.4.2 Matrices of form $A + xy'$

A further generalization of above is the result that if the $n \times n$ matrix $\boldsymbol{A}$ is non-singular and $\boldsymbol{x}$ and $\boldsymbol{y}$ are n-vectors such that $\boldsymbol{y}'\boldsymbol{A}^{-1}\boldsymbol{x} \neq -1$, then we have

$$(\boldsymbol{A} + \boldsymbol{x}\boldsymbol{y}')^{-1} = \boldsymbol{A}^{-1} - \frac{1}{1+\boldsymbol{y}'\boldsymbol{A}^{-1}\boldsymbol{x}}\boldsymbol{A}^{-1}\boldsymbol{x}\boldsymbol{y}'\boldsymbol{A}^{-1}:$$

because $(\boldsymbol{A} + \boldsymbol{x}\boldsymbol{y}')(\boldsymbol{A}^{-1} - \frac{1}{1+\boldsymbol{y}'\boldsymbol{A}^{-1}\boldsymbol{x}}\boldsymbol{A}^{-1}\boldsymbol{x}\boldsymbol{y}'\boldsymbol{A}^{-1})$

$$\begin{aligned} &= \boldsymbol{A}\boldsymbol{A}^{-1} - \frac{\boldsymbol{A}\boldsymbol{A}^{-1}\boldsymbol{x}\boldsymbol{y}'\boldsymbol{A}^{-1}}{1+\boldsymbol{y}'\boldsymbol{A}^{-1}\boldsymbol{x}} + \boldsymbol{x}\boldsymbol{y}'\boldsymbol{A}^{-1} - \frac{\boldsymbol{x}\boldsymbol{y}'\boldsymbol{A}^{-1}\boldsymbol{x}\boldsymbol{y}'\boldsymbol{A}^{-1}}{1+\boldsymbol{y}'\boldsymbol{A}^{-1}\boldsymbol{x}} \\ &= \mathbf{I_n} - \frac{\mathbf{I_n}\boldsymbol{x}\boldsymbol{y}'\boldsymbol{A}^{-1}}{1+\boldsymbol{y}'\boldsymbol{A}^{-1}\boldsymbol{x}} + \frac{\boldsymbol{x}\boldsymbol{y}'\boldsymbol{A}^{-1} + \boldsymbol{x}\boldsymbol{y}'\boldsymbol{A}^{-1}\boldsymbol{y}'\boldsymbol{A}^{-1}\boldsymbol{x} - \boldsymbol{x}\boldsymbol{y}'\boldsymbol{A}^{-1}\boldsymbol{x}\boldsymbol{y}'\boldsymbol{A}^{-1}}{1+\boldsymbol{y}'\boldsymbol{A}^{-1}\boldsymbol{x}} = \mathbf{I_n}. \end{aligned}$$

Numerical matrices of this form are not easy to recognise unless the matrix $\boldsymbol{A}$ is the identity matrix or a multiple of it. It is a little easier if additionally $\boldsymbol{x} = \boldsymbol{y}$. The main use of this result is that this form arises in various theoretical developments of methodology.

Example 5.2
If $A = a\mathbf{I_n}$ and $\boldsymbol{x} = \boldsymbol{y} = (x_1, x_2, \ldots, x_n)'$,

then $\boldsymbol{A} + \boldsymbol{x}\boldsymbol{y}' = \begin{pmatrix} a+x_1^2 & x_1x_2 & \cdots & x_1x_n \\ x_2x_1 & a+x_2^2 & \cdots & x_2x_n \\ \vdots & \vdots & \ddots & \vdots \\ x_nx_1 & x_nx_2 & \cdots & a+x_n^2 \end{pmatrix}$ which is symmetric.

For example, if $a = 2$ and $\boldsymbol{x} = (1,2,3)'$, then $\boldsymbol{A} + \boldsymbol{x}\boldsymbol{x}' = \begin{pmatrix} 3 & 2 & 3 \\ 2 & 6 & 6 \\ 3 & 6 & 11 \end{pmatrix}$.

Note that $\boldsymbol{A}^{-1} = {}^1\!/_2\mathbf{I_n}$, $\boldsymbol{x}'\boldsymbol{A}^{-1}\boldsymbol{x} = 7$ and so the formula gives the inverse as

$$\begin{pmatrix} 0.5 & 0 & 0 \\ 0 & 0.5 & 0 \\ 0 & 0 & 0.5 \end{pmatrix} - \frac{1}{(1+7)\times 2\times 2}\begin{pmatrix} 1 & 2 & 3 \\ 2 & 4 & 6 \\ 3 & 6 & 9 \end{pmatrix}$$
$$= \begin{pmatrix} 0.469 & -0.063 & -0.094 \\ -0.063 & 0.375 & -0.188 \\ -0.094 & -0.188 & 0.219 \end{pmatrix}.$$

Check:

```
> X<-matrix(c(3,2,3,2,6,6,3,6,11),3,3,byrow=T)
> X
     [,1] [,2] [,3]
[1,]    3    2    3
[2,]    2    6    6
[3,]    3    6   11
```

```
> solve(X)
         [,1]    [,2]     [,3]
[1,]  0.46875 -0.0625 -0.09375
[2,] -0.06250  0.3750 -0.18750
[3,] -0.09375 -0.1875  0.21875
```

5.5 Inverses of Partitioned Matrices

5.5.1 Some basic results

Consider the partitioned matrix $\begin{pmatrix} \boldsymbol{A} & \boldsymbol{B} \\ \boldsymbol{C} & \boldsymbol{D} \end{pmatrix}$ where the dimensions of the sub-matrices match suitably and non-singularity is assumed where necessary. Consider first some special cases where some of $\boldsymbol{A}$, ..., $\boldsymbol{D}$ are either $\boldsymbol{0}$ or identity matrices. Most of the results below can be demonstrated by direct multiplication.

Recall that $\begin{pmatrix} \boldsymbol{A} & \boldsymbol{B} \\ \boldsymbol{C} & \boldsymbol{D} \end{pmatrix}\begin{pmatrix} \boldsymbol{P} & \boldsymbol{Q} \\ \boldsymbol{R} & \boldsymbol{S} \end{pmatrix} = \begin{pmatrix} \boldsymbol{AP}+\boldsymbol{BR} & \boldsymbol{AQ}+\boldsymbol{BR} \\ \boldsymbol{CP}+\boldsymbol{DR} & \boldsymbol{CQ}+\boldsymbol{DS} \end{pmatrix}$.

(i) If $\boldsymbol{A}$ and $\boldsymbol{B}$ are square matrices, then [clearly]
$$\begin{pmatrix} \boldsymbol{A} & \boldsymbol{0} \\ \boldsymbol{0} & \boldsymbol{D} \end{pmatrix}^{-1} = \begin{pmatrix} \boldsymbol{A}^{-1} & \boldsymbol{0} \\ \boldsymbol{0} & \boldsymbol{D}^{-1} \end{pmatrix}.$$

(ii) $$\begin{pmatrix} \boldsymbol{0} & \boldsymbol{B} \\ \boldsymbol{C} & \boldsymbol{0} \end{pmatrix}^{-1} = \begin{pmatrix} \boldsymbol{0} & \boldsymbol{B}^{-1} \\ \boldsymbol{C}^{-1} & \boldsymbol{0} \end{pmatrix}.$$

(iii) $$\begin{pmatrix} \boldsymbol{0} & \boldsymbol{B} \\ \boldsymbol{B}^{-1} & \boldsymbol{0} \end{pmatrix}^{-1} = \begin{pmatrix} \boldsymbol{0} & \boldsymbol{B} \\ \boldsymbol{B}^{-1} & \boldsymbol{0} \end{pmatrix}.$$

(iv) $$\begin{pmatrix} \mathbf{I_m} & \boldsymbol{B} \\ \boldsymbol{0} & \mathbf{I_n} \end{pmatrix}^{-1} = \begin{pmatrix} \mathbf{I_m} & -\boldsymbol{B} \\ \boldsymbol{0} & \mathbf{I_n} \end{pmatrix}.$$

(v) $$\begin{pmatrix} \mathbf{I_m} & \boldsymbol{0} \\ \boldsymbol{C} & \mathbf{I_n} \end{pmatrix}^{-1} = \begin{pmatrix} \mathbf{I_m} & \boldsymbol{0} \\ -\boldsymbol{C} & \mathbf{I_n} \end{pmatrix}.$$

(vi) $$\begin{pmatrix} \boldsymbol{A} & \boldsymbol{B} \\ \boldsymbol{0} & \boldsymbol{D} \end{pmatrix}^{-1} = \begin{pmatrix} \boldsymbol{A}^{-1} & -\boldsymbol{A}^{-1}\boldsymbol{B}\boldsymbol{D}^{-1} \\ \boldsymbol{0} & \boldsymbol{D}^{-1} \end{pmatrix}.$$

(vii) $$\begin{pmatrix} \boldsymbol{A} & \boldsymbol{0} \\ \boldsymbol{C} & \boldsymbol{D} \end{pmatrix}^{-1} = \begin{pmatrix} \boldsymbol{A}^{-1} & -\boldsymbol{A}^{-1}\boldsymbol{B}\boldsymbol{D}^{-1} \\ -\boldsymbol{D}^{-1}\boldsymbol{D}\boldsymbol{A}^{-1} & \boldsymbol{D}^{-1} \end{pmatrix}.$$

(viii) $$\begin{pmatrix} \boldsymbol{A} & \mathbf{I_m} \\ \mathbf{I_n} & \boldsymbol{0} \end{pmatrix}^{-1} = \begin{pmatrix} \boldsymbol{0} & \mathbf{I_m} \\ \mathbf{I_n} & -\boldsymbol{A} \end{pmatrix}.$$

(ix) $\begin{pmatrix} \mathbf{0} & \mathbf{I_m} \\ \mathbf{I_n} & \boldsymbol{D} \end{pmatrix}^{-1} = \begin{pmatrix} -\boldsymbol{D} & \mathbf{I_m} \\ \mathbf{I_n} & \mathbf{0} \end{pmatrix}$.

(x) $\begin{pmatrix} \boldsymbol{A} & \boldsymbol{B} \\ \boldsymbol{C} & \mathbf{0} \end{pmatrix}^{-1} = \begin{pmatrix} \mathbf{0} & \boldsymbol{C}^{-1} \\ \boldsymbol{B}^{-1} & -\boldsymbol{B}^{-1}\boldsymbol{A}\boldsymbol{C}^{-1} \end{pmatrix}$.

(xi) $\begin{pmatrix} \mathbf{0} & \boldsymbol{B} \\ \boldsymbol{C} & \boldsymbol{D} \end{pmatrix}^{-1} = \begin{pmatrix} -\boldsymbol{C}^{-1}\boldsymbol{D}\boldsymbol{B}^{-1} & \boldsymbol{C}^{-1} \\ \boldsymbol{B}^{-1} & \mathbf{0} \end{pmatrix}$.

(xii) $\begin{pmatrix} \boldsymbol{A} & \boldsymbol{B} \\ \boldsymbol{C} & \boldsymbol{D} \end{pmatrix}^{-1} = \begin{pmatrix} \boldsymbol{A}^{-1} + \boldsymbol{A}^{-1}\boldsymbol{B}\boldsymbol{E}^{-1}\boldsymbol{C}\boldsymbol{A}^{-1} & -\boldsymbol{A}^{-1}\boldsymbol{B}\boldsymbol{E}^{-1} \\ -\boldsymbol{E}^{-1}\boldsymbol{C}\boldsymbol{A}^{-1} & \boldsymbol{E}^{-1} \end{pmatrix}$
where $\boldsymbol{E} = \boldsymbol{D} - \boldsymbol{C}\boldsymbol{A}^{-1}\boldsymbol{B}$.

(xiii) $\begin{pmatrix} \boldsymbol{A} & \boldsymbol{B} \\ \boldsymbol{C} & \boldsymbol{D} \end{pmatrix}^{-1} = \begin{pmatrix} \boldsymbol{F}^{-1} & -\boldsymbol{F}^{-1}\boldsymbol{B}\boldsymbol{D}^{-1} \\ -\boldsymbol{D}^{-1}\boldsymbol{C}\boldsymbol{F}^{-1} & \boldsymbol{D}^{-1} + \boldsymbol{D}^{-1}\boldsymbol{C}\boldsymbol{F}^{-1}\boldsymbol{B}\boldsymbol{D}^{-1} \end{pmatrix}$
where $\boldsymbol{F} = \boldsymbol{A} - \boldsymbol{B}\boldsymbol{D}^{-1}\boldsymbol{C}$.

5.5.1.1 Notes

The matrices $\boldsymbol{E}$ and $\boldsymbol{F}$ in (xii) and (xiii) above are termed the ***Schur complements*** of $\boldsymbol{A}$ and $\boldsymbol{D}$ respectively.

5.6 General Formulae

Let $\boldsymbol{A}$ be an $n \times n$ matrix (a_{ij}) and let $\boldsymbol{C} = (c_{ij})$ be the cofactor matrix of $\boldsymbol{A}$, so $\boldsymbol{C}' = \boldsymbol{A}^{\#}$, the adjoint of $\boldsymbol{A}$; see §4.1. Then expanding the determinant by row k we have $|\boldsymbol{A}| = \sum_{j=1}^{n} a_{kj}c_{kj}$ for any k and $|\boldsymbol{A}| = \sum_{j=1}^{n} a_{jk}c_{jk}$ for any column k, (these results follow from the definition of the cofactors c_{ij} in 4.1). Let $\boldsymbol{B}$ be the matrix obtained by replacing the kth row of $\boldsymbol{A}$ by a copy of row i, then $|\boldsymbol{B}| = 0$ since it has two identical rows. Thus, expanding $|\boldsymbol{B}|$ by this row we have $\sum_{j=1}^{n} a_{kj}c_{ij} = 0$ if $i \neq k$ and similarly $|\boldsymbol{A}| = \sum_{i=1}^{n} a_{ij}c_{ik} = 0$ if $k \neq j$, i.e., $\sum_{i=1}^{n} a_{ij}c_{ik} = \delta_{jk}|\boldsymbol{A}|$ where $\delta_{jk} = 1$ or 0 as $j = k$ or $j \neq k$. Similarly $\sum_{i=1}^{n} a_{ki}c_{ji} = \delta_{jk}|\boldsymbol{A}|$. Thus $\boldsymbol{A}\boldsymbol{C}' = \boldsymbol{C}'\boldsymbol{A} = |\boldsymbol{A}|\mathbf{I_n}$, i.e., $\boldsymbol{A}\boldsymbol{A}^{\#} = \boldsymbol{A}^{\#}\boldsymbol{A} = |\boldsymbol{A}|\mathbf{I_n}$, or $\boldsymbol{A}^{-1} = |\boldsymbol{A}|^{-1}\boldsymbol{A}^{\#}$.

Example 5.3

If $\boldsymbol{A} = \begin{pmatrix} a_{11} & a_{12} \\ a_{21} & a_{22} \end{pmatrix}$, then $|\boldsymbol{A}| = (a_{11}a_{22} - a_{12}a_{21})$ and $\boldsymbol{C} = \begin{pmatrix} a_{22} & -a_{21} \\ -a_{12} & a_{11} \end{pmatrix}$
so $\boldsymbol{A}^{\#} = \begin{pmatrix} a_{22} & -a_{12} \\ -a_{21} & a_{11} \end{pmatrix}$ so $\boldsymbol{A}^{-1} = \frac{1}{a_{11}a_{22} - a_{12}a_{21}} \begin{pmatrix} a_{22} & -a_{12} \\ -a_{21} & a_{11} \end{pmatrix}$.

5.7 Initial Applications Continued

5.7.1 Linear equations

In §3.4 we briefly considered systems of linear equations given by $\boldsymbol{y} = \boldsymbol{A}\boldsymbol{x}$. It is easy to see that if $\boldsymbol{A}$ is square and invertible, there is a unique solution for $\boldsymbol{x}$ given by $\boldsymbol{x} = \boldsymbol{A}^{-1}\boldsymbol{y}$. This is not the only situation where an exact solution exists, i.e., even if $\boldsymbol{A}$ is singular or non-square (and therefore no inverse $\boldsymbol{A}^{-1}$ exists) there may be a solution for $\boldsymbol{x}$, which may or may not be unique. This extension requires use of generalized inverses and is considered in some detail in §8.3.3.1.

5.7.2 Linear models

Returning to the simple linear model introduced in §2.12.2, $\boldsymbol{y} = \boldsymbol{X}\beta + \varepsilon$ where $\boldsymbol{y}$ and ε are $n \times 1$ vectors, $\boldsymbol{X}$ is a $n \times p$ design matrix and β is a $p \times 1$ vector of parameters where we noted that if there were a matrix $\boldsymbol{G}$ such that $\boldsymbol{G}\boldsymbol{X}\beta = \beta$, then we would have $\boldsymbol{G}\boldsymbol{y}$ (a linear function of the observations $\boldsymbol{y}$) as an unbiased estimator of β. If $\boldsymbol{X}$ is of full column rank p, then $\boldsymbol{X}'\boldsymbol{X}$ is a non-singular $p \times p$ matrix and so invertible. If we consider $\boldsymbol{G} = (\boldsymbol{X}'\boldsymbol{X})^{-1}\boldsymbol{X}'$, then $\boldsymbol{G}\boldsymbol{X}\beta = (\boldsymbol{X}'\boldsymbol{X})^{-1}\boldsymbol{X}'\boldsymbol{X}\beta = \beta$ and so $(\boldsymbol{X}'\boldsymbol{X})^{-1}\boldsymbol{X}'\boldsymbol{y}$ is an unbiased estimator of β. Again, this is not the only situation where we can obtain a useful estimator of β and we consider cases where $\boldsymbol{X}$ does not have full column rank in §8.3.3.1 using generalized inverses. In that section we also consider in what circumstances $(\boldsymbol{X}'\boldsymbol{X})^{-1}\boldsymbol{X}'\boldsymbol{y}$ is a least squares estimator of β and finally in §9.7.2.1 we show that if we make the assumption that the errors ε_i; $i = 1, \dots, n$ are independently and identically normally distributed, then this least squares solution is also the maximum likelihood estimator.

5.7.3 Multivariate methods

In §3.4 we mentioned that the rank of the $n \times p$ data matrix $\boldsymbol{X}'$ was important, in particular whether or not it had full column rank. This is because the rank of the $p \times p$ sample variance matrix $\boldsymbol{S} = \frac{1}{(n-1)}(\boldsymbol{X} - \overline{\boldsymbol{X}})(\boldsymbol{X} - \overline{\boldsymbol{X}})'$ is determined by the rank of $\boldsymbol{X}'$, so if $\rho(\boldsymbol{X}') = p$, then $\rho(\boldsymbol{S}) = p$ and so $\boldsymbol{S}$ is invertible. Certain exploratory multivariate analysis techniques such as linear discriminant analysis and canonical correlation analysis (discussed in Chapter 9) involve inversion of the sample variance $\boldsymbol{S}$ and so if $\boldsymbol{S}$ is singular, then these techniques are inapplicable as they stand and some form of dimensionality reduction is required. Other (primarily exploratory) multivariate methods such as principal component analysis and partial least squares (also discussed in Chapter 9) do not require inversion of the variance $\boldsymbol{S}$ and so can be useful in such situations. The most common practical reason for $\boldsymbol{S}$ to be singular (or equivalently $\boldsymbol{X}'$ not to be of full column rank) is that there are fewer observations than the number of dimensions, i.e., $n < p$.

5.8 Exercises

(1) Find the inverses, both 'by hand' and with **R**, of

(i) $A_1 = \begin{pmatrix} 12 & 7 \\ -4 & 6 \end{pmatrix}$.

(ii) $A_2 = \begin{pmatrix} 0 & 5 \\ -4 & 1 \end{pmatrix}$.

(iii) $A_3 = \begin{pmatrix} 1 & 5 \\ -4 & 0 \end{pmatrix}$.

(iv) $A_4 = \begin{pmatrix} 0 & 5 \\ -4 & 0 \end{pmatrix}$.

(v) $A_5 = \begin{pmatrix} 0 & 0 & 0 & 5 \\ 0 & 0 & 4 & 0 \\ 0 & 7 & 0 & 0 \\ 3 & 0 & 0 & 0 \end{pmatrix}$.

(2) Suppose $AB = BA$ and that A is a non-singular $n \times n$ matrix. Show that $A^{-1}B = BA^{-1}$.

(3) Suppose A is an $n \times n$ orthogonal matrix and B is $n \times n$. Show that AB is orthogonal if and only if B is orthogonal.

(4) Suppose X and Y are non-singular $n \times n$ matrices and all other matrices stated in this exercise are also non-singular.

(a) Show that $(\mathbf{I_n} + X)^{-1} = X^{-1}(\mathbf{I_n} + X^{-1})^{-1}$.
(b) Show that $(X + YY')^{-1}Y = X^{-1}Y(\mathbf{I_n} + Y'X^{-1}Y)^{-1}$.
(c) Show that $(X + Y)^{-1} = X^{-1}(X^{-1} + Y^{-1})^{-1}Y^{-1} = Y^{-1}(X^{-1} + Y^{-1})^{-1}X^{-1}$.
(d) Show that $X^{-1} + Y^{-1} = X^{-1}(X + Y)Y^{-1}$.
(e) Show that $X - X(X + Y)^{-1}X = Y - Y(X + Y)^{-1}Y$.
(f) Show that if $(X + Y)^{-1} = X^{-1} + Y^{-1}$, $XY^{-1}X = YX^{-1}Y$.
(g) Show that $(\mathbf{I_n} + XY)^{-1} = \mathbf{I_n} - X(\mathbf{I_n} + YX)^{-1}Y$.
(h) Show that $(\mathbf{I_n} + XY)^{-1}X = X(\mathbf{I_n} + YX)^{-1}$.

(5) Show that $(A + BCB')^{-1} = A^{-1} - A^{-1}B[C^{-1} + B'A^{-1}B]^{-1}B'A^{-1}$ where A and C are non-singular $m \times m$ and $n \times n$ matrices and B is $m \times n$.

(6) Show that $\begin{pmatrix} A & B \\ \mathbf{0} & D \end{pmatrix}^{-1} = \begin{pmatrix} A^{-1} & X \\ \mathbf{0} & D^{-1} \end{pmatrix}$ for a suitable X.

(7) Let $A = \begin{pmatrix} \mathbf{I_n} & \lambda \imath_n \\ \lambda \imath_n' & 1 \end{pmatrix}$.

(i) For what values of λ is $\boldsymbol{A}$ non-singular?

(ii) Find $\boldsymbol{A}^{-1}$ when it exists.

(8) Suppose $\boldsymbol{A}$ is skew-symmetric and non-singular.

(i) Show that $\boldsymbol{A}^{-1}$ is skew-symmetric.

(ii) Show that $(\mathbf{I_n} - \boldsymbol{A})(\mathbf{I_n} + \boldsymbol{A})^{-1}$ is orthogonal.

(9) Suppose that $\boldsymbol{AX} = \mathbf{0}$ and $\boldsymbol{A}$ is idempotent. Let $\boldsymbol{B} = (\boldsymbol{X} - \boldsymbol{A})^{-1}$. Prove that

(i) $\boldsymbol{XB} = \mathbf{I_n} - \boldsymbol{A}$.

(ii) $\boldsymbol{XBX} = \boldsymbol{X}$.

(iii) $\boldsymbol{XBA} = \mathbf{0}$.

(10) Suppose $\boldsymbol{A} = \mathbf{I_n} - 2\boldsymbol{xx}'$ with $\boldsymbol{x}'\boldsymbol{x} = 1$. Find $\boldsymbol{A}^{-1}$.

(11) Show that $\begin{pmatrix} \mathbf{0} & \boldsymbol{A} \\ \lambda & \boldsymbol{x}' \end{pmatrix}^{-1} = \frac{1}{\lambda}\begin{pmatrix} -\boldsymbol{x}'\boldsymbol{A}^{-1} & 1 \\ \lambda\boldsymbol{A}^{-1} & \mathbf{0} \end{pmatrix}$.

6

Eigenanalysis of Real Symmetric Matrices

6.1 Introduction and Definitions

Eigenvalues and eigenvectors of a matrix provide a fundamental characterisation of the matrix and are central to many of the theoretical results on matrices. They have a close connection to determinants and they provide a representation that permits definition of fractional powers of a matrix, i.e., square and cube roots etc. In most of this chapter the matrices considered are real symmetric matrices. This restriction is with statistical applications in mind. Details of extensions to non-symmetric and to complex matrices are readily available in the references given in §1.2.

In statistics, the eigenanalysis of the variance matrix is the basis of many statistical analyses. This matrix is necessarily symmetric and also it is positive semi-definite (see §2.9). Recall that the size of a matrix is reflected by its determinant (see §4.1.1, note (ii)). This, combined with the properties that the sum and product of the eigenvalues are equal to the trace and determinant of the matrix (see §6.4.4 below), means that the eigenvalues of a matrix are themselves key properties of a matrix. Further the eigenvectors associated with the eigenvalues, especially the dominant (largest) and the minor (smallest) values, give further information and provide interpretations of statistical interest, e.g., directions of dominant and minor variation in the case of the eigenanalysis of a variance matrix.

Principal component analysis is a fundamental tool of multivariate analysis and this involves projecting (or more exactly translating and rotating) the data onto the eigenvectors of the variance matrix. The advantages of doing this are explained later in this chapter in §6.5 and in fuller detail in §9.3. Other techniques of multivariate analysis such as linear discriminant analysis, canonical correlation analysis and partial least squares discussed in Chapter 9 all rest on the eigenanalysis of various matrices derived from the data.

Specific methods of statistical analysis are not the only areas where eigenanalysis plays a crucial role. In §6.7.1 and §8.2 we consider various matrix decompositions, i.e., expressing a matrix as a product of two or three matrices which have particular forms, for example the first and third matrices in a product of three factors might be orthogonal and the second term diagonal. Then in the later parts of §8.2 we show how these can be used for efficient calculation of determinants and inverses and in solving linear equations, all of which we have seen arise in many areas of statistical modelling and statistical analysis.

If S is a $n \times n$ matrix, then the ***eigenvalues*** of S are the roots of the ***characteristic equation*** of S, $|S - \lambda \mathbf{I_n}| = 0$. Some authors refer to eigenvalues as characteristic values or characteristic roots. The polynomial $p_S(\lambda) = |S - \lambda \mathbf{I_n}|$ is called the ***characteristic polynomial*** of S. This is a polynomial of degree n and so has n roots $\lambda_1, \lambda_2, \ldots, \lambda_n$ which are not necessarily distinct. Conventionally we order these so that $\lambda_1 \geq \lambda_2 \geq \ldots \geq \lambda_n$, provided, of course, that they are all real numbers.

It can be shown (e.g., Banerjee and Roy, 2014, Theorem 11.18), the ***Cayley-Hamilton theorem***, that S satisfies its own characteristic equation, i.e., $p_S(S) = \mathbf{0}$, where it is understood that any scalar μ in $p_S(\lambda)$ is replaced by $\mu \mathbf{I_n}$.

Example 6.1

(i) If $S = \begin{pmatrix} 1 & 4 \\ 9 & 1 \end{pmatrix}$, then the characteristic equations of S is

$$0 = |S - \lambda \mathbf{I_2}| = \begin{pmatrix} 1-\lambda & 4 \\ 9 & 1-\lambda \end{pmatrix} = (1-\lambda)^2 - 36 = \lambda^2 - 2\lambda - 35$$

$$= (\lambda - 7)(\lambda + 5) \text{ so } \lambda_1 = 7 \text{ and } \lambda_2 = -5.$$

(ii) If $S = \begin{pmatrix} 6 & 3 \\ 3 & 2 \end{pmatrix}$, then the characteristic equations of S is

$$0 = |S - \lambda \mathbf{I_2}| = \begin{pmatrix} 6-\lambda & 3 \\ 3 & 2-\lambda \end{pmatrix} = (6-\lambda)(2-\lambda) - 4 = \lambda^2 - 8\lambda + 8$$

$$= (\lambda - 4)^2 - 8 \text{ so } \lambda_1 = 4 + 2\sqrt{2} \text{ and } \lambda_2 = 4 - 2\sqrt{2}.$$

(iii) If $S = \begin{pmatrix} 2 & 1 & 1 \\ 1 & 2 & 1 \\ 1 & 1 & 2 \end{pmatrix}$, then the characteristic equations of S is

$$0 = |S - \lambda \mathbf{I_3}| = \lambda^3 - 6\lambda^2 + 9\lambda - 4 = (\lambda - 4)(\lambda - 1)^2$$

so $\lambda_1 = 4$ and $\lambda_2 = \lambda_3 = 1$ so the three eigenvalues of S are 4 (with multiplicity 1) and 1 (with multiplicity 2).

6.2 Eigenvectors

If λ is an eigenvalue of the $n \times n$ matrix S, then $|S - \lambda \mathbf{I_n}| = 0$ so $A = S - \lambda \mathbf{I_n}$ is a singular matrix and so there is a linear combination of the n columns of S equal to zero (i.e., the columns must be linearly dependent), i.e., there are constants $x_1, x_2, \ldots, x_n$, not all zero, such $A(x_1, x_2, \ldots, x_n)' = 0$, such that $Ax = 0$ or $Sx = \lambda x$. This last equation is termed an ***eigenequation***. The vector x is termed an ***eigenvector*** of S [corresponding to the eigenvalue λ]. The pair (x, λ) is termed an ***eigenpair*** of $\boldsymbol{S}$. Since there are n eigenvalues of $\boldsymbol{S}, \lambda_1, \lambda_2, \ldots, \lambda_n$ (not necessarily distinct) there are n eigenvectors $\boldsymbol{x}_1, \boldsymbol{x}_2, \ldots, \boldsymbol{x}_n$ corresponding to the n eigenvalues. To find the eigenvectors of a matrix $\boldsymbol{S}$ 'by hand' the first step is to find the eigenvalues $\lambda_1, \lambda_2, \ldots, \lambda_n$ by finding the roots of the n-degree polynomial $|\boldsymbol{S} - \lambda \mathbf{I_n}|$ and then for each λ_i in turn solving the simultaneous linear equations $\boldsymbol{S}\boldsymbol{x}_i = \lambda_i \boldsymbol{x}_i$ for $\boldsymbol{x}_i$.

6.2.1 General matrices

Strictly $\boldsymbol{x}$ is termed a *right eigenvector* if $\boldsymbol{S}\boldsymbol{x} = \lambda \boldsymbol{x}$ and a *left eigenvector* if $\boldsymbol{x}'\boldsymbol{S} = \lambda \boldsymbol{x}'$. Note that necessarily $\boldsymbol{S}$ must be a square matrix for an eigenvector to be defined. If $\boldsymbol{S}$ is symmetric (and therefore necessarily square), it is easily seen that left and right eigenvectors are identical with identical eigenvalues. Left and right eigenvectors of non-symmetric matrices have the same eigenvalues because $|\boldsymbol{S} - \lambda \mathbf{I_n}| = |(\boldsymbol{S} - \lambda \mathbf{I_n})'|$.

6.3 Implementation in R

The command for producing the eigenanalysis of a matrix is `eigen()`. This produces the eigenvalues and eigenvectors of a square matrix. If the matrix is not symmetric, the command produces the right eigenvectors. Left eigenvectors could be obtained by using `eigen()` on the transpose of the matrix. The eigenvalues of the matrix $\boldsymbol{S}$ are stored in the vector `eigen(S)$values` and the eigenvectors in the $n \times n$ matrix `eigen(S)$vectors`.

Example 6.2

(i)
```
> X<-matrix(c(1,4,9,1),
+ 2,2,byrow=T)
> X
     [,1] [,2]
[1,]    1    4
[2,]    9    1
```

```
> eigen(X)
$values
[1]  7 -5
$vectors
      [,1]   [,2]
[1,] 0.555 -0.555
[2,] 0.832  0.832
```

To verify the first eigenequation:

```
>  eigen(X)$values[1]* eigen(X)$vectors[,1]
[1] 3.88 5.82
>  X%*%eigen(X)$vectors[,1]
     [,1]
[1,] 3.88
[2,] 5.82
```

(ii)
```
> X<-matrix(c(2,1,1,1,2,1,1,
+ 1,2),3,3,byrow=T)
> X
     [,1] [,2] [,3]
[1,]    2    1    1
[2,]    1    2    1
[3,]    1    1    2
```

```
> eigen(X)
$values
[1] 4 1 1
$vectors
       [,1]   [,2]   [,3]
[1,] -0.577  0.816  0.000
[2,] -0.577 -0.408 -0.707
[3,] -0.577 -0.408  0.707
```

To verify the first eigenequation:

```
>  eigen(X)$values[1]* eigen(X)$vectors[,1]
[1] -2.31 -2.31 -2.31
>  X%*%eigen(X)$vectors[,1]
      [,1]
[1,] -2.31
[2,] -2.31
[3,] -2.31
```

(iii) To verify all the eigenequations simultaneously:

```
> options(digits=3)

> X<-matrix(c(2,1,1,1,2,1,1,
+ 1,2),3,3,byrow=T)

> one<-matrix((rep(1,3)),
+ 3,1)
> one
     [,1]
[1,]    1
[2,]    1
[3,]    1

> eigen(X)$values
[1] 4 1 1
```

```
> one%*%eigen(X)$values
     [,1] [,2] [,3]
[1,]    4    1    1
[2,]    4    1    1
[3,]    4    1    1
> X%*%eigen(X)$vectors
      [,1]   [,2]   [,3]
[1,] -2.31  0.000  0.816
[2,] -2.31 -0.707 -0.408
[3,] -2.31  0.707 -0.408
> one%*%eigen(X)$values*
+ eigen(X)$vectors
      [,1]   [,2]   [,3]
[1,] -2.31  0.000  0.816
[2,] -2.31 -0.707 -0.408
[3,] -2.31  0.707 -0.408
```

6.3.0.1 Notes on examples

(i) Note that the final multiplication in the last example is a triple product, the first multiplication is a standard matrix multiplication with %*% and the second is an element-by-element multiplication (i.e., a Hadamard product, see §8.4) of two

3×3 matrices with $*$. This 'trick' is one of the rare occasions when this form of multiplication is useful.

(ii) Use of the sum vector ι_3 to create a matrix with rows identical to a row vector is a useful 'trick'.

(iii) **R** treated the vector `eigen(X)$vectors` as a row vector because it was premultiplied by a 3×1 matrix ι_3 (see §2.10.1 and §2.11.2.7).

(iv) The matrices of eigenvectors in examples (ii) and (iii) above are different (the second and third are in different orders). This is because the corresponding two eigenvalues are identical and so it does not matter but **R** makes an arbitrary choice and the choice can be different in different sessions (even on the same installation). In fact there are many other possible choices of the second and third eigenvectors, e.g., $(0.408, -0.816, 0.408)'$ and $(0.707, 0.000, -0.707)'$, and again the choice is arbitrary and **R** may well make different choices on different occasions.

6.4 Properties of Eigenanalyses

6.4.1 Properties related to uniqueness

(i) If $\boldsymbol{x}$ is an eigenvector of $\boldsymbol{S}$, any scalar multiple $k\boldsymbol{x}$ is also an eigenvector because $\boldsymbol{S}(k\boldsymbol{x}) = \lambda k\boldsymbol{x}$. Usually (as in **R**) eigenvectors are normalized so that $\boldsymbol{x}'\boldsymbol{x} = 1$ which means that an eigenvector is determined up to its sign, i.e., the normalized eigenvector can be multiplied by -1 without altering its properties. Again **R** makes an arbitrary choice which may well not be consistent from one occasion to another. Some authorities prefer to resolve the ambiguity by always taking the first element to be positive (say). This would need specific checking and implementation in **R** on each occasion that an eigenanalysis-related function is used.

(ii) If λ_i and λ_j are distinct eigenvalues of $\boldsymbol{S}$ with eigenvectors $\boldsymbol{x}_i$ and $\boldsymbol{x}_j$, then $\boldsymbol{x}_i$ and $\boldsymbol{x}_j$ are distinct: suppose $\boldsymbol{x}_i = \boldsymbol{x}_j = \boldsymbol{x}$ say, then we have $\boldsymbol{S}\boldsymbol{x} = \lambda_i\boldsymbol{x}$ and $\boldsymbol{S}\boldsymbol{x} = \lambda_j\boldsymbol{x}$ so $(\lambda_I - \lambda_j)\boldsymbol{x} = 0$, so $\boldsymbol{x} = 0$ because $\lambda_i \neq \lambda_j$ which contradicts $\boldsymbol{x}$ being an eigenvector. Note that this is true whether or not S is symmetric.

(iii) If $\boldsymbol{x}_i$ and $\boldsymbol{x}_j$ are distinct eigenvectors of $\boldsymbol{S}$ with the same eigenvalue λ, then any linear combination of $\boldsymbol{x}_i$ and $\boldsymbol{x}_j$ is also an eigenvector of $\boldsymbol{S}$ since $\boldsymbol{S}\boldsymbol{x}_i = \lambda\boldsymbol{x}_i$ and $\boldsymbol{S}\boldsymbol{x}_j = \lambda\boldsymbol{x}_j$ then $\boldsymbol{S}(a_1\boldsymbol{x}_i + a_2\boldsymbol{x}_j) = \lambda(a_1\boldsymbol{x}_i + a_2\boldsymbol{x}_j)$. For example, the identity matrix $\mathbf{I}_n$ has every vector $\boldsymbol{x}$ as an eigenvector.

6.4.2 Properties related to symmetric matrices

(i) Suppose now that $\boldsymbol{S}$ is real and symmetric, i.e., $\boldsymbol{S} = \boldsymbol{S}'$, then the eigenvalues λ_j and eigenvectors $\boldsymbol{x}_j$ of $\boldsymbol{S}$ are real. To prove this, let $\lambda_j = \mu_j + i\nu_j, \boldsymbol{x}_j = \boldsymbol{y}_j + i\boldsymbol{z}_j$ (where here $i = \surd(-1)$). Equating real and imaginary parts of $\boldsymbol{Sx}_j = \lambda_j\boldsymbol{x}_j$ gives $\boldsymbol{Sy}_j = \mu_j\boldsymbol{y}_j - \nu_j\boldsymbol{z}_j$, and $\boldsymbol{Sz}_j = \mu_j\boldsymbol{y}_j + \nu_j\boldsymbol{z}_j$. Premultiplying the first equation by $\boldsymbol{z}'_j$ and the second by $\boldsymbol{y}'_j$ and noting $\boldsymbol{z}'_j\boldsymbol{Sy}_j = (\boldsymbol{z}'_j\boldsymbol{Sy}_j)'$ (since it's a scalar) $= \boldsymbol{y}'_j\boldsymbol{Sz}_j$ (since $\boldsymbol{S}$ is symmetric by presumption) and subtracting the two equations gives $\nu_j\boldsymbol{z}'_j\boldsymbol{z}_j + \nu_j\boldsymbol{y}'_j\boldsymbol{y}_j = 0$, so $\nu_j = 0$ because $\boldsymbol{z}'_j\boldsymbol{z}_j + \boldsymbol{y}'_j\boldsymbol{y}_j > 0$, i.e., λ_j is real.

(ii) If $\boldsymbol{S}$ is symmetric, eigenvectors corresponding to distinct eigenvalues are orthogonal because if $\boldsymbol{Sx}_i = \lambda_i\boldsymbol{x}_i$ and $\boldsymbol{Sx}_j = \lambda_j\boldsymbol{x}_j$, then $\boldsymbol{x}'_j\boldsymbol{Sx}_i = \lambda_i\boldsymbol{x}'_j\boldsymbol{x}_i$ and $\boldsymbol{x}'_i\boldsymbol{Sx}_j = \lambda_j\boldsymbol{x}'_i\boldsymbol{x}_j$ but $\boldsymbol{x}'_j\boldsymbol{Sx}_i = (\boldsymbol{x}'_j\boldsymbol{Sx}_i)'$ because it is a scalar (see §2.8.1) $(\boldsymbol{x}'_j\boldsymbol{Sx}_i)' = \boldsymbol{x}'_i\boldsymbol{S}'\boldsymbol{x}_j = \boldsymbol{x}'_i\boldsymbol{Sx}_j$ (see §2.11.2.5 and noting $\boldsymbol{S}$ is symmetric), so $(\lambda_i - \lambda_j)\boldsymbol{x}'_i\boldsymbol{x}_j = 0$ (noting $\boldsymbol{x}'_i\boldsymbol{x}_j = \boldsymbol{x}'_j\boldsymbol{x}_i$) and since $\lambda_i \neq \lambda_j$ we have $\boldsymbol{x}'_i\boldsymbol{x}_j = 0$ and thus $\boldsymbol{x}_i$ and $\boldsymbol{x}_j$ are orthogonal.

If $\boldsymbol{A}$ is not symmetric the eigenvalues may or may not be real. For example the matrix

$$\boldsymbol{A} = \begin{pmatrix} 0 & -1 \\ 1 & 0 \end{pmatrix}$$

is not symmetric and its eigenvalues are the roots of the characteristic equation $\lambda^2 + 1 = 0$ which are $\pm\sqrt{-1}$. **R** will handle complex arithmetic such as this:

```
> A<-matrix(c(0,-1,1,0),2,2)
> eigen(A)
$values
[1] 0+1i 0-1i

$vectors
             [,1]         [,2]
[1,] 0.707+0.000i 0.707+0.000i
[2,] 0.000+0.707i 0.000-0.707i
```

where i= $\sqrt{(}-1)$.

6.4.2.1 Eigenvectors of non-symmetric matrices

(i) If $\boldsymbol{x}_i$ is a right eigenvector of the $n \times n$ matrix $\boldsymbol{A}$ with eigenvalue λ_i, then $\boldsymbol{Ax}_i = \lambda_i\boldsymbol{x}_i$ so $\boldsymbol{x}'_i A' = \lambda_i\boldsymbol{x}'_i$ and so $\boldsymbol{x}'_i$ is a left eigenvector of $\boldsymbol{A}'$. Similarly, a right eigenvector $\boldsymbol{y}_i$ of $\boldsymbol{A}'$ is a left eigenvector of $\boldsymbol{A}$.

(ii) If $\boldsymbol{x}_i$ and $\boldsymbol{y}'_j$ are left and right eigenvectors of $\boldsymbol{A}$ corresponding to distinct eigenvalues λ_i and λ_j, then $\boldsymbol{x}_i$ and $\boldsymbol{y}_j$ are orthogonal because we have $\boldsymbol{Ax}_i = \lambda_i\boldsymbol{x}_i$ and $\boldsymbol{y}'_j\boldsymbol{A} = \lambda_j\boldsymbol{y}'_j$. Premultiplying the first by $\boldsymbol{y}'_j$ and postmultiplying the second by $\boldsymbol{x}_i$ and subtracting gives $(\lambda_i - \lambda_j)\boldsymbol{y}'_j\boldsymbol{x}_i = 0$ and so $\boldsymbol{y}_j$ and $\boldsymbol{x}_i$ are orthogonal for $i \neq j$. The vectors $\boldsymbol{x}_i$ and $\boldsymbol{y}_i$ can be standardized so that $\boldsymbol{x}'_i\boldsymbol{y}_i = 1$, in which case the left and right eigenvectors are said to be ***biorthogonal***. Note that neither the $\boldsymbol{x}_i$ nor the $\boldsymbol{y}_j$ are themselves orthogonal unless the matrix $\boldsymbol{A}$ is symmetric.

6.4.2.2 Illustration of biorthogonality

```
> X<-matrix(c(1,4,9,1),2,2,byrow=T); X
     [,1] [,2]
[1,]    1    4
[2,]    9    1
```

```
> eigen(X)
$values
[1]  7 -5
$vectors
      [,1]   [,2]
[1,] 0.555 -0.555
[2,] 0.832  0.832
```

```
> eigen(t(X))
$values
[1]  7 -5
$vectors
      [,1]   [,2]
[1,] 0.832 -0.832
[2,] 0.555  0.555
```

```
> t(eigen(X)$vectors[,1])%*%eigen(t(X))$vectors[,2]
         [,1]
[1,] -4.5e-17
```

6.4.3 Properties related to functions of matrices

(i) If $\boldsymbol{x}$ and λ are an eigenvector and an eigenvalue of $\boldsymbol{S}$ and if $\boldsymbol{S}$ is non-singular, then $\boldsymbol{x}$ is an eigenvector of $\boldsymbol{S}^{-1}$ with eigenvalue λ^{-1} since if $\boldsymbol{Sx} = \lambda\boldsymbol{x}$ then $\boldsymbol{S}^{-1}\boldsymbol{Sx} = \lambda\boldsymbol{S}^{-1}\boldsymbol{x}$ so $\boldsymbol{S}^{-1}\boldsymbol{x} = \lambda^{-1}\boldsymbol{x}$ showing $\boldsymbol{x}$ is an eigenvector of $\boldsymbol{S}^{-1}$ with eigenvalue λ^{-1}.

(ii) If $\boldsymbol{x}$ and λ are an eigenpair of $\boldsymbol{S}$, $\boldsymbol{x}$ is an eigenvector of $\boldsymbol{S}^k$ with eigenvalue λ^k since $\boldsymbol{S}^k\boldsymbol{x} = \boldsymbol{S}^{k-1}(\boldsymbol{Sx}) = \boldsymbol{S}^{k-1}(\lambda\boldsymbol{x}) = \ldots = \lambda^k\boldsymbol{x}$.

(iii) If $\boldsymbol{x}$ and λ are an eigenpair of $\boldsymbol{S}$, $\boldsymbol{x}$ and $(a - b\lambda)$ are an eigenvector and eigenvalue of $a\mathbf{I_n} - b\boldsymbol{S}$ since $(a\mathbf{I_n} - b\boldsymbol{S})\boldsymbol{x} = a\boldsymbol{x} - b\lambda\boldsymbol{x} = (a - b\lambda)\boldsymbol{x}$.

(iv) If $\boldsymbol{S}$ is $n \times m$ and $\boldsymbol{T}$ is $m \times n$ where $n \geq m$, then $\boldsymbol{ST}$ and $\boldsymbol{TS}$ have the same non-zero eigenvalues. The eigenvalues of ST are the roots of $|\boldsymbol{ST} - \lambda\mathbf{I_n}| = 0$ but $|\boldsymbol{ST} - \lambda\mathbf{I_n}| = (-\lambda)^{n-m}|\boldsymbol{TS} - \lambda\mathbf{I_m}|$, see §4.6.2. Note that this implies that $\boldsymbol{ST}$ has at most $n - m$ non-zero eigenvalues.

(v) If $\boldsymbol{x}$ and λ are an eigenpair of $\boldsymbol{ST}$, then $\boldsymbol{Tx}$ is an eigenvector of $\boldsymbol{TS}$ corresponding to eigenvalue λ because we have $\boldsymbol{STx} = \lambda\boldsymbol{x}$ so $\boldsymbol{TS}(\boldsymbol{Tx}) = \lambda(\boldsymbol{Tx})$.

(vi) If $\boldsymbol{X}$ is an $m \times n$ matrix, then $\boldsymbol{XX}'$ and $\boldsymbol{X}'\boldsymbol{X}$ have the same non-zero eigenvalues. This follows directly from (iv) above.

(vii) If $\boldsymbol{x}$ and λ are an eigenpair of $\boldsymbol{S}$ and $\boldsymbol{T}$ is a non-singular $n \times n$ matrix, then λ is an eigenvalue of $\boldsymbol{TST}^{-1}$ corresponding to eigenvector $\boldsymbol{Tx}$ because if $\boldsymbol{Sx} = \lambda\boldsymbol{x}$ then $(\boldsymbol{TST}^{-1})\boldsymbol{Tx} = \lambda\boldsymbol{Tx}$. $\boldsymbol{S}$ and $\boldsymbol{TST}^{-1}$ are similar matrices (see §2.5.9).

6.4.4 Properties related to determinants

(i) If S is diagonal [or triangular] , then the eigenvalues are the diagonal elements since $S-\lambda\mathbf{I_n}$ is diagonal [or triangular] and the determinant of a diagonal [or triangular] matrix is the product of its diagonal elements (see §4.3.2).

(ii) S is non-singular if and only if all of its eigenvalues are non-zero since $0=|S-\lambda\mathbf{I_n}|=|S|$ if $\lambda=0$ and if $|S|=0$, then $\lambda=0$ satisfies $|S-\lambda\mathbf{I_n}|=0$ and so is an eigenvalue of S.

(iii) If μ is **not** an eigenvalue of S, then $S-\mu\mathbf{I_n}$ is non-singular since if $|S-\mu\mathbf{I_n}|=0$ then μ would be an eigenvalue of S.

(iv) If S has eigenvalues $\lambda_1,\lambda_2,\ldots,\lambda_n$, then $|S|=\prod_{i=1}^n\lambda_i$ because the λ_i are the n roots of $|S-\lambda\mathbf{I_n}|=0$, so $|S-\lambda\mathbf{I_n}|=(\lambda_1-\lambda)(\lambda_2-\lambda)\ldots(\lambda_n-\lambda)$ for any value of λ and putting $\lambda=0$ gives the result.

(v) If S has eigenvalues $\lambda_1,\lambda_2,\ldots,\lambda_n$, then $\text{tr}(S)=\sum_{i=1}^n\lambda_i$, comparing the coefficients of λ^{n-1} in $|S-\lambda\mathbf{I_n}|=(\lambda_1-\lambda)(\lambda_2-\lambda)\ldots(\lambda_n-\lambda)$.

(vi) If S has eigenvalues $\lambda_1,\lambda_2,\ldots,\lambda_n$, then $\text{tr}(S^k)=\sum_{i=1}^n\lambda_i^k$ which follows from (v) and §6.4.3(ii).

6.4.5 Properties related to diagonalisation

(i) If X is an $n\times n$ matrix with distinct eigenvalues $\lambda_1>\lambda_2>\ldots>\lambda_n$, then there exists a non-singular matrix T and diagonal matrix Λ such that $T^{-1}XT=\Lambda$ and the elements of Λ are the λ_i. If the eigenvectors of X are $x_i, i=1,2,\ldots,n$ then $Xx_i=\lambda_i x_i, i=1,2,\ldots,n$ and if $T=(x_1,x_2,\ldots,x_n)$ (i.e., the matrix composed of the n eigenvectors as columns), then T is non-singular since the eigenvectors are linearly independent. Further $XT=(\lambda_1x_1,\lambda_2x_2,\ldots,\lambda_nx_n)=\Lambda T$ where $\Lambda=\text{diag}(\lambda_1,\lambda_2,\ldots,\lambda_n)$ so, multiplying by T^{-1} we have $T^{-1}XT=\Lambda$.

(ii) If X is an $n\times n$ matrix with distinct eigenvalues $\lambda_1>\lambda_2>\ldots>\lambda_n$ and Y commutes with X, then $T^{-1}YT=M$ for some diagonal matrix M because if $Xx_i=\lambda_i x_i$, then $YXx_i=\lambda_i Yx_i$ so $X(Yx_i)=\lambda_i(Yx_i)$ showing that Yx_i is another eigenvector of X corresponding to λ_i but the λ_i are distinct so Yx_i must be a scalar multiple of x_i, i.e., $Yx_i=\mu_i x_i$ for some scalar μ_i. Thus x_i is an eigenvector of Y and thus $YT=TM$ where $M=\text{diag}(\mu_1,\mu_2,\ldots,\mu_n)$.

(iii) If S is a symmetric matrix, then there exists an orthogonal matrix T and a diagonal matrix Λ such that $T'ST=\Lambda$. This result requires a substantial proof in the general case. In the case where the eigenvalues of S are distinct, it follows from (i) since by choosing the eigenvectors to be normalised to ensure $x_i'x_i=1$ we can ensure T is orthogonal so $T^{-1}=T'$. In the general case where there are some multiple eigenvalues (possibly some of which may be zero) we need to choose k orthogonal eigenvectors corresponding to an eigenvalue with multiplicity k. The most straightforward proof that this is possible is by induction and can be found in the references cited.

(iv) If $\boldsymbol{S}$ is a symmetric matrix, we can write $\boldsymbol{S}$ as $\boldsymbol{T}\Lambda\boldsymbol{T}' = \boldsymbol{S}$ where Λ is the diagonal matrix of eigenvalues of $\boldsymbol{S}$ and $\boldsymbol{T}$ is the matrix of eigenvectors.

(v) If $\boldsymbol{X}$ and $\boldsymbol{Y}$ are both symmetric and if $\boldsymbol{X}$ and $\boldsymbol{Y}$ commute, then result (ii) above can be generalized so there are diagonal matrices Λ and $\boldsymbol{M}$ and a matrix $\boldsymbol{T}$ such that $\boldsymbol{T}'\boldsymbol{X}\boldsymbol{T} = \Lambda$ and $\boldsymbol{T}'\boldsymbol{Y}\boldsymbol{T} = \boldsymbol{M}$. If we have $\boldsymbol{T}'\boldsymbol{X}\boldsymbol{T} = \Lambda$ and $\boldsymbol{T}'\boldsymbol{Y}\boldsymbol{T} = \boldsymbol{M}$, then $\boldsymbol{XY} = \boldsymbol{T}\Lambda\boldsymbol{T}'\boldsymbol{TMT}' = \boldsymbol{T}\Lambda\boldsymbol{MT}' = \boldsymbol{TM}\Lambda\boldsymbol{T}' = \boldsymbol{TMT}'\boldsymbol{T}\Lambda\boldsymbol{T}' = \boldsymbol{YX}$, noting diagonal matrices commute. The proof of the converse is more difficult and is not given here. In the particular case that the eigenvalues are distinct, the result follows from arguments similar to that in (ii) noting that the eigenvectors in $\boldsymbol{T}$ can be chosen to be orthogonal.

6.4.6 Properties related to values of eigenvalues

(i) (Bounds for a Rayleigh quotient $\boldsymbol{x}'\boldsymbol{Sx}/\boldsymbol{x}'\boldsymbol{x}$). If $\boldsymbol{S}$ is an $n \times n$ symmetric matrix with eigenvalues $\lambda_1 \geq \lambda_2 \geq \ldots \geq \lambda_n$ and $\boldsymbol{x}$ is any vector, then $\lambda_1 \geq \frac{\boldsymbol{x}'\boldsymbol{Sx}}{\boldsymbol{x}'\boldsymbol{x}} \geq \lambda_n$. This follows by noting that $\boldsymbol{x}'\boldsymbol{Sx} = \boldsymbol{x}'\boldsymbol{T}\Lambda\boldsymbol{T}'\boldsymbol{x} = \boldsymbol{y}'\Lambda\boldsymbol{y} = \sum_j \lambda_j y_j^2$, where $\boldsymbol{y} = \boldsymbol{T}'\boldsymbol{x}$ so $\lambda_1 \sum_j y_j^2 \geq \boldsymbol{x}'\boldsymbol{Sx} \geq \lambda_n \sum_j y_j^2$ and $\sum_j y_j^2 = \boldsymbol{y}'\boldsymbol{y} = \boldsymbol{x}'\boldsymbol{TT}'\boldsymbol{x} = \boldsymbol{x}'\boldsymbol{x}$ since $\boldsymbol{T}$ is orthogonal.

(ii) $\boldsymbol{S}$ is positive definite (i.e., $\boldsymbol{S} > 0$) if and only if all the eigenvalues of $\boldsymbol{S}$ are strictly positive: if all the λ_i are positive, then in particular $\lambda_n > 0$ so $\boldsymbol{x}'\boldsymbol{Sx} \geq \lambda_n \sum_j y_j^2 > 0$ for any $\boldsymbol{x}$ and thus $\boldsymbol{S} > 0$. Conversely, if $\boldsymbol{S} > 0$, then $\boldsymbol{x}'\boldsymbol{Sx} > 0$ for any $\boldsymbol{x}$. In particular we have $\boldsymbol{Sx}_n = \lambda_n \boldsymbol{x}_n$ so $\boldsymbol{x}'\boldsymbol{Sx}_n = \lambda_n \boldsymbol{x}_n'\boldsymbol{x}_n > 0$ so $\lambda_n > 0$ since $\boldsymbol{x}_n'\boldsymbol{x}_n > 0$.

(iii) $\boldsymbol{S}$ is positive semi-definite if and only if all the eigenvalues of $\boldsymbol{S}$ are non-negative. The proof of this is similar to that in (ii) above.

(iv) If $\boldsymbol{S}$ is positive definite, then it is non-singular since its determinant (equal to the product of its eigenvalues) is strictly positive.

(v) If $\boldsymbol{S}$ is positive semi-definite and non-singular, then it must be positive definite since its determinant (equal to the product of its eigenvalues) is strictly positive and so all its eigenvalues must be strictly positive.

6.4.7 Rank and non-zero eigenvalues

The rank of a symmetric matrix $\boldsymbol{S}$ is equal to the number of non-zero eigenvalues. We have $\rho(\boldsymbol{T}'\boldsymbol{ST}) = \rho(\Lambda) =$ number of non-zero diagonal elements of Λ, noting §3.3.2.1 and §3.1.1. Note that this result is not true in general for non-symmetric matrices but it can be shown that the number of non-zero eigenvalues cannot exceed the rank of the matrix. For any matrix $\boldsymbol{A}$ we have $\rho(\boldsymbol{A}) = \rho(\boldsymbol{AA}')$; see §3.1; = number of non-zero eigenvalues of $\boldsymbol{AA}'$ since $\boldsymbol{AA}'$ is symmetric.

6.4.7.1 Example of non-symmetric matrix

Consider the matrix $A = \begin{pmatrix} 0 & 1 & 0 & \cdots & 0 \\ 0 & 0 & 1 & \cdots & 0 \\ 0 & 0 & \ddots & \ddots & 0 \\ \vdots & \vdots & \ddots & \ddots & 1 \\ 0 & 0 & \cdots & 0 & 0 \end{pmatrix}$ which has all eigenvalues equal to zero since it is a triangular matrix with all diagonal elements zero but it is clearly of rank $n-1$ since the first $n-1$ rows are linearly independent and the last row is entirely composed of zeroes. Matrices of this form are termed ***Jordan matrices***.

6.5 A Key Statistical Application: PCA

In the introduction to this chapter we mentioned that a prime statistical application of eigenanalysis is the exploratory technique of multivariate analysis known as principal component analysis (PCA). If $\boldsymbol{X}'$ is a $n \times p$ data matrix and $\text{var}(\boldsymbol{X}') = \boldsymbol{S}$ (where $\boldsymbol{S}$ is a $p \times p$ symmetric positive semi-definite matrix) and $\boldsymbol{S}$ has eigenvectors $\boldsymbol{a}_i$, $i = 1, \ldots, p$ with $\boldsymbol{A}$ the matrix whose i^{th} column is $\boldsymbol{a}_i$, then the transformation to principal components is given by $\boldsymbol{Y}' = (\boldsymbol{X} - \overline{\boldsymbol{X}})'\boldsymbol{A}$. Since $\boldsymbol{S}$ is symmetric, the columns of $\boldsymbol{A}$, $\boldsymbol{a}_i$, are orthogonal so $\boldsymbol{A}$ is orthogonal so multiplying the data matrix $\boldsymbol{X}'$ by $\boldsymbol{A}$ is a rotation (with a possible reflection, depending on whether $|\boldsymbol{A}| = +1$ or -1; see §4.4). A change of location (or translation) by subtracting the overall mean $\overline{\boldsymbol{X}}'$ followed by a rotation does not alter any intrinsic statistical property of the data, i.e., the transformed data $\boldsymbol{Y}'$ have essentially the same overall statistical properties as the original data $\boldsymbol{X}'$.

However, what is gained is that the first component of the data $\boldsymbol{Y}'$, $\boldsymbol{y}_1' = (\boldsymbol{X} - \overline{\boldsymbol{X}})'\boldsymbol{a}_1$, which is a linear combination of the original observations, has the maximum possible variance amongst all such linear combinations. This is not obvious and requires some constrained optimization involving calculus to prove so the details of the proof are deferred to §9.3. In this sense $\boldsymbol{y}_1'$ is the most important component of the data. Further, it can be shown that the second component of $\boldsymbol{Y}'$ given by $\boldsymbol{y}_2' = (\boldsymbol{X} - \overline{\boldsymbol{X}})'\boldsymbol{a}_2$ is the second most important component, having the maximum possible variance amongst all such linear combinations of the data subject to the constraint of being orthogonal with the first.

Since statistical information is measured by variance, the first implication of this important property is that the major statistical features of the data (such as dividing into subgroups etc.) are likely to be exhibited in the first few components of $\boldsymbol{Y}'$, $\boldsymbol{y}_1, \boldsymbol{y}_2, \ldots, \boldsymbol{y}_k$, where k is chosen so that the variances of $\boldsymbol{y}_j'$ are negligible for $j > k$. It can be much easier to examine just k components (e.g., with pairwise scatterplots) than it is if the search is amongst a much larger number p. In this sense we could say we have reduced the dimensionality of the data from p to k. This can be a considerable saving in effort if p is large and k is relatively small (less than 10 say, or at least appreciably smaller than p).

In §5.7.3 we mentioned that some multivariate techniques were impossible to perform on data where $n < p$, i.e., there are fewer observations than dimensions. One solution is to perform a PCA to reduce the dimensionality from p to k and perform the subsequent analyses on just the reduced data matrix composed of the first k columns of $\boldsymbol{Y}'$.

Similarly in linear models such as $\boldsymbol{y} = \boldsymbol{X}\boldsymbol{\beta} + \boldsymbol{\varepsilon}$, if the design matrix $\boldsymbol{X}$ is not of full column rank, one solution is to consider the principal components of $\boldsymbol{X}$ and investigate the regression of $\boldsymbol{y}$ on the first k principal components of $\boldsymbol{X}$, choosing k so that this reduced design matrix is of full column rank, a technique known as ***principal component regression***.

6.6 Matrix Exponential

If $\boldsymbol{X}$ is a $n \times n$ square matrix, then the ***exponential of X*** is defined to be

$$e^{\boldsymbol{X}} = \exp(\boldsymbol{X}) = 1 + \boldsymbol{X} + \frac{\boldsymbol{X}^2}{2!} + \frac{\boldsymbol{X}^3}{3!} + \ldots = \sum_{r=1}^{\infty} \frac{\boldsymbol{X}^r}{r!}.$$

If $\boldsymbol{X}$ has distinct eigenvalues, then $\boldsymbol{X} = \boldsymbol{T}\Lambda\boldsymbol{T}^{-1}$, §6.4.5, where Λ is a diagonal matrix of the eigenvalues of $\boldsymbol{X}$. Then

$$\exp(\boldsymbol{X}) = \sum_{r=1}^{\infty} \frac{\boldsymbol{X}^r}{r!} = \boldsymbol{T} \begin{pmatrix} e^{\lambda_1} & 0 & \cdots & 0 \\ 0 & e^{\lambda_2} & \cdots & 0 \\ \vdots & \vdots & \ddots & \vdots \\ 0 & 0 & \cdots & e^{\lambda_n} \end{pmatrix} \boldsymbol{T}^{-1}.$$

If $\boldsymbol{A}$ is not similar to a diagonal matrix, then it can still be shown (e.g., §11.4, Banerjee and Roy, 2014) that the exponential series above converges and so $\exp(\boldsymbol{X})$ is properly defined. It is straightforward to establish basic elementary properties of the matrix exponential function and this is left to the exercises.

6.7 Decompositions

6.7.1 Spectral decomposition of a symmetric matrix

If $\boldsymbol{S}$ is a symmetric matrix with eigenvalues $\lambda_1 \geq \lambda_2 \geq \ldots \geq \lambda_n$ and eigenvectors $(\boldsymbol{x}_1, \boldsymbol{x}_2, \ldots, \boldsymbol{x}_n)$ we have $\boldsymbol{S} = \boldsymbol{T}\Lambda\boldsymbol{T}'$, with $\boldsymbol{T}'\boldsymbol{T} = \mathbf{I_n}$ (§6.4.5(iii)). This is known as the ***spectral decomposition*** of $\boldsymbol{S}$. It is often expressed in the form of a sum of rank 1 matrices: If $\boldsymbol{T} = (\boldsymbol{x}_1, \boldsymbol{x}_2, \ldots, \boldsymbol{x}_n)$, then $\boldsymbol{T}\Lambda = (\lambda_1\boldsymbol{x}_1, \lambda_2\boldsymbol{x}_2, \ldots, \lambda_n\boldsymbol{x}_n)$ and so

$\boldsymbol{T\Lambda T'} = \sum_j \lambda_j \boldsymbol{x}_j\boldsymbol{x}'_j$. Each of the matrices $\boldsymbol{x}_j\boldsymbol{x}'_j$ is of rank 1. If there are r non-zero eigenvalues (so $\lambda_{r+1} = \ldots\ldots = \lambda_n = 0$), then the summation is over r terms.

6.7.1.1 Square root of positive semi-definite matrices

If $\boldsymbol{S} \geq 0$, then all of its eigenvalues are non-negative, i.e., $\lambda_i \geq 0$ and also we have $\boldsymbol{S} = \boldsymbol{T\Lambda T'}$, with $\boldsymbol{T'T} = \mathbf{I_n}$ and $\Lambda = \text{diag}(\lambda_1, \lambda_2, \ldots, \lambda_n)$. Define $\Lambda^{1/2}$ by $\Lambda^{1/2} = \text{diag}(\lambda_1^{1/2}, \lambda_2^{1/2}, \ldots, \lambda_n^{1/2})$. Then let $\boldsymbol{S}^{1/2} = \boldsymbol{T}\Lambda^{1/2}\boldsymbol{T'}$ and $(\boldsymbol{S}^{1/2})^2 = \boldsymbol{T}\Lambda^{1/2}\boldsymbol{T'T}\Lambda^{1/2}\boldsymbol{T'} = \boldsymbol{T}\Lambda^{1/2}\Lambda^{1/2}\boldsymbol{T'} = \boldsymbol{T\Lambda T'} = \boldsymbol{S}$ and so $\boldsymbol{S}^{1/2}$ is a square root of $\boldsymbol{S}$. Note that there are many other matrices $\boldsymbol{Q}$ such that $\boldsymbol{Q}^2 = \boldsymbol{S}$ but $\boldsymbol{S}^{1/2}$ is the only one such that $\boldsymbol{T}$ is orthogonal. Other [positive] powers of $\boldsymbol{S}$ can be defined similarly. If $\boldsymbol{S}$ is positive definite, then negative powers can also be defined in a similar way.

6.7.2 Singular value decomposition (svd) of an $m \times n$ matrix

If $\boldsymbol{A}$ is an $m \times n$ matrix with $\rho(\boldsymbol{A}) = r \leq \min(m,n)$, then there are orthogonal matrices $\boldsymbol{U}$ and $\boldsymbol{V}$ and a diagonal matrix Λ with positive diagonal elements such that $\boldsymbol{A} = \boldsymbol{U}\Lambda^{1/2}\boldsymbol{V'}$. The elements $\Lambda^{1/2}$ are called the ***singular values*** of the matrix $\boldsymbol{A}$.

6.7.2.1 Proof of svd

To prove this, first note that $\boldsymbol{AA'}$ and $\boldsymbol{A'A}$ are both positive semi-definite since $\boldsymbol{x'AA'x} = (\boldsymbol{A'x})'(\boldsymbol{A'x}) \geq 0$ (likewise $\boldsymbol{A'A}$) and they have the same non-zero eigenvalues; see §6.4.3(vi), $\lambda_1, \lambda_2, \ldots, \lambda_r$ say. If we define $\boldsymbol{U}$ to be the $m \times r$ matrix of eigenvectors of $\boldsymbol{AA'}$ corresponding to the non-zero eigenvalues, so $\boldsymbol{AA'U} = \boldsymbol{U}\Lambda$ and the $m \times r$ matrix $\boldsymbol{U}_2$ to be chosen so that it is orthogonal and $\boldsymbol{AA'U}_2 = O$, i.e., the columns of $\boldsymbol{U}_2$ corresponds to the zero eigenvalues of $\boldsymbol{AA'}$, then $\boldsymbol{UU'} + \boldsymbol{U}_2\boldsymbol{U}'_2 = \mathbf{I_m}$. Define $\boldsymbol{V} = \boldsymbol{A'U}\Lambda^{-1/2}$. Then we have $\boldsymbol{A'AV} = \boldsymbol{V}\Lambda$ and $\boldsymbol{V'V} = \mathbf{I_r}$. Since $\boldsymbol{AA'U}_2 = O$ we must have $\boldsymbol{A'U}_2 = O$; see §2.13.1; so $\boldsymbol{A} = \mathbf{I_m}\boldsymbol{A} = (\boldsymbol{UU'} + \boldsymbol{U}_2\boldsymbol{U}'_2)\boldsymbol{A} = \boldsymbol{UU'A} = \boldsymbol{U}\mathbf{I_r}\boldsymbol{U'A} = \boldsymbol{U}\Lambda^{1/2}\Lambda^{-1/2}\boldsymbol{U'A} = \boldsymbol{U}\Lambda^{1/2}\boldsymbol{V'}$.

6.7.2.2 Note

Note that $\boldsymbol{U}$ and $\boldsymbol{V}$ are eigenvectors of the symmetric matrices $\boldsymbol{AA'}$ and $\boldsymbol{A'A}$, each with eigenvalues Λ. These are termed the ***left singular vectors*** and ***right singular vectors*** of $\boldsymbol{A}$.

6.7.3 Implementation of svd in R

The command for producing the singular value decomposition of a matrix is `svd()`. This produces the singular values and left and right singular vectors of a matrix. The singular values of a $m \times n$ matrix $\boldsymbol{A}$ are held in a vector `svd(A)$d` of length $\min(m,n)$, the left singular vectors are stored in a matrix `svd(A)$u` of dimensions $(\min(m,n), m)$ and the right singular vectors in a matrix `svd(A)$v` of dimensions $(\min(m,n), m)$. Notice that if $\boldsymbol{A}$ has rank $r < \min(m,n)$, the matrices $\boldsymbol{U}$ and $\boldsymbol{V}$

produced by **R** contain singular vectors corresponding to zero singular values (i.e., eigenvectors of matrices AA' and $A'A$ corresponding to zero eigenvalues). These are not actually required for the singular value decomposition but they can easily be discarded by extracting just the columns of `svd(A)$u` and `svd{A}$v` corresponding to the non-zero eigenvalues, i.e., using `svd(A)$u[,1:r]` and `svd(A)$v[,1:r]`. This is illustrated in the last two examples below.

Example 6.3

```
> options(digits=3)
```

(i) A symmetric matrix

```
> S<-matrix(c(2,1,1,1,2,1,1,
+ 1,3),3,3)
> S
     [,1] [,2] [,3]
[1,]    2    1    1
[2,]    1    2    1
[3,]    1    1    3
```

```
> eigen(S)
$values
[1] 4.41 1.59 1.00
$vectors
       [,1]   [,2]   [,3]
[1,] -0.500 -0.500  0.707
[2,] -0.500 -0.500 -0.707
[3,] -0.707  0.707      0
```

```
> # Check on spectral
> # decomposition of S
> eigen(S)$vectors%*%
+ diag(eigen(S)$values)
+ %*%t(eigen(S)$vectors)
     [,1] [,2] [,3]
[1,]    2    1    1
[2,]    1    2    1
[3,]    1    1    3
> svd(S)
$d
[1] 4.41 1.59 1.00
$u
       [,1]   [,2]   [,3]
[1,] -0.500  0.500  0.707
[2,] -0.500  0.500 -0.707
[3,] -0.707 -0.707      0
```

```
$v
       [,1]   [,2]   [,3]
[1,] -0.500  0.500  0.707
[2,] -0.500  0.500 -0.707
[3,] -0.707 -0.707      0

> # Check on svd of S:
> svd(S)$u%*%diag(svd(S)$
+ d)%*%t(svd(S)$v)
     [,1] [,2] [,3]
[1,]    2    1    1
[2,]    1    2    1
[3,]    1    1    3
```

```
> eigen(S%*%t(S))

$values
[1] 19.49  2.51  1.00

$vectors
        [,1]   [,2]   [,3]
[1,] -0.500  0.500  0.707
[2,] -0.500  0.500 -0.707
[3,] -0.707 -0.707      0
```

```
> eigen(t(S)%*%S)

$values
[1] 19.49  2.51  1.00

$vectors
        [,1]   [,2]   [,3]
[1,] -0.500  0.500  0.707
[2,] -0.500  0.500 -0.707
[3,] -0.707 -0.707      0
```

Note that in this case the eigenvectors of $\boldsymbol{S}$, $\boldsymbol{SS}'$ and $\boldsymbol{S}'\boldsymbol{S}$ are identical and the eigenvectors are the same as both of the $\boldsymbol{U}$ and $\boldsymbol{V}$ matrices of the singular value decomposition. This is because $\boldsymbol{S}$ is symmetric. The eigenvalues of $\boldsymbol{SS}'$ and $\boldsymbol{S}'\boldsymbol{S}$ are the squares of the eigenvalues of $\boldsymbol{S}$ and the singular values produced by the svd. For this reason some texts rather misleadingly refer to the eigenvalues of *symmetric* matrices as singular values but this is only true for symmetric matrices. The next example illustrates that eigenvalues and singular values of square matrices are in general diffferent. Eigenvalues are not defined for matrices which are not square but singular values always exist for both square and non-square matrices.

(ii) A non-symmetric matrix

```
> A<-matrix(c(2,1,1,1,2,1,4,
+ 2,3),3,3)
> A
     [,1] [,2] [,3]
[1,]    2    1    4
[2,]    1    2    2
[3,]    1    1    3
> eigen(A)
$values
[1] 5.45 1.00 0.55

$vectors
      [,1]   [,2]   [,3]
[1,] 0.716  0.707 -0.925
[2,] 0.494 -0.707  0.268
[3,] 0.494      0  0.268
```

```
> svd(A)
$d
[1] 6.265 1.269 0.377

$u
        [,1]   [,2]   [,3]
[1,] -0.725  0.458 -0.514
[2,] -0.444 -0.882 -0.160
[3,] -0.526  0.113  0.843

$v
        [,1]   [,2]    [,3]
[1,] -0.386  0.117 -0.9149
[2,] -0.342 -0.940  0.0245
[3,] -0.857  0.322  0.4028
```

```
> # Check on svd of A:
> svd(A)$u%*%diag(svd(A)$d)%*%t(svd(A)$v)
     [,1] [,2] [,3]
[1,]    2    1    4
[2,]    1    2    2
[3,]    1    1    3
>
```

```
> eigen(A%*%t(A))
$values
[1] 39.248  1.610  0.142

$vectors
       [,1]   [,2]   [,3]
[1,] -0.725  0.458  0.514
[2,] -0.444 -0.882  0.160
[3,] -0.526  0.113 -0.843
```

```
> eigen(t(A)%*%A)
$values
[1] 39.248  1.610  0.142

$vectors
       [,1]   [,2]    [,3]
[1,] -0.386  0.117  0.9149
[2,] -0.342 -0.940 -0.0245
[3,] -0.857  0.322 -0.4028
```

Note that in this example the eigenvalues of $\boldsymbol{A}$ are different from the singular values which are the square roots of the eigenvalues of $\boldsymbol{AA'}$ and $\boldsymbol{A'A}$. The eigenvectors of $\boldsymbol{AA'}$ are identical to the $\boldsymbol{U}$ matrix of the singular value decomposition and those of $\boldsymbol{A'A}$ are identical to the $\boldsymbol{V}$ matrix of the svd.

(iii) A non-square matrix of full rank

```
> B<-matrix(c(1,2,3,4,5,6)
+ ,2,3)
> B
     [,1] [,2] [,3]
[1,]    1    3    5
[2,]    2    4    6
> svd(B)
$d
[1] 9.526 0.514
```

```
$u
       [,1]   [,2]
[1,] -0.620 -0.785
[2,] -0.785  0.620
$v
       [,1]   [,2]
[1,] -0.230  0.883
[2,] -0.525  0.241
[3,] -0.820 -0.402
```

```
> eigen(B%*%t(B))
$values
[1] 90.735  0.265

$vectors
      [,1]   [,2]
[1,] 0.620 -0.785
[2,] 0.785  0.620
```

```
> eigen(t(B)%*%B)
$values
[1]  9.07e+01  2.65e-01 0

$vectors
       [,1]   [,2]   [,3]
[1,] -0.230  0.883  0.408
[2,] -0.525  0.241 -0.816
[3,] -0.820 -0.402  0.408
```

```
> # Check on svd of B
> svd(B)$u%*%diag(svd(B)$d)%*%t(svd(B)$v)
     [,1] [,2] [,3]
[1,]    1    3    5
[2,]    2    4    6
>
```

Note that here the 2×3 matrix $\boldsymbol{B}$ is of full rank so there are no zero singular values. The matrix $\boldsymbol{BB}'$ is 2×2 and has rank 2 and so both eigenvalues are non-zero but the 3×3 matrix $\boldsymbol{B}'\boldsymbol{B}$ has rank 2 and so has one zero eigenvalue.

(iv) A non-square matrix not of full rank

```
> X<-matrix(c(3,2,1,4,2,0,5,
+ 2,-1,-1,0,1),4,3,byrow=T)
> X
     [,1] [,2] [,3]
[1,]    3    2    1
[2,]    4    2    0
[3,]    5    2   -1
[4,]   -1    0    1
> svd(X)
```

```
$d
[1] 7.9  1.90  0.0
```

Note one zero singular value so final eigenvector is arbitrary, though harmless and irrelevant.

```
$u
      [,1]  [,2]   [,3]
[1,] -0.44  0.68  0.069
[2,] -0.56  0.12  0.504
[3,] -0.68 -0.44 -0.573
[4,]  0.12  0.56 -0.642

$v
       [,1]  [,2]  [,3]
[1,] -0.902 -0.14  0.41
[2,] -0.428  0.39 -0.82
[3,]  0.045  0.91  0.41
> U<-svd(X)$u
> V<-svd(X)$v
> D<-svd(X)$d
> # Check on SVD of X
> U%*%diag(D)%*%t(V)
     [,1]    [,2]     [,3]
[1,]    3 2.0e+00  1.0e+00
[2,]    4 2.0e+00  2.1e-16
[3,]    5 2.0e+00 -1.0e+00
[4,]   -1 2.5e-16  1.0e+00
```

```
> # Now use only non-zero
> # eigenvectors & values
>
> UU<-U[,1:2]
>
> VV<-V[,1:2]
>
> DD<-diag(D[1:2])
>
> # Check on SVD of X with
> # this reduced set
>
> UU%*%DD%*%t(VV)
     [,1]    [,2]     [,3]
[1,]    3 2.0e+00  1.0e+00
[2,]    4 2.0e+00  1.4e-16
[3,]    5 2.0e+00 -1.0e+00
[4,]   -1 5.6e-17  1.0e+00
```

This example illustrates that retaining the singular vectors corresponding to zero singular values does not affect the decomposition in the sense that we still have $\boldsymbol{UDV}' = \boldsymbol{X}$ even if $\boldsymbol{U}$ and $\boldsymbol{V}$ include the arbitrary columns corresponding to zero singular values. Consequently, it may be that there is little practical value in discarding these columns (as performed with $\boldsymbol{UU}$, $\boldsymbol{VV}$ and $\boldsymbol{DD}$), especially if there are only very few as is the case in this example. It could be sensible to do so, however, if the original matrix is large (many tens of rows and columns or more, say) but of low rank. Carrying several tens of irrelevant dimensions in $\boldsymbol{U}$ and $\boldsymbol{V}$ could decrease numerical accuracy by introducing more rounding errors as well as adding to the computational task of matrix manipulations.

(v) A 3×3 Jordan matrix of rank 2

```
> J<-matrix(c(0,1,0,0,0,1,
+ 0,0,0),3,3)
> J
     [,1] [,2] [,3]
[1,]    0    0    0
[2,]    1    0    0
[3,]    0    1    0
```

```
> eigen(J)
$values
[1] 0 0 0
$vectors
     [,1]  [,2] [,3]
[1,]    0     0    0
[2,]    0     0    0
[3,]    1    -1    0
```

```
> svd(J)
$d
[1] 1 1 0
$u
     [,1] [,2] [,3]
[1,]    0    0    1
[2,]    0   -1    0
[3,]   -1    0    0
$v
     [,1] [,2] [,3]
[1,]    0   -1    0
[2,]   -1    0    0
[3,]    0    0    1
```

```
> U<-svd(J)$u[,1:2]
> V<-svd(J)$v[,1:2]
> D<-diag(svd(J)$d[1:2])
> U%*%D%*%t(V)
     [,1] [,2] [,3]
[1,]    0    0    0
[2,]    1    0    0
[3,]    0    1    0
```

6.8 Eigenanlysis of Matrices with Special Structures

To show that a vector $\boldsymbol{x}$ and scalar λ are an eigenvector and eigenvalue of a matrix $\boldsymbol{A}$, it is only necessary to demonstrate that $\boldsymbol{Ax} = \lambda\boldsymbol{x}$. Sometimes the matrix $\boldsymbol{A}$ has a particular structure that can be manipulated, perhaps using the useful tricks indicated in §2.8. Sometimes this may be deceptively simple and not obvious without experience. Particular results relate to rank 1 matrices and matrices composed as the sum of a rank 1 matrix with a scalar multiple of the identity matrix.

6.8.1 The rank one matrix $\boldsymbol{xx}'$

If $\boldsymbol{x}$ is a vector of length n, then $\boldsymbol{xx}'$ is an $n \times n$ matrix of rank 1 (since $\rho(\boldsymbol{xx}') < \min(\rho(\boldsymbol{x}), \rho(\boldsymbol{x}')) = 1$). We have $\boldsymbol{xx}'\boldsymbol{x} = \boldsymbol{x}(\boldsymbol{x}'\boldsymbol{x}) = (\boldsymbol{x}'\boldsymbol{x})\boldsymbol{x}$, noting that $(\boldsymbol{x}'\boldsymbol{x})$ is a scalar and so commutes with the vector $\boldsymbol{x}$. This is in the form $\boldsymbol{Ax} = \lambda\boldsymbol{x}$ with $\boldsymbol{A} = \boldsymbol{xx}'$ and $\lambda = (\boldsymbol{x}'\boldsymbol{x})$ and so $\boldsymbol{x}$ is an eigenvector of $\boldsymbol{xx}'$ with eigenvalue $(\boldsymbol{x}'\boldsymbol{x})$. Since $\boldsymbol{xx}'$ is symmetric and of rank 1 it has only one non-zero eigenvalue.

6.8.2 The rank one matrix $\boldsymbol{Sxx}'$

If $\boldsymbol{x}$ is a vector of length n, then $\boldsymbol{Sxx}'$ is an $n \times n$ matrix of rank 1 (since $\rho(\boldsymbol{Sxx}') < \min(\rho(\boldsymbol{S}), \rho(\boldsymbol{xx}')) = 1$). We have $\boldsymbol{Sxx}'\boldsymbol{Sx} = \boldsymbol{Sx}(\boldsymbol{x}'\boldsymbol{Sx}) = (\boldsymbol{x}'\boldsymbol{Sx})\boldsymbol{Sx}$, noting that $(\boldsymbol{x}'\boldsymbol{Sx})$ is a scalar and so commutes with the vector $\boldsymbol{Sx}$. This is in the form $\boldsymbol{Ax} = \lambda\boldsymbol{x}$ with $\boldsymbol{A} = \boldsymbol{Sxx}'$ and $\lambda = (\boldsymbol{x}'\boldsymbol{Sx})$ and so $\boldsymbol{Sx}$ is an eigenvector of $\boldsymbol{Sxx}'$ with eigenvalue $(\boldsymbol{x}'\boldsymbol{Sx})$. Since $\boldsymbol{Sxx}'$ is of rank 1 it has only one non-zero eigenvalue (see §6.4.7).

6.8.3 The matrix $a\mathbf{I}_n + b\boldsymbol{xy}'$

$(a\mathbf{I}_n + b\boldsymbol{xy}')\boldsymbol{x} = a\mathbf{I}_n\boldsymbol{x} + b\boldsymbol{xy}'\boldsymbol{x} = a\boldsymbol{x} + b(\boldsymbol{y}'\boldsymbol{x})\boldsymbol{x}$ (noting that $\boldsymbol{y}'\boldsymbol{x}$ is a scalar and so commutes with $\boldsymbol{x}$). Thus $(a\mathbf{I}_n + b\boldsymbol{xy}')\boldsymbol{x} = (a + b\boldsymbol{y}'\boldsymbol{x})\boldsymbol{x}$ and $\boldsymbol{x}$ is an eigenvector of $a\mathbf{I}_n + b\boldsymbol{xy}'$ with eigenvalue $(a + b\boldsymbol{y}'\boldsymbol{x})$. The rank of $a\mathbf{I}_n + b\boldsymbol{xy}'$ is not in general 1 (e.g., consider $a = 1$ and $b = 0$) and so will in general have other non-zero eigenvalues and corresponding non-trivial eigenvectors.

To find the other eigenvalues, consider $|a\mathbf{I}_n + b\boldsymbol{xy}' - \lambda\mathbf{I}_n| = |(a-\lambda)\mathbf{I}_n + b\boldsymbol{xy}'| = (a-\lambda)^n|\mathbf{I}_n + b\boldsymbol{xy}'/(a-\lambda)| = (a-\lambda)^n|1 + b\boldsymbol{y}'\boldsymbol{x}/(a-\lambda)|$; see §4.6.2; $= (a-\lambda)^{n-1}(a + b\boldsymbol{y}'\boldsymbol{x} - \lambda)$ and so the other eigenvalues are a with multiplicity $n-1$. Note that if $\boldsymbol{x} = \boldsymbol{y}$ and $a \neq 0$ and $a + b\boldsymbol{y}'\boldsymbol{x} \neq 0$, the matrix is symmetric with n non-zero eigenvalues and so is of full rank and thus non-singular.

6.9 Summary of Key Results

$n \times n$ matrix A with eigenvalues $\lambda_1, \lambda_2, \ldots, \lambda_n$ and [right] eigenvectors $x_1, x_2, \ldots, x_n$.

(1) $\sum_{i=1}^{n} \lambda_i = \text{trace}(A)$.

(2) $\prod_{i=1}^{n} \lambda_i = |A|$.

(3) A and CAC^{-1} have identical eigenvalues for C non-singular.

(4) Eigenvectors of CAC^{-1} are Cx_i.

(5) AB and BA have identical non-zero eigenvalues.

(6) Eigenvectors of $BA = B$ times those of AB.

(7) If A is symmetric, the eigenvalues of A are real.

(8) If A is symmetric, the eigenvectors corresponding to distinct eigenvalues are orthogonal.

(9) The single non-zero eigenvalue of the $n \times n$ rank 1 matrix xx' is $x'x$ with corresponding eigenvector x.

(10) The single non-zero eigenvalue of the $n \times n$ rank 1 matrix Sxx' is $x'Sx$ with corresponding eigenvector Sx.

6.10 Exercises

(1) Let $X = \begin{pmatrix} 0 & 0 & 6 \\ 1/2 & 0 & 0 \\ 0 & 1/3 & 0 \end{pmatrix}$.

(i) Show that $|X| = 1$.

(ii) Show that $\lambda = 1$ is an eigenvalue of X.

(iii) Show that $X^3 = \mathbf{I_n}$ but $X^2 \neq \mathbf{I_n}$.

(2) Find the eigenvalues of the $n \times n$ matrix $\begin{pmatrix} 1 & \rho & \rho & \cdots & \rho \\ \rho & 1 & \rho & \cdots & \rho \\ \vdots & \vdots & \ddots & \vdots & \vdots \\ \rho & \cdots & \cdots & \ddots & \rho \\ \rho & \rho & \cdots & \rho & 1 \end{pmatrix}$ and show that one eigenvector is proportional to $\imath_{\mathbf{n}}$.

(3) If $\boldsymbol{X}$ has eigenvalues $2\lambda,\ \lambda+\alpha$ and $\lambda+3\alpha$ and $|\boldsymbol{X}| = 80$ and $\mathrm{tr}(\boldsymbol{X}) = 16$ find the eigenvalues of $\boldsymbol{X}$.

(4) If λ is an eigenvalue of $\boldsymbol{X}$ show that $\lambda+\alpha$ is an eigenvalue of $\boldsymbol{X}+\alpha\mathbf{I_n}$.

(5) If $\boldsymbol{x}$ is a $n \times 1$ vector find the spectral decomposition of $\boldsymbol{xx}'$.

(6) If $\boldsymbol{x}$ is a $n \times 1$ vector and $\boldsymbol{S}$ is a symmetric $n \times n$ matrix find the spectral decomposition of $\boldsymbol{xSx}'$.

(7) If $\boldsymbol{T} = k\boldsymbol{S}$ where k is a scalar, show that $\boldsymbol{T}$ and $\boldsymbol{S}$ have identical eigenvectors and the eigenvalues of $\boldsymbol{T}$ are obtained by multiplying those of $\boldsymbol{S}$ by k.

(8) Find the square root of the matrix

$$\boldsymbol{S} = \begin{pmatrix} 2 & 1 & 1 \\ 1 & 2 & 1 \\ 1 & 1 & 3 \end{pmatrix}.$$

(9) If $\boldsymbol{X}$ is 2×2, show that $|\boldsymbol{X}+\mathbf{I_n}| = |\boldsymbol{X}|+1$ if and only if $\mathrm{tr}(\boldsymbol{A}) = 0$.

(10) Suppose $\boldsymbol{X}$ is $n \times n$ matrix with distinct eigenvalues.

(a) Show that $\exp(\alpha\boldsymbol{X})\exp(\beta\boldsymbol{X}) = \exp((\alpha+\beta)\boldsymbol{X})$.
(b) Show that $\boldsymbol{X}\exp(\boldsymbol{X}) = \exp(\boldsymbol{X})\boldsymbol{X}$.
(c) Show that $\left(\exp(\boldsymbol{X})\right)^r = \exp(r\boldsymbol{X})$ where r is an integer and $r > 0$.
(d) Show that $\left(\exp(\boldsymbol{X})\right)^{-1} = \exp(-\boldsymbol{X})$.
(e) Show that $\left(\exp(\boldsymbol{X})\right)' = \exp(\boldsymbol{X}')$.
(f) Show that $|\exp(\boldsymbol{X})| = \exp(\mathrm{tr}(\boldsymbol{X}))$.

(11) If $\boldsymbol{X}$ and $\boldsymbol{Y}$ are $n \times n$ matrices, each with distinct eigenvalues, and $\boldsymbol{XY} = \boldsymbol{YX}$, show that $\exp(\boldsymbol{X}+\boldsymbol{Y}) = \exp(\boldsymbol{X})\exp\boldsymbol{Y}) = \exp(\boldsymbol{Y})\exp\boldsymbol{X})$.

(12) Suppose $\boldsymbol{A}$ is idempotent and symmetric,

(i) Show that the eigenvalues of $\boldsymbol{A}$ are 1 or 0.
(ii) Show that $\mathrm{tr}(\boldsymbol{A}) = \rho(\boldsymbol{A})$.
(iii) Show that $\rho(\mathbf{I_n} - \boldsymbol{A}) = n - \rho(\boldsymbol{A})$.

(13) Where is the fallacy in the 'deceptively obvious proof' of the Cayley-Hamilton theorem, $p_A(\lambda) = |\boldsymbol{A} - \lambda\mathbf{I_n}|$ so $p_A(\boldsymbol{A}) = |\boldsymbol{A} - \boldsymbol{A}\mathbf{I_n}| = |\boldsymbol{A} - \boldsymbol{A}| = 0$?

(14) If $\boldsymbol{A}$ is nilpotent show that all eigenvalues of $\boldsymbol{A}$ are 0.

(15) If all eigenvalues of $\boldsymbol{A}$ are 0, show that $\boldsymbol{A}$ is nilpotent.

(16) Suppose$\boldsymbol{A}$ is a real $n \times n$ skew-symmetric matrix.

(i) Show that the only real eigenvalue of $\boldsymbol{A}$ is 0.
(ii) Show that if n is odd, then $\boldsymbol{A}$ has at least one eigenvalue of 0.

7

Vector and Matrix Calculus

7.1 Introduction

This section considers various simple cases of differentiation of scalar-valued, vector-valued and matrix-valued functions of scalars, vectors and matrices with respect to scalars, vectors and matrices. For example, the quadratic form $\boldsymbol{x}'\boldsymbol{A}\boldsymbol{x}$ is a scalar function of a vector $\boldsymbol{x}$, $\boldsymbol{x}\boldsymbol{x}'$ is a matrix function of a vector $\boldsymbol{x}$, $\text{tr}(\boldsymbol{X})$ and $|\boldsymbol{X}|$ are scalar functions of a matrix $\boldsymbol{X}$ and $\boldsymbol{A}\boldsymbol{x}$ is a vector function of a vector $\boldsymbol{x}$. Not all combinations of these will be covered here.

Broadly, the procedure consists of [partially] differentiating each element of the function with respect to each element of the arguments and arranging the results in a vector or matrix as appropriate. We use partial differentiation if the argument is not a scalar; this is equivalent to differentiation with respect to the individual elements. Thus the result of differentiating a scalar with respect to a vector [matrix] will consist of the vector [matrix] of partial derivatives of the scalar with respect to the elements of the vector [matrix]. Differentiating a vector-valued function of a vector argument with respect to another vector will result in a matrix where the individual elements are the partial derivatives of each element of the function with respect to each element of the vector argument. Generally, differentiating an $m \times n$ matrix with respect to a $p \times q$ matrix can be defined and will result in a matrix of dimension $mp \times nq$. We consider only the cases where not only is one of m and p equal to 1 but also one of n and q equals 1. Other cases can be handled with the use of Kronecker products and vec operators (Chapter 8) and some of the exercises of this chapter illustrate some of the simpler cases requiring these further ideas.

Most of the basic results can be obtained by expressing the functions of the vectors in terms of the individual elements and expanding, for example, inner products of vectors as sums of products of individual elements. It may help understanding to write out explicitly the cases $n = 1$ and $n = 2$. Many of the basic rules of scalar calculus of single variables (e.g., differentiation of products etc.) carry through recognisably to the vector and matrix cases.

7.2 Differentiation of a Scalar with Respect to a Vector

If $\boldsymbol{x}$ is a vector of length n and $f = f(\boldsymbol{x})$ is a scalar function of $\boldsymbol{x}$ then $\frac{\partial f}{\partial \boldsymbol{x}}$ is ***defined*** to be the vector

$$\frac{\partial f}{\partial \boldsymbol{x}} = \begin{pmatrix} \frac{\partial f}{\partial x_1} \\ \frac{\partial f}{\partial x_2} \\ \vdots \\ \frac{\partial f}{\partial x_n} \end{pmatrix}.$$

7.2.1 Derivative of $\boldsymbol{a}'\boldsymbol{x}$

$f(\boldsymbol{x}) = \boldsymbol{a}'\boldsymbol{x} = \sum_j a_j x_j$ so $\frac{\partial f}{\partial x_j} = \frac{\partial(\sum_j a_j x_j)}{\partial x_j} = a_j$ so

$$\frac{\partial f}{\partial \boldsymbol{x}} = \frac{\partial \boldsymbol{a}'\boldsymbol{x}}{\partial \boldsymbol{x}} = \begin{pmatrix} a_1 \\ a_2 \\ \vdots \\ a_n \end{pmatrix} = \boldsymbol{a}.$$

7.2.2 Derivative of $\boldsymbol{x}'\boldsymbol{x}$

$f(\boldsymbol{x}) = \boldsymbol{x}'\boldsymbol{x} = \sum_j x_j^2$ so $\frac{\partial f}{\partial x_j} = \frac{\partial(\sum_j x_j^2)}{\partial x_j} = 2x_j$ so

$$\frac{\partial f}{\partial \boldsymbol{x}} = \frac{\partial \boldsymbol{x}'\boldsymbol{x}}{\partial \boldsymbol{x}} = \begin{pmatrix} 2x_1 \\ 2x_2 \\ \vdots \\ 2x_n \end{pmatrix} = 2\boldsymbol{x}.$$

7.2.3 Derivative of quadratic forms $\boldsymbol{x}'\boldsymbol{S}\boldsymbol{x}$

We can, without loss of generality, take $\boldsymbol{S}$ to be symmetric (see §2.9 on Page 36). First consider the special cases of $n = 1$ and $n = 2$:

Case $n = 1$: $\boldsymbol{x} = (x_1)$, $\boldsymbol{S} = (s_{11})$, $f(\boldsymbol{x}) = x_1 s_{11} x_1 = x_1^2 s_{11}$ so $\frac{\partial f}{\partial \boldsymbol{x}} = 2s_{11}x_1 = 2\boldsymbol{S}\boldsymbol{x}$.

Case $n = 2$: i.e., $\boldsymbol{x} = (x_1.x_2)'$, $\boldsymbol{S} = \begin{pmatrix} s_{11} & s_{12} \\ s_{12} & s_{22} \end{pmatrix}$ then $\boldsymbol{x}'\boldsymbol{S}\boldsymbol{x} = x_1^2 + 2x_1x_2s_{12} + x_2^2 s_{22}$ so $\frac{\partial f}{\partial \boldsymbol{x}} = (\frac{\partial f}{\partial x_1}, \frac{\partial f}{\partial x_2})' = ((2x_1s_{11} + 2x_2s_{12}), (2x_1s_{12} + 2x_2s_{22}))' = 2\boldsymbol{S}\boldsymbol{x}$.

General case: $f(\boldsymbol{x}) = \boldsymbol{x}'\boldsymbol{S}\boldsymbol{x} = \sum_k \sum_j x_k s_{kj} x_j = \sum_k \sum_{j, j\neq k} x_k s_{kj} x_j + \sum_j s_{jj} x_j^2$ so

$$\begin{aligned}\frac{\partial f}{\partial x_i} &= \frac{\partial(\sum_k \sum_{j,\ j\neq k} x_k s_{kj} x_j + \sum_j s_{jj} x_j^2)}{\partial x_i} = \sum_{j, j\neq i} s_{ij} x_j + \sum_{k, k\neq i} s_{ik} x_k + 2 s_{ii} x_i \\ &= s_{ii} x_i + \sum_{j, j\neq i} s_{ij} x_j + \sum_{k, k\neq i} s_{ik} x_k + s_{ii} x_i = 2\boldsymbol{S}\boldsymbol{x}.\end{aligned}$$

Noting §2.9, clearly if $\boldsymbol{A}$ is not symmetric then the derivative of $\boldsymbol{x}'\boldsymbol{A}\boldsymbol{x}$ is $(\boldsymbol{A}+\boldsymbol{A}')x$.

7.3 Differentiation of a Scalar with Respect to a Matrix

If $\boldsymbol{X}$ is an $m \times n$ matrix (x_{ij}) and $f = f(\boldsymbol{X})$ is a scalar valued function of $\boldsymbol{X}$ then $\frac{\partial f}{\partial \boldsymbol{X}}$ is ***defined*** to be the $m \times n$ matrix

$$\frac{\partial f}{\partial \boldsymbol{X}} = \begin{pmatrix} \frac{\partial f}{\partial x_{11}} & \frac{\partial f}{\partial x_{12}} & \cdots & \frac{\partial f}{\partial x_{1n}} \\ \frac{\partial f}{\partial x_{21}} & \frac{\partial f}{\partial x_{22}} & \cdots & \frac{\partial f}{\partial x_{2n}} \\ \vdots & \vdots & \ddots & \vdots \\ \frac{\partial f}{\partial x_{m1}} & \frac{\partial f}{\partial x_{m2}} & \cdots & \frac{\partial f}{\partial x_{mn}} \end{pmatrix} = \left(\frac{\partial f}{\partial x_{ij}}\right).$$

Special care needs to be taken with this definition since the x_{ij} may not be functionally independent. For example, if the matrix $\boldsymbol{X}$ is symmetric so that $x_{ij} = x_{ji}$ then $\frac{\partial f}{\partial x_{ij}}$ may be different from the value obtained in the non-symmetric case. Symmetry is the most common situation in statistical applications where this arises but skew-symmetric matrices and also matrices of other special structures (diagonal, triangular, banded etc.) need careful handling. In the cases considered below, it is to be understood that there is no other functional relationship between the elements other than symmetry where that case is declared.

7.3.1 Derivative of tr(X)

If $\boldsymbol{X}$ is $n \times n$ then $f(\boldsymbol{X}) = \mathrm{tr}(\boldsymbol{X}) = \sum_k x_{kk}$ so $\frac{\partial f}{\partial x_{ij}} = 0$ if $i \neq j$ and 1 if $i = j$. Thus $\frac{\partial f}{\partial \boldsymbol{X}} = \mathbf{I_n}$.

7.3.2 Derivative of $a'Xa$ when X is not symmetric

If $\boldsymbol{X}$ is an $n \times n$ matrix then $\boldsymbol{a}'\boldsymbol{X}\boldsymbol{a} = \sum_i \sum_j a_i a_j x_{ij}$ so assuming $x_{ij} \neq x_{ji}$, $\frac{\partial(\boldsymbol{a}'\boldsymbol{X}\boldsymbol{a})}{\partial x_{ij}} = a_i a_j$. Thus $\frac{\partial(\boldsymbol{a}'\boldsymbol{X}\boldsymbol{a})}{\partial \boldsymbol{X}} = \boldsymbol{a}\boldsymbol{a}'$ provided $\boldsymbol{X}$ is not symmetric.

7.3.3 Derivative of $a'Xa$ when X is symmetric

If X is a symmetric $n \times n$ matrix and a is a vector of length n then

$$a'Xa = 2\sum_{i=1}^{n}\sum_{j=1}^{i-1} a_i a_j x_{ij} + \sum_{i=1}^{n} a_i^2 x_{ii},$$

so $\frac{\partial(a'Xa)}{\partial x_{ij}} = 2a_i a_j$ if $i \neq j$ and $\frac{\partial(a'Xa)}{\partial x_{ii}} = a_i^2 \;\; (i = j)$.
Thus $\frac{\partial(a'Xa)}{\partial X} = 2aa' - \text{diag}(\text{diag}(aa'))$ if X is symmetric.

7.3.4 Derivative of tr(XA) when X is not symmetric

If X and A are $n \times n$ matrices then $\text{tr}(XA) = \sum_i \sum_j x_{ij} a_{ji}$ so $\frac{\partial(\text{tr}(XA))}{\partial x_{ij}} = a_{ji}$ thus $\frac{\partial(\text{tr}(XA))}{\partial X} = A'$ provided X is not symmetric.

7.3.5 Derivative of tr(XA) when X is symmetric

If X and A are $n \times n$ matrices and X is symmetric then

$$\text{tr}(XA) = \sum_{i=1}^{n}\sum_{j=1}^{i-1} x_{ij}(a_{ij} + a_{ji}) + \sum_{i=1}^{n} x_{ii} a_{ii}$$

so $\frac{\partial(\text{tr}(XA))}{\partial x_{ij}} = a_{ij} + a_{ji}$ if $i \neq j$ and $\frac{\partial(\text{tr}(XA))}{\partial x_{ij}} = a_{ii}$ (if $i = j$).
Thus $\frac{\partial(\text{tr}(XA))}{\partial X} = A + A' - \text{diag}(\text{diag}(A))$ if X is symmetric.

7.3.6 Derivative of tr($A'XA$)

If X and A are $n \times n$ matrices then $A'XA$ is $n \times n$. Since $\text{tr}(A'XA) = \text{tr}(XAA')$ (see §2.4 on Page 29) and since AA' is symmetric, we have $\frac{\partial \text{tr}(A'XA)}{\partial X} = 2AA' - \text{diag}(\text{diag}(A))$ or AA' according as X is symmetric or non-symmetric (by §7.3.5 and §7.3.4).

7.3.7 Derivative of $|X|$ when X is not symmetric

If X is $n \times n$ and $f = f(X) = |X|$ then $\frac{\partial f}{\partial X} = |X|\,(X^{-1})'$. First consider the case $n = 2$: $X = (x_{ij}), f(X) = |X| = x_{11}x_{22} - x_{12}x_{21}$. So

$$\frac{\partial f}{\partial X} = \begin{pmatrix} x_{22} & -x_{21} \\ -x_{12} & x_{11} \end{pmatrix} = |X|\,(X^{-1})'.$$

7.3.7.1 General proof in non-symmetric case

Generally we have $|X| = \sum_{j=1}^{n} x_{jk} c_{jk}$ for any row k where c_{jk} is the cofactor of x_{jk} (§4.1 on Page 60) so $\frac{\partial f}{\partial x_{ij}} = c_{ij}$ and thus $\frac{\partial f}{\partial X} = (c_{ij}) = \left(X^{\#}\right)'$ where $X^{\#}$ is the adjoint matrix of $\boldsymbol{X}$. Since $\boldsymbol{X}^{-1} = |\boldsymbol{X}|^{-1}\boldsymbol{X}^{\#}$ (§5.6 on Page 79) we have $\boldsymbol{X}^{\#} = |\boldsymbol{X}|\left(\boldsymbol{X}^{-1}\right)'$ and thus $\frac{\partial f}{\partial \boldsymbol{X}} = |\boldsymbol{X}|\,(\boldsymbol{X}^{-1})'$.

7.3.8 Derivative of $|\boldsymbol{X}|$ when $\boldsymbol{X}$ is symmetric

If $\boldsymbol{X}$ is $n \times n$ and $f = f(\boldsymbol{X}) = |\boldsymbol{X}|$ then $\frac{\partial f}{\partial \boldsymbol{X}} = |\boldsymbol{X}|\{2\boldsymbol{X}^{-1} - \text{diag}(\text{diag}(\boldsymbol{X}^{-1}))\}$. First consider the case $n = 2:\;\; \boldsymbol{X} = (x_{ij}), f(\boldsymbol{X}) = |\boldsymbol{X}| = x_{11}x_{22} - x_{12}^2$. So

$$\frac{\partial f}{\partial \boldsymbol{X}} = \begin{pmatrix} x_{22} & -2x_{21} \\ -2x_{21} & x_{11} \end{pmatrix} = |\boldsymbol{X}|\{2\boldsymbol{X}^{-1} - \text{diag}(\text{diag}(\boldsymbol{X}^{-1}))\}.$$

7.3.8.1 General proof in symmetric case

Generally we have $|\boldsymbol{X}| = \sum_{j=1}^{n} x_{jk}c_{jk}$ for any row k where c_{jk} is the cofactor of x_{jk} (§4.1 on Page 60) so $\frac{\partial f}{\partial x_{ij}} = 2c_{ij}$ for $i \neq j$, and c_{ii} for $i = j$. Thus $\frac{\partial f}{\partial \boldsymbol{X}} = 2(c_{ij}) - \text{diag}(c_{ij}) = 2\boldsymbol{X}^{\#} - \text{diag}\left(\text{diag}\left(\boldsymbol{X}^{\#}\right)\right)$ where $\boldsymbol{X}^{\#}$ is the adjoint matrix of $\boldsymbol{X}$. Since $\boldsymbol{X}^{-1} = |\boldsymbol{X}|^{-1}\boldsymbol{X}^{\#}$ (§5.6 on Page 79) we have $\boldsymbol{X}^{\#} = |\boldsymbol{X}|\left(\boldsymbol{X}^{-1}\right)$ and thus $\frac{\partial f}{\partial \boldsymbol{X}} = |\boldsymbol{X}|\{2\boldsymbol{X}^{-1} - \text{diag}(\text{diag}(\boldsymbol{X}^{-1}))\}$.

7.3.9 Derivative of $|\boldsymbol{X}|^r$

If $f = f(\boldsymbol{X}) = |\boldsymbol{X}|^r$ then $\frac{\partial f}{\partial \boldsymbol{X}} = r|\boldsymbol{X}|^{r-1}\frac{\partial f}{\partial |\boldsymbol{X}|} = r|\boldsymbol{X}|^{r-1}|\boldsymbol{X}|\left(\boldsymbol{X}^{-1}\right)' = r|\boldsymbol{X}|^r\left(\boldsymbol{X}^{-1}\right)'$ provided $\boldsymbol{X}$ is not symmetric and $r \neq 0$.
When $\boldsymbol{X}$ is symmetric then clearly $\frac{\partial f}{\partial \boldsymbol{X}} = r|\boldsymbol{X}|^r\{2\boldsymbol{X}^{-1} - \text{diag}(\text{diag}(\boldsymbol{X}^{-1}))\}$.

7.3.10 Derivative of $\log(|\boldsymbol{X}|)$

If $f = f(\boldsymbol{X}) = \log(|\boldsymbol{X}|)$ then $\frac{\partial f}{\partial \boldsymbol{X}} = |\boldsymbol{X}|^{-1}\frac{\partial f}{\partial |\boldsymbol{X}|} = |\boldsymbol{X}|^{-1}|\boldsymbol{X}|\left(\boldsymbol{X}^{-1}\right)' = \left(\boldsymbol{X}^{-1}\right)'$ provided $\boldsymbol{X}$ is not symmetric.
When $\boldsymbol{X}$ is symmetric then clearly $\frac{\partial f}{\partial \boldsymbol{X}} = 2\boldsymbol{X}^{-1} - \text{diag}(\text{diag}(\boldsymbol{X}^{-1}))$.

7.4 Differentiation of a Vector with Respect to a Vector

If $\boldsymbol{x}$ is a vector of length n and $f = f(\boldsymbol{x})$ is a vector function of length m then $\frac{\partial f}{\partial \boldsymbol{x}}$ is ***defined*** to be the $m \times n$ matrix

$$\frac{\partial f}{\partial \boldsymbol{x}} = \begin{pmatrix} \frac{\partial f_1}{\partial x_1} & \frac{\partial f_1}{\partial x_2} & \cdots & \frac{\partial f_1}{\partial x_n} \\ \frac{\partial f_2}{\partial x_1} & \frac{\partial f_2}{\partial x_2} & \cdots & \frac{\partial f_2}{\partial x_n} \\ \vdots & \vdots & \ddots & \vdots \\ \frac{\partial f_m}{\partial x_1} & \frac{\partial f_m}{\partial x_2} & \cdots & \frac{\partial f_m}{\partial x_n} \end{pmatrix} = \left(\frac{\partial f_i}{\partial x_j}\right).$$

If $\boldsymbol{y} = f(\boldsymbol{x})$ then the determinant of $\frac{\partial \boldsymbol{x}}{\partial \boldsymbol{y}}$, $|\frac{\partial x_i}{\partial y_j}|$ is known as the Jacobean of the transformation $\boldsymbol{x} \rightarrow \boldsymbol{y}$ and is used when changing an integral with respect to $\boldsymbol{x}$ to one with respect to $\boldsymbol{y}$; see §9.2.1.

7.4.1 Derivative of Ax

If $\boldsymbol{A}$ is $m \times n$ then $f = f(\boldsymbol{x}) = \boldsymbol{A}\boldsymbol{x} = (\sum_j a_{1j}x_j, \sum_j a_{2j}x_j, \ldots, \sum_j a_{mj}x_j)'$, so

$$\frac{\partial f}{\partial \boldsymbol{x}} = \begin{pmatrix} \frac{\partial f_1}{\partial x_1} & \frac{\partial f_1}{\partial x_2} & \cdots & \frac{\partial f_1}{\partial x_n} \\ \frac{\partial f_2}{\partial x_1} & \frac{\partial f_2}{\partial x_2} & \cdots & \frac{\partial f_2}{\partial x_n} \\ \vdots & \vdots & \ddots & \vdots \\ \frac{\partial f_m}{\partial x_1} & \frac{\partial f_m}{\partial x_2} & \cdots & \frac{\partial f_m}{\partial x_n} \end{pmatrix} = \begin{pmatrix} a_{11} & a_{12} & \cdots & a_{1n} \\ a_{21} & a_{22} & \cdots & a_{2n} \\ \vdots & \vdots & \ddots & \vdots \\ a_{m1} & a_{m2} & \cdots & a_{mn} \end{pmatrix} = \boldsymbol{A}.$$

7.5 Differentiation of a Matrix with Respect to a Scalar

If the elements of a $m \times n$ matrix $\boldsymbol{A}$ are functions of a scalar x, i.e., $\boldsymbol{A} = (a_{ij}) = (a_{ij}(x))$, the derivative with respect to x is simply the matrix of derivatives of $a_{ij}(x)$ with respect to x.

7.5.1 $\boldsymbol{A} = (1, x, x^2, x^3, \ldots, x^{n-1})$

If $\boldsymbol{A} = \boldsymbol{A}(x) = (1, x, x^2, x^3, \ldots, x^{n-1})$, the derivative of $\boldsymbol{A}$ with respect to x is $(0, 1, 2x, 3x^2, x^3, \ldots, (n-1)x^{n-2})$.

7.5.2 Autocorrelation matrix

$$\text{If } \boldsymbol{A}(x) = \begin{pmatrix} 1 & x & x^2 & \cdots & x^{n-1} \\ x & 1 & x & x^2 & x^{n-2} \\ x^2 & x & \ddots & \cdots & \vdots \\ \vdots & \vdots & \cdots & \ddots & \vdots \\ x^{n-1} & \cdots & x^2 & x & 1 \end{pmatrix}$$

$$\text{then } \frac{\partial \boldsymbol{A}}{\partial x} = \begin{pmatrix} 0 & 1 & 2x & \cdots & (n-1)x^{n-2} \\ 1 & 0 & 1 & 2x & (n-2)x^{n-3} \\ x^2 & x & \ddots & \cdots & \vdots \\ \vdots & \vdots & \cdots & \ddots & x \\ (n-1)x^{n-2} & \cdots & 2x & 1 & 0 \end{pmatrix}.$$

7.6 Use of Eigenanalysis in Constrained Optimization

A powerful procedure for optimizing a function involving quadratic forms subject to a quadratic constraint rests on reducing the problem to solving an eigenequation (see §6.2). This is analogous to solving a polynomial equation where showing that the solution is one of the roots of a polynomial almost solves the problem completely but an extra step or argument is need to determine which particular root is required. In the same way an extra step is required to determine which particular eigenpair that satisfies the eigenequation is required. In outline, the procedure is as follows (where optimization is to be performed with respect to $\boldsymbol{x}$):

(1) Introduce a Lagrange multiplier to incorporate the constraint in a new objective function.

(2) Differentiate with respect to $\boldsymbol{x}$ and set the derivative equal to zero.

(3) Recognise this is an eigenequation with the Lagrange multiplier as eigenvalue.

(4) Deduce that there are ONLY a limited number of possible values for this eigenvalue (all of which can be calculated numerically).

(5) Use some extra step to determine which eigenvalue gives the desired optimum (typically using the constraint somewhere).

Sometimes this procedure can be used when there is no explicit constraint but where it can be seen that the problem is invariant to scalar multiplication of $\boldsymbol{x}$ (for example a ratio of quadratic forms in $\boldsymbol{x}$) and so a scalar constraint can be introduced without affecting the solution to the problem. This is illustrated in the examples below.

Example 7

(i) To optimize $\boldsymbol{x}'\boldsymbol{S}\boldsymbol{x}$ subject to $\boldsymbol{x}'\boldsymbol{x} = 1$ where $\boldsymbol{S}$ is symmetric.

Let $\Omega = \boldsymbol{x}'\boldsymbol{S}\boldsymbol{x} - \lambda(\boldsymbol{x}'\boldsymbol{x} - 1)$ where λ is a Lagrange multiplier.

$$\frac{\partial \Omega}{\partial \boldsymbol{x}} = 2\boldsymbol{S}\boldsymbol{x} - 2\lambda\boldsymbol{x} \text{ (see §7.2.3) so we require } \boldsymbol{S}\boldsymbol{x} = \lambda\boldsymbol{x}.$$

Thus we require $\boldsymbol{x}$ to be an eigenvector of $\boldsymbol{S}$ corresponding to eigenvalue λ. Since $\boldsymbol{S}\boldsymbol{x} = \lambda\boldsymbol{x}$, we have $\boldsymbol{x}'\boldsymbol{S}\boldsymbol{x} = \lambda\boldsymbol{x}'\boldsymbol{x} = \lambda$ (since $\boldsymbol{x}'\boldsymbol{x} = 1$). So the maximum and minimum values of $\boldsymbol{x}'\boldsymbol{S}\boldsymbol{x}$ are obtained by taking $\boldsymbol{x}$ to be the eigenvectors corresponding to the largest and smallest eigenvalues of $\boldsymbol{S}$.

(ii) To optimize $\frac{\boldsymbol{x}'\boldsymbol{S}\boldsymbol{x}}{\boldsymbol{x}'\boldsymbol{A}\boldsymbol{x}}$ where $\boldsymbol{S}$ and $\boldsymbol{A}$ are symmetric and $\boldsymbol{A}$ is non-singular.

Since the ratio is invariant with respect to scalar multiplication of $\boldsymbol{x}$, the problem is not affected by imposing a scalar constraint on $\boldsymbol{x}$. The most convenient is to require $\boldsymbol{x}'\boldsymbol{A}\boldsymbol{x} = 1$ and optimize $\boldsymbol{x}'\boldsymbol{S}\boldsymbol{x}$ subject to the constraint $\boldsymbol{x}'\boldsymbol{A}\boldsymbol{x} = 1$. Define $\Omega = \boldsymbol{x}'\boldsymbol{S}\boldsymbol{x} - \lambda(\boldsymbol{x}'\boldsymbol{A}\boldsymbol{x} - 1)$ where λ is a Lagrange multiplier.

$$\frac{\partial \Omega}{\partial \boldsymbol{x}} = 2\boldsymbol{Sx} - 2\lambda\boldsymbol{Ax} \text{ so we require } \boldsymbol{Sx} = \lambda\boldsymbol{Ax} \text{ or } \boldsymbol{A}^{-1}\boldsymbol{Sx} = \lambda\boldsymbol{x}.$$

Thus we require $\boldsymbol{x}$ to be an [right] eigenvector of $\boldsymbol{A}^{-1}\boldsymbol{S}$ corresponding to eigenvalue λ. Since $\boldsymbol{A}^{-1}\boldsymbol{Sx} = \lambda\boldsymbol{x}$, we have $\boldsymbol{x}'\boldsymbol{Sx} = \lambda\boldsymbol{x}'\boldsymbol{Ax} = \lambda$ (since $\boldsymbol{x}'\boldsymbol{Ax} = 1$). So the maximum and minimum values of $\boldsymbol{x}'\boldsymbol{Sx}/\boldsymbol{x}'\boldsymbol{Ax}$ are obtained by taking $\boldsymbol{x}$ to be the [right] eigenvectors corresponding to the largest and smallest eigenvalues of $\boldsymbol{A}^{-1}\boldsymbol{S}$.

(iii) To maximize $\boldsymbol{x}'\beta(\beta'\boldsymbol{XX}'\beta)^{-1}\beta'\boldsymbol{x}$ with respect to the $p \times 1$ vector β.

Since $\beta'\boldsymbol{XX}'\beta$ is $1 \times p \times p \times n \times n \times p \times 1 \equiv 1$, i.e., a scalar, we need to maximize $\boldsymbol{x}'\beta\beta'\boldsymbol{x}/\beta'\boldsymbol{XX}'\beta = \beta'\boldsymbol{xx}'\beta/\beta'\boldsymbol{XX}'\beta$, noting $\beta'\boldsymbol{x}$ is $1 \times p \times p \times 1 \equiv 1$, also a scalar and so commutes with $\boldsymbol{x}'\beta$. This is in the form of the previous example with $\boldsymbol{S} = \boldsymbol{xx}'$ and $\boldsymbol{A} = \boldsymbol{XX}'$ so the maximum value is given when β is the [right] eigenvector of $(\boldsymbol{XX}')^{-1}\boldsymbol{xx}'$ corresponding to its largest eigenvalue. Since $\rho((\boldsymbol{XX}')^{-1}\boldsymbol{xx}') \leq \rho(\boldsymbol{x}) = 1$ this has only one non-zero eigenvalue which is $\boldsymbol{x}'(\boldsymbol{XX}')^{-1}\boldsymbol{x}$ with [right] eigenvector $(\boldsymbol{XX}')^{-1}\boldsymbol{x}$ (because $(\boldsymbol{XX}')^{-1}\boldsymbol{xx}'(\boldsymbol{XX}')^{-1}\boldsymbol{x} = \boldsymbol{x}'(\boldsymbol{XX}')^{-1}\boldsymbol{x}(\boldsymbol{XX}')^{-1}\boldsymbol{x}$ noting that the scalar $\boldsymbol{x}'(\boldsymbol{XX}')^{-1}\boldsymbol{x}$ commutes with the other terms.

(iv) To optimize $\Phi = \frac{\boldsymbol{x}'\boldsymbol{Ay}}{\sqrt{\boldsymbol{x}'\boldsymbol{Sxy}'\boldsymbol{Ty}}}$ with respect to $\boldsymbol{x}$ and $\boldsymbol{y}$, where $\boldsymbol{x}$ and $\boldsymbol{y}$ are $p \times 1$ and $q \times 1$ vectors, $\boldsymbol{A}$ is $p \times q$, $\boldsymbol{S}$ is $p \times p$ and $\boldsymbol{T}$ is $q \times q$ and $\boldsymbol{S}$ and $\boldsymbol{T}$ are both symmetric and non-singular.

First, note that Φ is invariant to scalar multiplication of both $\boldsymbol{x}$ and $\boldsymbol{y}$ and so the problem is unaltered if scale constraints are imposed on both $\boldsymbol{x}$ and $\boldsymbol{y}$. The most convenient constraints to take are $\boldsymbol{x}'\boldsymbol{Sx} = \boldsymbol{y}'\boldsymbol{Ty} = 1$, so define $\Omega = \boldsymbol{x}'\boldsymbol{Ay} - \lambda(\boldsymbol{x}'\boldsymbol{Sx} - 1) - \mu(\boldsymbol{y}'\boldsymbol{Ty} - 1)$ where λ and μ are Lagrange multipliers. Differentiating with respect to $\boldsymbol{x}$ and to $\boldsymbol{y}$ and setting the derivatives equal to zero shows that we require $\boldsymbol{Ay} - 2\lambda\boldsymbol{Sx} = 0$ and $\boldsymbol{A}'\boldsymbol{x} - 2\mu\boldsymbol{Ty} = 0$ (see §7.2.1). Premultiplying the first by $\boldsymbol{x}'$ and postmultiplying the transpose of the second by $\boldsymbol{y}$ and recalling the constraints show that $2\lambda = 2\mu = \Phi$. Premultiplying the first equation by $\boldsymbol{S}^{-1}\boldsymbol{x}'\boldsymbol{Ay}$ gives $\boldsymbol{S}^{-1}\boldsymbol{x}'\boldsymbol{AyAy} - \boldsymbol{S}^{-1}\boldsymbol{x}'\boldsymbol{Ayx}'\boldsymbol{AySx} = 0$, i.e., $\boldsymbol{S}^{-1}\boldsymbol{x}'\boldsymbol{AyAy} - (\boldsymbol{x}'\boldsymbol{Ay})^2\boldsymbol{x} = 0$ (noting $\boldsymbol{x}'\boldsymbol{Ay}$ is a scalar). Postmultiplying the transpose of the second by $\boldsymbol{T}^{-1}\boldsymbol{A}'\boldsymbol{S}^{-1}$ gives $\boldsymbol{x}'\boldsymbol{AT}^{-1}\boldsymbol{A}'\boldsymbol{S}^{-1} - \boldsymbol{x}'\boldsymbol{Ayy}'\boldsymbol{TT}^{-1}\boldsymbol{A}'\boldsymbol{S}^{-1} = 0$, i.e., (taking the transpose) $\boldsymbol{S}^{-1}\boldsymbol{x}'\boldsymbol{AyAy} = \boldsymbol{S}^{-1}\boldsymbol{AT}^{-1}\boldsymbol{A}'\boldsymbol{x}$. Substituting this in the preceding equation gives $\boldsymbol{S}^{-1}\boldsymbol{AT}^{-1}\boldsymbol{A}'\boldsymbol{x} = (\boldsymbol{x}'\boldsymbol{Ay})^2\boldsymbol{x}$. Thus we require $\boldsymbol{x}$ to be an eigenvector of $\boldsymbol{S}^{-1}\boldsymbol{AT}^{-1}\boldsymbol{A}'$ corresponding to eigenvalue $(\boldsymbol{x}'\boldsymbol{Ay})^2 = \Phi^2$. Clearly we need to take the largest eigenvalue to maximize Φ for any value of $\boldsymbol{y}$. A similar argument shows that to maximize Φ for any value of $\boldsymbol{x}$ we need to take the eigenvector of $\boldsymbol{T}^{-1}\boldsymbol{A}'\boldsymbol{S}^{-1}\boldsymbol{A}$ corresponding to its largest eigenvalue $(\boldsymbol{x}'\boldsymbol{Ay})^2 = \Phi^2$. Thus to maximize Φ with respect to both $\boldsymbol{x}$ and $\boldsymbol{y}$, these should be taken as the eigenvectors of $\boldsymbol{S}^{-1}\boldsymbol{AT}^{-1}\boldsymbol{A}'$ and $\boldsymbol{T}^{-1}\boldsymbol{A}'\boldsymbol{S}^{-1}\boldsymbol{A}$ corresponding to their largest eigenvalues.

7.7 Exercises

(1) If $\boldsymbol{x} = (x_1, x_2, \ldots, x_n)'$ and $f(\boldsymbol{x}) = \boldsymbol{x}$ find $\frac{\partial f}{\partial \boldsymbol{x}}$.

(2) If $\boldsymbol{A}$, $\boldsymbol{X}$ and $\boldsymbol{B}$ are $m \times n$, $n \times p$ and $p \times m$ matrices (where $n \neq p$) find $\frac{\partial \text{tr}(\boldsymbol{AXB})}{\partial \boldsymbol{X}}$.

(3) If $\boldsymbol{A}$, $\boldsymbol{X}$ and $\boldsymbol{B}$ are $n \times n$ matrices find $\frac{\partial \text{tr}(\boldsymbol{AXB})}{\partial \boldsymbol{X}}$.

(4) If $\boldsymbol{A}$, $\boldsymbol{X}$ and $\boldsymbol{B}$ are $n \times n$ matrices find $\frac{\partial \text{tr}(\boldsymbol{AX'B})}{\partial \boldsymbol{X}}$.

(5) Find $\frac{\partial \text{tr}(\boldsymbol{X}^2)}{\partial \boldsymbol{X}}$.

(6) If $\boldsymbol{X}$ is a $n \times n$ matrix with elements x_{ij} and is not symmetric (i.e., $x_{ij} \neq x_{ji}$) show that $\frac{\partial \boldsymbol{X}^{-1}}{\partial x_{ij}} = -\boldsymbol{X}^{-1}\Delta_{[ij]}\boldsymbol{X}^{-1}$ where $\Delta_{[ij]}$ is a $n \times n$ matrix with $\Delta_{rs} = \delta_{ri}\delta_{sj}$, where δ is Kronecker's delta function, i.e., the elements of Δ are all zero except in the $(i, j)^{th}$ place where there is a 1.

(7) If $\boldsymbol{X}$ is a $n \times n$ matrix with elements x_{ij} and is symmetric ($x_{ij} = x_{ji}$) show that $\frac{\partial \boldsymbol{X}^{-1}}{\partial x_{ij}} = -\boldsymbol{X}^{-1}(\Delta_{[ij]} + \Delta_{[ji]})\boldsymbol{X}^{-1}$.

(8) Find the maximum value of $\boldsymbol{x}'\boldsymbol{Xaa}'\boldsymbol{X}'\boldsymbol{x}/\boldsymbol{x}'\boldsymbol{XX}'\boldsymbol{x}$.

(9) Find the maximum value of $\boldsymbol{x}'\boldsymbol{A}'\boldsymbol{By}$ with respect to $\boldsymbol{x}$ and $\boldsymbol{y}$ subject to the constraints $\boldsymbol{x}'\boldsymbol{x} = \boldsymbol{y}'\boldsymbol{y} = 1$ where $\boldsymbol{A}$ and $\boldsymbol{B}$ are $n \times p$ and $n \times m$ matrices.

(10) Show that $\frac{\partial \boldsymbol{AB}}{\partial x} = \boldsymbol{A}\frac{\partial \boldsymbol{B}}{\partial x} + \frac{\partial \boldsymbol{A}}{\partial x}\boldsymbol{B}$.

(11) Show that $\frac{\partial \boldsymbol{A} \odot \boldsymbol{B}}{\partial x}. = \boldsymbol{A} \odot \frac{\partial \boldsymbol{B}}{\partial x} + \frac{\partial \boldsymbol{A}}{\partial x} \odot \boldsymbol{B}$ where $\odot$ indicates the Hadamard product (see §8.4).

(12) Show that $\frac{\partial \boldsymbol{A} \otimes \boldsymbol{B}}{\partial x} = \boldsymbol{A} \otimes \frac{\partial \boldsymbol{B}}{\partial x} + \frac{\partial \boldsymbol{A}}{\partial x} \otimes \boldsymbol{B}$ where $\otimes$ indicates the Kronecker product (see §8.5).

8

Further Topics

8.1 Introduction

This section provides an introduction to a few topics in matrix algebra which are unlikely to be considered in an early undergraduate level course on linear algebra but can be of immense practical use in certain statistical applications. The next section is a collation of various further decompositions of matrices into factors of particular structures. The first of these is restricted to matrices of full column rank and the others are restricted to square matrices. These have application both in developing methodological techniques and in efficient numerical calculation of solutions of linear equations, eigenanalyses, determinants and inverses of non-singular matrices.

Section 3.2.2 considered the factorization of a $m \times n$ matrix of rank r into factors $m \times r$ and $r \times n$, both of rank r. Spectral and singular value decompositions which rest on eigenanalyses of symmetric matrices (symmetric $\boldsymbol{S}$ and $\boldsymbol{AA}'$ in §6.7.1 and §6.7.2 respectively) were considered in some detail in Chapter 6 but the decompositions considered here do not rest on eigenanalyses. Details of the derivations of these are beyond the scope of this text but guidance on their implementation in **R** is given.

The second topic is generalized inverses which extend the definition of an inverse beyond that of a non-singular square matrix. This has application in the analysis of linear models, particularly models where the design matrix is not of full rank. The third and fourth consider different forms of the product of two matrices, the Hadamard and Kronecker products. The latter has application in the formulation and analysis of mixed fixed and random effects linear models.

8.2 Further Matrix Decompositions

8.2.1 *QR* Decomposition

If $\boldsymbol{A}$ is a $m \times n$ matrix of full column rank (i.e., $\rho(\boldsymbol{A}) = n, \quad n \leq m$) and ***if*** $\boldsymbol{A}$ can be factored into $\boldsymbol{A}$=$\boldsymbol{QR}$ with $\boldsymbol{Q}$ a $m \times n$ matrix with orthonormal columns and $\boldsymbol{R}$ $n \times n$ upper triangular, (see §8.2, Banerjee and Roy, 2014) then $\boldsymbol{QR}$ is said to be the $\boldsymbol{QR}$ decomposition of $\boldsymbol{A}$. The factorization is unique. Since $n = \rho(\boldsymbol{X}) = \rho(\boldsymbol{QR}) \leq \min(\rho(\boldsymbol{Q}), \rho(\boldsymbol{R})) \leq n$ we have $\rho(\boldsymbol{Q}) = \rho(\boldsymbol{R}) = n$, i.e., $\boldsymbol{Q}$ has full column rank and $\boldsymbol{R}$ is non-singular. Since $\boldsymbol{Q}$ has orthonormal columns we have $\boldsymbol{Q}'\boldsymbol{Q} = \mathbf{I_n}$ but in general $\boldsymbol{QQ}' \neq \mathbf{I_m}$ unless $m = n$ in which case $\boldsymbol{Q}$ is a non-singular orthogonal matrix.

8.2.1.1 Implementation in R

The **R** function for obtaining a $\boldsymbol{QR}$ decomposition of a matrix $\boldsymbol{A}$ is `qr(A)`. It is in the base system so no special packages have to be loaded. This function produces a $m \times n$ matrix (in `qr(A)$qr`) containing the factor $\boldsymbol{R}$ in the upper triangle and the lower triangle contains information on the factor $\boldsymbol{Q}$ in compact form. The actual factors can be extracted from this using two auxiliary functions `qr.R(.)` and `qr.Q(.)`. The function also returns the rank of the matrix (in `qr(A)$rank`) which is a convenient alternative to the function `rk(.)` in the package `fBasics` mentioned in §3.1.1. Thus the sequence of commands needed is:

```
qra<-qr(A)          ### perform QR decomposition
QA<-qr.Q(qra)       ### obtain Q factor in QA
RA<-qr.R(qra)       ### obtain R factor in RA
```

The function `qr(.)` will actually produce a factorization even if $\boldsymbol{A}$ is not of full column rank or if $m < n$. If $m > n$ but if $\rho(\boldsymbol{A}) < n$ then the $\boldsymbol{R}$ factor will be a singular triangular matrix, i.e., with some zero diagonal elements.

Example 8.1:

(i) A 4×3 matrix of full column rank

```
> set.seed(137)
> options(digits=2)
> A<-matrix(c(sample(1:12)),
+ 4,3)
> A
     [,1] [,2] [,3]
[1,]    8    3   12
[2,]    5    9    1
[3,]   10    6    2
[4,]    7    4   11
> qra<-qr(A)
> qra
```

```
$qr
       [,1]    [,2]  [,3]
[1,] -15.43 -10.177 -12.8
[2,]   0.32  -6.199   4.8
[3,]   0.65   0.061   9.1
[4,]   0.45   0.010  -0.4
```

Note calculation of $\rho(\boldsymbol{A})$:

```
$rank
[1] 3
```

```
$qraux
[1] 1.5 2.0 1.9
$pivot
[1] 1 2 3
attr(,"class")
[1] "qr"
> QA<-qr.Q(qra);RA<-qr.R(qra)
> QA;RA
```

```
[,1]    [,2]  [,3]
[1,] -0.52  0.367  0.40
[2,] -0.32 -0.920  0.14
[3,] -0.65  0.096 -0.75
[4,] -0.45  0.100  0.52
     [,1]  [,2]  [,3]
[1,]  -15 -10.2 -12.8
[2,]    0  -6.2   4.8
[3,]    0   0.0   9.1
```

```
> QA%*%RA
     [,1] [,2] [,3]
[1,]    8    3   12
[2,]    5    9    1
[3,]   10    6    2
[4,]    7    4   11
```

So $\boldsymbol{A} = \boldsymbol{QR}$.

```
> A
     [,1] [,2] [,3]
[1,]    8    3   12
[2,]    5    9    1
[3,]   10    6    2
[4,]    7    4   11
```

```
> QA%*%t(QA)
      [,1]  [,2]  [,3]  [,4]
[1,]  0.56 -0.12  0.08  0.48
[2,] -0.12  0.97  0.02  0.13
[3,]  0.08  0.02  0.99 -0.08
[4,]  0.48  0.13 -0.08  0.48
```

```
> t(QA)%*%QA
    ,1] [,2] [,3]
[1,] 1    0    0
[2,] 0    1    0
[3,] 0    0    1
```

So $\boldsymbol{Q}'\boldsymbol{Q} = \mathbf{I}_3$ but $\boldsymbol{QQ}' \neq \mathbf{I}_4$.

(ii) A 4×3 matrix of rank 2.

```
> X<-matrix(c(3,2,1,4,2,0,5,
+ 2,-1,-1,0,1),4,3,byrow=T)
> X
     [,1] [,2] [,3]
[1,]    3    2    1
[2,]    4    2    0
[3,]    5    2   -1
[4,]   -1    0    1
> qrx<-qr(X)
```

```
> qrx
$qr
      [,1]  [,2]     [,3]
[1,] -7.14 -3.36  4.2e-01
[2,]  0.56  0.84  1.7e+00
[3,]  0.70  0.77  7.0e-16
[4,] -0.14 -0.63 -9.5e-01
$rank
[1] 2
```

```
> QX<-qr.Q(qrx); QX
      [,1]  [,2]  [,3]
[1,] -0.42  0.70  0.15
[2,] -0.56  0.14 -0.71
[3,] -0.70 -0.42  0.56
[4,]  0.14  0.56  0.41
```

```
> RX<-qr.R(qrx); RX
     [,1]  [,2]    [,3]
[1,] -7.1 -3.36 4.2e-01
[2,]  0.0  0.84 1.7e+00
[3,]  0.0  0.00 7.0e-16
```

Note 0 entry on diagonal of $\boldsymbol{R}$ so $\boldsymbol{R}$ is singular and $\rho(\boldsymbol{R}) = 2$.

```
> QX%*%RX
     [,1]     [,2]     [,3]
[1,]    3  2.0e+00  1.0e+00
[2,]    4  2.0e+00 -8.8e-16
[3,]    5  2.0e+00 -1.0e+00
[4,]   -1 -1.7e-16  1.0e+00
```

So $X = QR$.

```
> X
     [,1] [,2] [,3]
[1,]    3    2    1
[2,]    4    2    0
[3,]    5    2   -1
[4,]   -1    0    1
```

```
> QX%*%t(QX)
      [,1]    [,2]   [,3]   [,4]
[1,] 0.689  0.23  0.083   0.39
[2,] 0.228  0.83 -0.061  -0.29
[3,] 0.083 -0.06  0.978  -0.11
[4,] 0.394 -0.29 -0.105   0.50
```

```
> t(QX)%*%QX
     ,1] [,2] [,3]
[1,] 1    0    0
[2,] 0    1    0
[3,] 0    0    1
```

So $Q'Q = \mathbf{I}_3$ but $QQ' \neq \mathbf{I}_4$.

(iii) A 3×4 matrix of rank 2 (i.e., $m < n$ and not of full rank).

```
> X<-matrix(c(3,2,1,4,2,0,5,
+ 2,1,2,-4,2),3,4,byrow=T)
> X
     [,1] [,2] [,3] [,4]
[1,]    3    2    1    4
[2,]    2    0    5    2
[3,]    1    2   -4    2
> qrx<-qr(X)
```

```
> qrx
$qr
     [,1]   [,2]   [,3] [,4]
[1,]-3.74  -2.14  -2.4 -4.80
[2,] 0.53   1.85  -6.0  0.93
[3,] 0.27  -0.75     0     0
$rank
[1] 2
```

```
> QX<-qr.Q(qrx) ; QX
      [,1]   [,2]   [,3]
[1,] -0.80  0.15 -0.58
[2,] -0.53 -0.62  0.58
[3,] -0.27  0.77  0.58
```

```
> RX<-qr.R(qrx); RX
     [,1] [,2]  [,3]  [,4]
[1,] -3.7 -2.1  -2.4 -4.80
[2,]  0.0  1.9  -6.0  0.93
[3,]  0.0  0.0     0     0
```

```
> QX%*%RX
     [,1] [,2] [,3] [,4]
[1,]    3    2    1    4
[2,]    2    0    5    2
[3,]    1    2   -4    2
```

```
> X
     [,1] [,2] [,3] [,4]
[1,]    3    2    1    4
[2,]    2    0    5    2
[3,]    1    2   -4    2
```

So $X = QR$.

```
> t(QX)%*%QX
     [,1] [,2] [,3]
[1,]    1    0    0
[2,]    0    1    0
[3,]    0    0    1
```

```
> QX%*%t(QX)
     [,1] [,2] [,3]
[1,]    1    0    0
[2,]    0    1    0
[3,]    0    0    1
```

So $\boldsymbol{Q}'\boldsymbol{Q} = \mathbf{I}_3$ and in this case $\boldsymbol{Q}\boldsymbol{Q}' = \mathbf{I}_4$.

8.2.2 *LU* and *LDU* decompositions

If $\boldsymbol{A}$ is a $n \times n$ square matrix and ***if*** $\boldsymbol{A}$ can be factored into $\boldsymbol{A} = \boldsymbol{LU}$ with $\boldsymbol{L}$ lower triangular and all diagonal entries equal to 1 and $\boldsymbol{U}$ upper triangular, then $\boldsymbol{LU}$ is said to be the $\boldsymbol{LU}$ decomposition of $\boldsymbol{A}$. Note that it may not be possible to obtain an $\boldsymbol{LU}$ decomposition of $\boldsymbol{A}$ but it can be shown (e.g., §3.3, Banerjee and Roy, 2014) that there is a permutation matrix $\boldsymbol{P}$ such that $\boldsymbol{A} = \boldsymbol{PLU}$ or equivalently $\boldsymbol{P}'\boldsymbol{A} = \boldsymbol{LU}$ recalling that all permutation matrices are orthogonal; see §2.5.5. The effect of premultiplying $\boldsymbol{A}$ by a permutation matrix $\boldsymbol{P}'$ is to re-order the rows of $\boldsymbol{A}$; see §2.5.5. Further, it can be shown that this factorization is unique. Some authors (e.g., Banerjee and Roy, 2014) restrict the definition of a $\boldsymbol{LU}$ decomposition to non-singular matrices, in which case all the diagonal entries of $\boldsymbol{U}$ are non-zero.

If $\boldsymbol{A}$ is non-singular then all the diagonal entries of $\boldsymbol{U}$ are non-zero and so it is possible to obtain a $\boldsymbol{LDU}$ decomposition where the diagonal elements of $\boldsymbol{U}$ are all equal to 1 (as are those of $\boldsymbol{L}$) from a $\boldsymbol{LU}$ decomposition by extracting the diagonal of $\boldsymbol{U}$ into a diagonal matrix $\boldsymbol{D}$ and premultiplying $\boldsymbol{U}$ by $\boldsymbol{D}^{-1}$, with $\boldsymbol{A}$ non-singular all the diagonal entries of $\boldsymbol{D}$ are be non-zero and so $\boldsymbol{D}$ is non-singular.

8.2.2.1 Implementation in R

R provides a function `lu(A)` for obtaining the $\boldsymbol{LU}$ decomposition of a matrix $\boldsymbol{A}$. It is in the library `Matrix` which must be loaded by the command `library(Matrix)` (or using the `Packages` drop-down menu) before using the function. Note the capitalisation of `Matrix`. This produces a permutation matrix $\boldsymbol{P}$ such that $\boldsymbol{A} = \boldsymbol{PLU}$ or equivalently $\boldsymbol{P}'\boldsymbol{A} = \boldsymbol{LU}$.

The `Matrix` library is included when **R** is first installed and so does not have to be installed specially. The library contains a collection of functions written specifically for handling very large sparse matrices, though these functions will also operate on dense matrices of modest size. Sparse matrices are ones where most of the entries are zero and dense matrices are ones where most entries are non-zero. Because the functions are designed for handling sparse very large matrices if the results of the functions are matrices, they are stored in a compact form and so an auxiliary function is needed to convert them to ordinary matrices. Typically this function is `expand(.)`. After use of `expand(.)` the $\boldsymbol{L}$, $\boldsymbol{U}$ factors and the permutation matrix $\boldsymbol{P}$ are stored in values accessed by `$L`, `$U` and `$P` respectively. Thus the sequence of commands is:

```
library(Matrix)       ### load the library Matrix
ALU<-lu(A)            ### Obtain the LU deocmposition for A
                      ###  (permuted if necessary by P)
E<-expand(ALU)        ### Expand the results of lu(A) to
                      ###   ordinary matrix form
L<-E$L                ### Obtain the L factor
U<-E$U                ### Obtain the U factor
P<-E$P                ### Obtain the P permutation matrix
L%*%U ; t(P)%*%A         ### Check that P'A=LU
P%*%L%*%U ; A         ### Check that A=PLU
```

There is no ready-made function in **R** library `Matrix` for obtaining the ***LDU*** decomposition but it can be constructed in a few extra lines to the above code as follows (where the matrix `V` holds the matrix `U` with diagonal elements set to 1):

```
D<-diag(diag(U))      ### Obtain the diagonal matrix of
                      ###   diagonal elements of U
V<-solve(D)%*%U       ### ensure diagonal elements of U are 1
P%*%L%*%D%*%V ; A     ### Check that A=PLDV
```

Example 8.2:

(i) ***LU*** decomposition of a 4×4 arbitrary matrix

```
> library(Matrix)
> options(digits=2)
> set.seed(137)
> A<-matrix(c(sample(-10:10,
+ 16)),4,4)
> A
     [,1] [,2] [,3] [,4]
[1,]    3   -4    9    4
[2,]   -2    5   -7    6
[3,]    7    8   -3   -6
[4,]   10    0    2   -9
```

```
> ALU<-lu(A)
> E<-expand(ALU)
> L<-E$L
> U<-E$U
> P<-E$P
>P
4 x 4 sparse Matrix of
 class "pMatrix"
[1,] . . | .
[2,] . . . |
[3,] . | . .
[4,] | . . .
```

```
>L; U
4 x 4 Matrix of class
"dtrMatrix" (unitriangular)
     [,1]  [,2]  [,3]  [,4]
[1,]  1.00     .     .     .
[2,]  0.70  1.00     .     .
[3,]  0.30 -0.50  1.00     .
[4,] -0.20  0.62 -0.62  1.00
```

```
4 x 4 Matrix of
class "dtrMatrix"
     [,1] [,2] [,3] [,4]
[1,] 10.0  0.0  2.0 -9.0
[2,]    .  8.0 -4.4  0.3
[3,]    .    .  6.2  6.8
[4,]    .    .    .  8.3
```

```
> L%*%U ; P%*%A
4 x 4 Matrix of
class "dgeMatrix"
     [,1] [,2] [,3] [,4]
[1,]   10    0    2   -9
[2,]    7    8   -3   -6
[3,]    3   -4    9    4
[4,]   -2    5   -7    6
```

```
     [,1] [,2] [,3] [,4]
[1,]   10    0    2   -9
[2,]    7    8   -3   -6
[3,]    3   -4    9    4
[4,]   -2    5   -7    6
```

So ***P'A=LU***.

```
> P%*%L%*%U ; A
4 x 4 Matrix of
class "dgeMatrix"
     [,1] [,2] [,3] [,4]
[1,]    3   -4    9    4
[2,]   -2    5   -7    6
[3,]    7    8   -3   -6
[4,]   10    0    2   -9
```

```
     [,1] [,2] [,3] [,4]
[1,]    3   -4    9    4
[2,]   -2    5   -7    6
[3,]    7    8   -3   -6
[4,]   10    0    2   -9
```

So ***A=PLU***.

(ii) Example (i) continued to obtain the ***LDU*** decomposition

```
> D<-diag(diag(U))
> V<-solve(D)%*%U
> U; V; D
4 x 4 Matrix of
class "dtrMatrix"
     [,1] [,2] [,3] [,4]
[1,] 10.0  0.0  2.0 -9.0
[2,]    .  8.0 -4.4  0.3
[3,]    .    .  6.2  6.8
[4,]    .    .    .  8.3
4 x 4 Matrix of
class "dgeMatrix"
     [,1] [,2]  [,3]   [,4]
[1,]    1    0  0.20 -0.900
[2,]    0    1 -0.55  0.038
[3,]    0    0  1.00  1.105
[4,]    0    0  0.00  1.000
     [,1] [,2] [,3] [,4]
[1,]   10    0  0.0  0.0
[2,]    0    8  0.0  0.0
[3,]    0    0  6.2  0.0
[4,]    0    0  0.0  8.3
```

```
> P%*%L%*%D%*%V
4 x 4 Matrix of class
  "dgeMatrix"
     [,1] [,2] [,3] [,4]
[1,]    3   -4    9    4
[2,]   -2    5   -7    6
[3,]    7    8   -3   -6
[4,]   10    0    2   -9
>
> A
     [,1] [,2] [,3] [,4]
[1,]    3   -4    9    4
[2,]   -2    5   -7    6
[3,]    7    8   -3   -6
[4,]   10    0    2   -9
```

So $\boldsymbol{A} = \boldsymbol{PLDU}$.

8.2.3 Cholesky decomposition

The Cholesky decomposition of a symmetric and positive semi-definite $n \times n$ matrix $\boldsymbol{A}$ is $\boldsymbol{A} = \boldsymbol{TT}'$, with $\boldsymbol{T}$ lower triangular. We have $\boldsymbol{A} = \boldsymbol{LDU}$ with the diagonal elements of $\boldsymbol{D}$ non-negative since $\boldsymbol{A}$ is positive semi-definite. Since $\boldsymbol{A}$ is symmetric we have $\boldsymbol{LDU} = \boldsymbol{A} = \boldsymbol{A}' = \boldsymbol{U}'\boldsymbol{D}'\boldsymbol{L}' = \boldsymbol{U}'\boldsymbol{DL}'$ and since the $\boldsymbol{LDU}$ decomposition is unique we have $\boldsymbol{U} = \boldsymbol{L}'$ so $\boldsymbol{A} = \boldsymbol{LDL}' = (\boldsymbol{LD}^{1/2})(\boldsymbol{LD}^{1/2})' = \boldsymbol{TT}'$ with $\boldsymbol{T}$ lower triangular.

8.2.3.1 Implementation in R

The **R** function for obtaining a Cholesky decomposition of a matrix $\boldsymbol{A}$ is `chol(A)`. It is in the base system so no special packages have to be loaded. Additionally, there is a function `Cholesky(A)` in the library `Matrix` written to handle sparse matrices. Here details are given only of the use of `chol(A)`. Care must be taken when using `chol(A)` because **R** does not check whether $\boldsymbol{A}$ is symmetric, nor does it check that $\boldsymbol{A}$ is positive semi-definite. Further, **R** actually returns an upper triangular matrix rather than a lower triangular even though the usual statement of the decomposition is in terms of lower triangular matrices. That is if $\boldsymbol{A}$ is positive definite it produces an upper triangular matrix $\boldsymbol{U}$ such that $\boldsymbol{U}'\boldsymbol{U} = \boldsymbol{A}$, i.e., $\boldsymbol{L} = \boldsymbol{U}'$. If $\boldsymbol{A}$ is not positive semi-definite then **R** will still produce results but these will be meaningless, i.e., it is not true that `t(chol(A))chol(A)=A`. If $\boldsymbol{A}$ is not symmetric then **R** will produce a decomposition of the symmetric matrix obtained by reflecting the upper triangle of $\boldsymbol{A}$ about the main diagonal, provided this is positive semi-definite. If $\boldsymbol{A}$ is positive semi-definite but not positive definite then it is possible that the upper triangular matrix $\boldsymbol{U}$ is such that $\boldsymbol{U}'\boldsymbol{U} = \boldsymbol{AP}$ for some permutation matrix $\boldsymbol{P} \neq \mathbf{I_n}$. To obtain the permutation $\boldsymbol{P}$ it is necessary to call the function with an extra parameter `chol(A,pivot=TRUE)` and then use additional steps. Details of this are given in the **R** help system (see `help(chol)`) and are illustrated in the exercises below.

Thus before using `chol(A)` it is advisable to do two preliminary checks:

```
A-t(A)            ### This should be 0 if A is symmetric
eigen(A)$values   ### These should all be non-negative.
                  ### If any are 0 then use chol(A,pivot=TRUE)
```

Example 8.3:

(i) A 3×3 positive definite symmetric matrix

```
> options(digits=2)
> A<-matrix(c(8,3,5,3,7,2,5
+ ,2,9),3,3)
> A
     [,1] [,2] [,3]
[1,]    8    3    5
[2,]    3    7    2
[3,]    5    2    9
```

```
> t(A)-A
     [,1] [,2] [,3]
[1,]    0    0    0
[2,]    0    0    0
[3,]    0    0    0
```

So $\boldsymbol{A}$ is symmetric.

```
> eigen(A)$values
[1] 15.1  5.8  3.2
```

So A is positive definite.

```
> U<-chol(A)
> t(U)
     [,1]  [,2] [,3]
[1,]  2.8 0.000  0.0
[2,]  1.1 2.424  0.0
[3,]  1.8 0.052  2.4
```

Take $T = U'$ and check $TT' = A$:

```
> t(U)%*%U
     [,1] [,2] [,3]
[1,]    8    3    5
[2,]    3    7    2
[3,]    5    2    9
```

8.2.4 Schur decomposition

The Schur decomposition of a $n \times n$ matrix A is $Q'AQ = T$ where Q is orthogonal and T is upper triangular with diagonal elements equal to the eigenvalues of A. A proof is provided in more advanced texts, (see §11.5, Banerjee and Roy, 2014). Since Q is orthogonal $Q^{-1} = Q'$ so A and T are similar matrices (§2.5.9) and so have identical eigenvalues (§6.4.3). Thus every square matrix is similar to a triangular matrix. The eigenvalues of a triangular matrix are given by its diagonal elements; see §6.4.4.

8.2.4.1 Implementation in R

R provides a function `Schur(A)` for obtaining the Schur decomposition of a matrix A. It is in the library `Matrix` which must be loaded by the command `library(Matrix)` (or using the `Packages` drop-down menu) before using the function.

Example 8.4:

(i) Schur decomposition of a 3×3 arbitrary matrix

```
> library(Matrix)
> options(digits=2)
> set.seed(999)
> A<-matrix(c(sample(-10:10,
+ 9)),3,3)
> A
     [,1] [,2] [,3]
[1,]   -2    5   -1
[2,]    1    3    7
[3,]   -9    8   -5
> T<-Schur(A)$T
> diag(T);eigen(A)$values
[1] -12.0   3.1   4.9
[1] -12.0   4.9   3.1
```

```
> Q<-Schur(A)$Q
> t(Q)%*%A%*%Q; T
      [,1]  [,2]  [,3]
[1,] -12.0   2.5  -5.7
[2,]   0.0   3.1   6.5
[3,]   0.0   0.0   4.9

     [,1] [,2] [,3]
[1,]  -12  2.5 -5.7
[2,]    0  3.1  6.5
[3,]    0  0.0  4.9
```

Note that the diagonal elements of $\boldsymbol{T}$ are the eigenvalues of $\boldsymbol{A}$ though given in a different order and $\boldsymbol{Q}'\boldsymbol{A}\boldsymbol{Q} = T$ up to negligible rounding errors.

8.2.5 Application to obtaining eigenvalues

If $\boldsymbol{A}$ is a square matrix then the ***QR*-algorithm** for finding the eigenvalues of $\boldsymbol{A}$ is as follows: define $\boldsymbol{A}_0 = \boldsymbol{A}$ and $\boldsymbol{A}_{k+1} = \boldsymbol{R}_k\boldsymbol{Q}_k$ where $\boldsymbol{A}_k = \boldsymbol{Q}_k\boldsymbol{R}_k$ is the $\boldsymbol{QR}$ decomposition of $\boldsymbol{A}_k$. Since $\boldsymbol{A}_{k+1} = \boldsymbol{R}_k\boldsymbol{Q}_k = \boldsymbol{Q}_k^{-1}\boldsymbol{Q}_k\boldsymbol{R}_k\boldsymbol{Q}_k = \boldsymbol{Q}_k^{-1}\boldsymbol{A}_k\boldsymbol{Q}_k$ all the $\boldsymbol{A}_k$ are similar and so have identical eigenvalues. It can be shown that the sequence $\boldsymbol{A}_k$ converges to an upper triangular matrix $\boldsymbol{T}$ whose eigenvalues are given by the diagonal elements which must therefore be the eigenvalues of $\boldsymbol{A}$.

8.2.6 Application to finding determinants

If $\boldsymbol{A}$ is a $n \times n$ non-singular square matrix with $\boldsymbol{QR}$ decomposition $\boldsymbol{A} = \boldsymbol{QR}$ with $\boldsymbol{Q}$ orthogonal and $\boldsymbol{R}$ upper triangular then $|\boldsymbol{A}| = |\boldsymbol{Q}||\boldsymbol{R}| = |\boldsymbol{R}| = \pm\prod_i r_{ii}$ recalling that the determinant of an orthogonal matrix is ± 1 (§4.4) and the determinant of a triangular matrix is the product of its diagonal elements (§4.4). If $\boldsymbol{A} = \boldsymbol{PLU}$ then $|\boldsymbol{A}| = (-1)^s \prod_i l_{ii} \prod_i u_{ii}$ where s is the number of row interchanges needed to convert the permutation matrix $\boldsymbol{P}$ to the identity matrix $\mathbf{I_n}$. If $\boldsymbol{A}$ is positive semi-definite and $\boldsymbol{A} = \boldsymbol{TT}'$ is the Cholesky decomposition of $\boldsymbol{A}$ then $|\boldsymbol{A}| = \prod_i e_{ii}^2$.

8.2.7 Applications of decompositions to solving $\boldsymbol{Ax} = \boldsymbol{y}$

First, note that if $\boldsymbol{A}$ happens to be a $n \times n$ square lower triangular matrix then it is easy to solve the equation $\boldsymbol{Ax} = \boldsymbol{y}$ without inverting $\boldsymbol{A}$ by ***forward substitution***: since $\boldsymbol{A}$ is lower triangular the first equation involves only x_1 so this can be substituted into the second to obtain x_2 and then both values into the third, and so on. Similarly if $\boldsymbol{A}$ is upper triangular the equation is easily solved by ***back substitution***, starting with x_n. If $\boldsymbol{A}$ is not triangular then the equation $\boldsymbol{Ax} = \boldsymbol{y}$ may be solved in two stages by decomposing $\boldsymbol{A}$ into factors, at least one of which is triangular.

If $\boldsymbol{A}$ is a $n \times n$ square matrix then $\boldsymbol{A} = \boldsymbol{PLU}$ for some permutation matrix $\boldsymbol{P}$ and lower and upper triangular matrices $\boldsymbol{L}$ and $\boldsymbol{U}$. If $\boldsymbol{Ax} = \boldsymbol{y}$ then $\boldsymbol{PLUx} = \boldsymbol{y}$ so $\boldsymbol{LUx} = \boldsymbol{P}'\boldsymbol{y}$ (noting $\boldsymbol{P}$ is orthogonal). Let $\boldsymbol{z} = \boldsymbol{Ux}$ and solve $\boldsymbol{Lz} = \boldsymbol{P}'\boldsymbol{y}$ for $\boldsymbol{z}$ by forward substitution and then solve $\boldsymbol{Ux} = \boldsymbol{z}$ for $\boldsymbol{x}$ by backward substitution. Once the decomposition of $\boldsymbol{A}$ has been found all subsequent steps involve no matrix inversions or multiplications and so giving substantial computation savings if the equation has to be solved for several different values of $\boldsymbol{y}$.

If $\boldsymbol{A}$ is symmetric and positive definite then $\boldsymbol{A} = \boldsymbol{TT}'$ with $\boldsymbol{T}$ lower triangular is the Cholesky decomposition and so $\boldsymbol{Ax} = \boldsymbol{y}$ can be solved in two stages as with the $\boldsymbol{LU}$ decomposition.

If $\boldsymbol{A}$ is $m \times n$ with $m > n$ and of full column rank with $\boldsymbol{QR}$ decomposition $\boldsymbol{A} = \boldsymbol{QR}$ and if $\boldsymbol{Ax} = \boldsymbol{y}$, we have $\boldsymbol{QRx} = \boldsymbol{y}$ so $\boldsymbol{Rx} = \boldsymbol{Q}'\boldsymbol{y}$ (recalling $\boldsymbol{Q}'\boldsymbol{Q} = \mathbf{I_n}$) and this can be solved by back substitution. This may be an exact solution or it may be an approximate solution in which case it is the least squares solution. This is discussed further in §8.3.3.1.

If $\boldsymbol{A}$ is $m \times n$ with $m < n$ and of full row rank then $\boldsymbol{A}'$ is of full column rank. Let the $\boldsymbol{QR}$ decomposition of $\boldsymbol{A}'$ be $\boldsymbol{A}' = \boldsymbol{QR}$. Then it can be shown (see 8.3.3.1 on Page 131) that a solution to $\boldsymbol{Ax} = \boldsymbol{y}$ is given by $\boldsymbol{xQ}(\boldsymbol{R}')^{-1}\boldsymbol{y}$. Since $\boldsymbol{R}'$ is lower triangular $(\boldsymbol{R}')^{-1}\boldsymbol{y}$ can be calculated by forward substitution.

8.2.8 Application to matrix inversion

The inverse of a square $n \times n$ matrix $\boldsymbol{A}$ is a solution $\boldsymbol{X}$ of the set of equations $\boldsymbol{AX} = \mathbf{I_n}$ so any of the methods descussed above can be used to solve the equations $\boldsymbol{Ax}_i = \boldsymbol{e}_i$ for $i = 1, 2, \ldots, n$ obtaining the columns of $\boldsymbol{X}$ in turn.

8.3 Generalized Inverses

So far in this text we have only considered inverses of square non-singular matrices. In this section we relax these conditions to consider matrices which are square and singular and also non-square matrices. The aim is to define matrices which have many of the properties of an inverse and so allow the extension of many of the techniques and results in statistics which involve inversion of a square non-singular matrix to such cases.

For example, if $\boldsymbol{A}$ is a non-singular $n \times n$ matrix then the [unique] solution of the linear equation $\boldsymbol{Ax} = \boldsymbol{b}$ is $\boldsymbol{x} = \boldsymbol{A}^{-1}\boldsymbol{b}$ where $\boldsymbol{A}^{-1}$ is the ordinary inverse of $\boldsymbol{A}$ defined in Chapter 5. If $\boldsymbol{A}$ is singular or if $\boldsymbol{A}$ is $m \times n$ with $m \neq n$ then we can consider solutions of the linear equation $\boldsymbol{Ax} = \boldsymbol{b}$ in terms of generalized inverses.

We consider first the ***Moore–Penrose inverse*** which is defined by requiring the MP inverse to satisfy four properties. These ensure that this type of inverse is unique and that if the matrix should happen to be square and non-singular then it is the ordinary inverse. We follow this by considering matrices satisfying only the first of the Moore–Penrose conditions, these are termed ***generalized inverses***. This single condition does not uniquely define a matrix and there may be arbitrarily many satisfying the condition but nevertheless they can play a useful role in statistical methodology. In particular, many of the key results needed for investigating exact and approximate solutions to the linear equation $\boldsymbol{Ax} = \boldsymbol{b}$ (including least squares solutions) depend only upon the first of the Moore–Penrose conditions and so they can be expressed in terms of generalized inverses although in practice they would be calculated using the Moore–Penrose inverse, not least since **R** has a ready-made function for this but does not provide any easy facility for calculating the arbitrarily many generalized inverses.

8.3.1 Moore–Penrose inverse

Recall that the inverse $\boldsymbol{A}^{-1}$ of a non-singular $n \times n$ matrix $\boldsymbol{A}$ is defined to be that matrix satisfying $\boldsymbol{A}^{-1}\boldsymbol{A} = \boldsymbol{A}\boldsymbol{A}^{-1} = \mathbf{I_n}$. It is easy to verify that the matrix $\boldsymbol{M} = \boldsymbol{A}^{-1}$ satisfies the following four properties:

(i) $\boldsymbol{AMA} = \boldsymbol{A}$

(ii) $\boldsymbol{MAM} = \boldsymbol{M}$

(iii) $(\boldsymbol{AM})' = \boldsymbol{AM}$, i.e., $\boldsymbol{AM}$ is symmetric

(iv) $(\boldsymbol{MA})' = \boldsymbol{MA}$, i.e., $\boldsymbol{MA}$ is symmetric

If $\boldsymbol{A}$ is any matrix, not necessarily non-singular or square, then a matrix $\boldsymbol{M}$ satisfying these four conditions is termed the Moore–Penrose inverse of $\boldsymbol{A}$ and we denote it as $\boldsymbol{A}^{+}$. Note that we do not include the most obvious condition satisfied by $\boldsymbol{A}^{-1}$ that $\boldsymbol{A}\boldsymbol{A}^{-1} = \boldsymbol{A}^{-1}\boldsymbol{A} = \mathbf{I_n}$. We show below that $\boldsymbol{A}^{+}$ is uniquely defined by these four conditions and so it is justified to refer to ***the*** Moore–Penrose inverse rather than **a** Moore–Penrose inverse. However, first we show that there is indeed a matrix $\boldsymbol{A}^{+}$ satisfying these four conditions and so it does make sense to consider whether it is unique or not.

To do this we need a result from an earlier chapter using the singular value decomposition of $\boldsymbol{A}$ (see §6.7.2 on Page 94): if $\boldsymbol{A}$ is $m \times n$ of rank r then there exist $m \times r$ and $n \times r$ orthogonal matrices $\boldsymbol{U}$ and $\boldsymbol{V}$ and a $r \times r$ diagonal matrix Λ with positive diagonal elements such that $\boldsymbol{A} = \boldsymbol{U}\Lambda^{1/2}\boldsymbol{V}'$. Define $\boldsymbol{M} = \boldsymbol{V}\Lambda^{-1/2}\boldsymbol{U}'$, then we have to demonstrate that $\boldsymbol{M}$ satisfies each of the four Moore–Penrose conditions. $\boldsymbol{AMA} = \boldsymbol{U}\Lambda^{1/2}\boldsymbol{V}'\boldsymbol{V}\Lambda^{-1/2}\boldsymbol{U}'\boldsymbol{U}\Lambda^{1/2}\boldsymbol{V}' = \boldsymbol{U}\Lambda^{1/2}\boldsymbol{V}' = \boldsymbol{A}$ since $\boldsymbol{V}'\boldsymbol{V} = \boldsymbol{U}'\boldsymbol{U} = \mathbf{I_r}$ since both $\boldsymbol{U}$ and $\boldsymbol{V}$ are orthogonal. Similarly $\boldsymbol{MAM} = \boldsymbol{M}$. $(\boldsymbol{AM})' = \boldsymbol{U}\Lambda\boldsymbol{V}'\boldsymbol{V}\Lambda^{-1/2}\boldsymbol{U}' = \boldsymbol{UU}'$ which is symmetric. Similarly $(\boldsymbol{MA})' = \boldsymbol{VV}'$ which is symmetric and so $\boldsymbol{M}$ satisfies all four of the conditions and so is a Moore–Penrose inverse of $\boldsymbol{A}$.

To show $\boldsymbol{A}^{+}$ is unique, suppose there are two matrices $\boldsymbol{M}$ and $\boldsymbol{P}$ that satisfy the four conditions. We note that $\boldsymbol{AM} = \boldsymbol{AM}' = \boldsymbol{M}'\boldsymbol{A}' = \boldsymbol{M}'\boldsymbol{APA}' = \boldsymbol{M}'\boldsymbol{A}'\boldsymbol{P}'\boldsymbol{A}' = (\boldsymbol{AM})'(\boldsymbol{AP})' = \boldsymbol{AMAP} = \boldsymbol{AC}$, using just conditions (i) and (iii). Also we have $\boldsymbol{MA} = \boldsymbol{MA}' = \boldsymbol{A}'\boldsymbol{M}' = \boldsymbol{APA}'\boldsymbol{M}' = \boldsymbol{A}'\boldsymbol{P}'\boldsymbol{A}'\boldsymbol{M}' = (\boldsymbol{PA})'(\boldsymbol{MA})' = \boldsymbol{PAMAP} = \boldsymbol{PA}$ using just conditions (i) and (iv). Then, using condition (ii) $\boldsymbol{M} = \boldsymbol{MAM} = \boldsymbol{BAP} = \boldsymbol{PAP} = \boldsymbol{P}$.

It follows immediately that if $\boldsymbol{A}$ is a square non-singular matrix then $\boldsymbol{A}^{+} = \boldsymbol{A}^{-1}$ since we have already seen that $\boldsymbol{A}^{-1}$ satisfies the four Moore–Penrose conditions.

We have $\rho(\boldsymbol{A}) = \rho(\boldsymbol{A}^{+})$ because $\rho(\boldsymbol{A}) = \rho(\boldsymbol{AA}^{+}\boldsymbol{A}) \leq \rho(\boldsymbol{A}^{+}) = \rho(\boldsymbol{A}^{+}\boldsymbol{AA}^{+}) \leq \rho(\boldsymbol{A})$, using MP conditions (i) and (ii). Similarly $\rho(\boldsymbol{A}) = \rho(\boldsymbol{AA}^{+}) = \rho(\boldsymbol{A}^{+}\boldsymbol{A})$.

We leave as exercises the results that $(\boldsymbol{A}^{+})^{+} = \boldsymbol{A}$, $(\boldsymbol{A}')^{+} = (\boldsymbol{A}^{+})'$, $(\boldsymbol{AA}^{+})^{+} = \boldsymbol{AA}^{+}$ and $(\boldsymbol{A}^{+}\boldsymbol{A})^{+} = \boldsymbol{A}^{+}\boldsymbol{A}$. All can be verified by checking that the four MR conditions are satisfied.

If $\boldsymbol{A}$ is a $m \times n$ matrix then further key results are

(i) If $\rho(\boldsymbol{A}) = m$ (i.e., is of full row rank) then $\boldsymbol{A}^{+} = \boldsymbol{A}'(\boldsymbol{AA}')^{-1}$ and so $\boldsymbol{AA}^{+} = \mathbf{I_m}$.

(ii) If $\rho(\boldsymbol{A}) = n$ (i.e., is of full column rank) then $\boldsymbol{A}^{+} = (\boldsymbol{A}'\boldsymbol{A})^{-1}\boldsymbol{A}'$ and so $\boldsymbol{A}^{+}\boldsymbol{A} = \mathbf{I_n}$.

(iii) If $\rho(\boldsymbol{A}) = 1$ then $\boldsymbol{A}^+ = (\text{tr}(\boldsymbol{AA}'))^{-1}\boldsymbol{A}'$.

(iv) If $\boldsymbol{B}$ is a $n \times p$ matrix and $\rho(\boldsymbol{A}) = n$ and $\rho(\boldsymbol{B}) = n$ then $(\boldsymbol{AB})^+ = \boldsymbol{B}^+\boldsymbol{A}^+$.

The proofs of these results are not given here but can be found in many texts, in particular in Abadir and Magnus (2005). Note that (i) and (ii) show that it is not possible for both $\boldsymbol{AA}^+ = \mathbf{I_m}$ and $\boldsymbol{A}^+\boldsymbol{A} = \mathbf{I_n}$ to be true simultaneously unless $m = n$, in which case $\boldsymbol{A}$ is non-singular and $\boldsymbol{A}^+ = \boldsymbol{A}^{-1}$.

8.3.2 Moore–Penrose inverses in R

R provides a function `ginv(.)` for calculating the Moore–Penrose inverse of a matrix. It is in the `MASS` library so the MASS library needs to be loaded before using the function by the command `library(MASS)`. This library is automatically included when **R** is first installed and so does not need to be installed separately. There is also an equivalent function `MPinv(.)` in the `gnm` library which does have to be installed separately (this library provides facilities for fitting non-linear generalized linear models).

Example 8.5:

(i) A non-square matrix not of full rank

```
> library(MASS)
> options(digits=2)
>  A<-matrix(c(3,2,1,4,2,0,5,
+ 2,-1,-1,0,1),4,3,byrow=T)
> A
     [,1] [,2] [,3]
[1,]    3    2    1
[2,]    4    2    0
[3,]    5    2   -1
[4,]   -1    0    1
```

```
>
> M<-ginv(A)
> M
     [,1] [,2]  [,3]  [,4]
[1,] 0    0.06  0.11 -0.06
[2,] 0.20 0.06 -0.06  0.11
[3,] 0.30 0.06 -0.22  0.28
>
```

To check that the four Moore–Penrose conditions are satisfied:

```
> A%*%M%*%A
     [,1]  [,2]   [,3]
[1,]    3 2e+00  1e+00
[2,]    4 2e+00  3e-16
[3,]    5 2e+00 -1e+00
[4,]   -1 2e-16  1e+00
```

(i) so $\boldsymbol{AMA} = \boldsymbol{A}$,

```
> M%*%A%*%M
      [,1] [,2]  [,3]  [,4]
[1,] 2e-17 0.06  0.11 -0.06
[2,] 2e-01 0.06 -0.06  0.11
[3,] 3e-01 0.06 -0.22  0.28
```

(ii) so $\boldsymbol{MAM} = \boldsymbol{M}$,

```
> M%*%A
     [,1] [,2] [,3]
[1,]  0.8  0.3 -0.2
[2,]  0.3  0.3  0.3
[3,] -0.2  0.3  0.8
```

(iii) so $(\boldsymbol{AM})' = \boldsymbol{AM}$, i.e., $\boldsymbol{AM}$ is symmetric,

```
> A%*%M
      [,1]  [,2]   [,3]   [,4]
[1,] 7e-01 3e-01 -1e-16  3e-01
[2,] 3e-01 3e-01  3e-01  1e-16
[3,] 1e-16 3e-01  7e-01 -3e-01
[4,] 3e-01 8e-17 -3e-01  3e-01
>
```

(iv) so $(\boldsymbol{MA})' = \boldsymbol{MA}$, i.e., $\boldsymbol{MA}$ is symmetric (within negligible rounding errors).

(ii) A non-square matrix of full row rank

```
> options(digits=2)
> library(MASS)
> B<-matrix(c(1,2,3,4,5,6)
+ ,2,3)
> B
     [,1] [,2] [,3]
[1,]    1    3    5
[2,]    2    4    6
> M<-ginv(B)
> M
      [,1]  [,2]
[1,] -1.33  1.08
[2,] -0.33  0.33
[3,]  0.67 -0.42
```

```
> M%*%B
      [,1] [,2]  [,3]
[1,]  0.83 0.33 -0.17
[2,]  0.33 0.33  0.33
[3,] -0.17 0.33  0.83

> B%*%M
        [,1] [,2]
[1,] 1.0e+00    0
[2,] 2.7e-15    1
```

So $\boldsymbol{B}^+\boldsymbol{B} \neq \mathbf{I}_3$ but $\boldsymbol{BB}^+ = \mathbf{I}_2$.

(iii) A non-square matrix of full column rank

```
> library(MASS)
> C<-matrix(c(1,2,3,5,1,5,6,4,5,3,1,4),4,3); C
     [,1] [,2] [,3]
[1,]    1    1    5
[2,]    2    5    3
[3,]    3    6    1
[4,]    5    4    4
> M<-ginv(C)
> M
        [,1]   [,2]   [,3]   [,4]
[1,] -0.1336 -0.198 -0.028  0.323
[2,] -0.0072  0.154  0.133 -0.140
[3,]  0.1930  0.064 -0.071 -0.022
> M%*%C
         [,1]     [,2]     [,3]
[1,]  1.0e+00 -2.2e-16  4.4e-16
[2,] -1.1e-16  1.0e+00 -1.1e-16
[3,] -2.8e-17  5.6e-17  1.0e+00
> C%*%M
       [,1]  [,2]  [,3]   [,4]
[1,]  0.824  0.28 -0.25  0.075
[2,]  0.276  0.57  0.39 -0.118
[3,] -0.251  0.39  0.64  0.108
[4,]  0.075 -0.12  0.11  0.968
```

So $\boldsymbol{C}^{+}\boldsymbol{C} = \mathbf{I}_3$ but $\boldsymbol{C}\boldsymbol{C}^{+} \neq \mathbf{I}_4$.

(iv) A column vector

```
> library(MASS)
> D<-matrix(c(1,3,7)); D
     [,1]
[1,]    1
[2,]    3
[3,]    7
> M<-ginv(D); M
      [,1]  [,2] [,3]
[1,] 0.017 0.051 0.12
```

```
> D%*%M
      [,1]  [,2] [,3]
[1,] 0.017 0.051 0.12
[2,] 0.051 0.153 0.36
[3,] 0.119 0.356 0.83
> M%*%D
     [,1]
[1,]    1
> sum(diag(D%*%t(D)))*M
     [,1] [,2] [,3]
[1,]    1    3    7
```

So $\text{tr}(\boldsymbol{D}\boldsymbol{D}')\boldsymbol{D}^{+} = \boldsymbol{D}'$, i.e., $\boldsymbol{D}^{+} = [\text{tr}(\boldsymbol{D}\boldsymbol{D}')]^{-1}\boldsymbol{D}'$.

8.3.3 Generalized inverse

A matrix $\boldsymbol{G}$ is termed a generalized inverse of the $m \times n$ matrix $\boldsymbol{A}$ if it satisfies the first of the Moore–Penrose conditions: $\boldsymbol{AGA} = \boldsymbol{A}$. It is usually denoted by $\boldsymbol{A}^-$ and necessarily is a $n \times m$ matrix. The Moore–Penrose inverse of $\boldsymbol{A}$, $\boldsymbol{A}^+$ is clearly **a** generalized inverse of $\boldsymbol{A}$ and we know that the Moore–Penrose inverse is unique but the proof of this required an appeal to each of the four MP conditions (see §8.3.1). This suggests that a generalized inverse satisfying only the first condition is not necessarily unique. This is indeed the case and it can be shown (see Abadir and Magnus (2005) §10.5) that any generalized inverse can be written in the form $\boldsymbol{A}^- = \boldsymbol{A}^+ + \boldsymbol{Q} - \boldsymbol{A}^+\boldsymbol{AQAA}^+$ where $\boldsymbol{Q}$ is any $n \times m$ matrix, i.e., $\boldsymbol{Q}$ is arbitrary. It is easily verified that $\boldsymbol{AA}^-\boldsymbol{A} = \boldsymbol{A}$, recalling that $\boldsymbol{AA}^+\boldsymbol{A} = \boldsymbol{A}$ (the first of the MP conditions). Note that taking $\boldsymbol{Q} = \boldsymbol{A}^+$ gives $\boldsymbol{A}^- = \boldsymbol{A}^+$ as the generalized inverse, recalling that $\boldsymbol{A}^+\boldsymbol{AA}^+ = \boldsymbol{A}^+$ (the second of the MP conditions).

R has no ready-made function to produce generalized inverses but the form above can be used, together with `ginv(.)` in the `MASS` library, to produce a generalized inverse (which of course will be different for different choices of $\boldsymbol{Q}$).

The primary role of generalized inverses is in discussing solutions of the system of linear equations in $\boldsymbol{x}$, $\boldsymbol{Ax} = \boldsymbol{y}$. There may be many solutions for $\boldsymbol{x}$ if $\boldsymbol{A}$ is singular or non-square.

Example 8.6:

(i) A non-square matrix not of full rank

```
> library(MASS)
> options(digits=2)
>  A<-matrix(c(3,2,1,4,2,0,5,
+ 2,-1,-1,0,1),4,3,byrow=T)
> A
     [,1] [,2] [,3]
[1,]    3    2    1
[2,]    4    2    0
[3,]    5    2   -1
[4,]   -1    0    1
> ### First need the
> ### MP-Inverse of A
> M<-ginv(A)
> M
        [,1]  [,2]  [,3]  [,4]
[1,] 6.9e-18 0.06  0.11 -0.06
[2,] 1.7e-01 0.06 -0.06  0.11
[3,] 3.3e-01 0.06 -0.22  0.28
```

```
> ### Next generate an
> ### arbitrary 3x4
> ### matrix Q
> set.seed(137)
> Q<-matrix(c(sample(1:12
+ ,replace=T)),3,4)
> Q
     [,1] [,2] [,3] [,4]
[1,]    8   10   11    4
[2,]    5    5   10    9
[3,]   11   12    8   10
> ### Now calculate
> ### Generalized Inverse
> G<-M+Q-M%*%A%*%Q%*%A%*%M
> G
      [,1] [,2] [,3] [,4]
[1,] -0.56  1.5  2.6  3.9
[2,] -5.72 -3.8  3.1  7.1
[3,] -1.89  2.8  2.6  6.3
```

```
> ### check G satisfies
> ### the first condition
> A%*%G%*%A
     [,1]     [,2]     [,3]
[1,]    3  2.0e+00  1.0e+00
[2,]    4  2.0e+00 -6.4e-14
[3,]    5  2.0e+00 -1.0e+00
[4,]   -1 -2.7e-14  1.0e+00
>
> ### Next generate a
> ### different arbitrary
> ### 3x4 matrix Q
> set.seed(163)
> Q<-matrix(c(sample(1:12
+ ,replace=T)),3,4)
> Q
     [,1] [,2] [,3] [,4]
[1,]    4    6    2    9
[2,]    9    8    6    4
[3,]    7    1    5    9
>
> ### Now calculate
> ### Generalized Inverse
> G<-M+Q-M%*%A%*%Q%*%A%*%M
> G
      [,1]  [,2] [,3] [,4]
[1,] -4.39  0.89 0.17 5.72
[2,]  0.61  2.72 3.83 0.89
[3,] -1.39 -4.44 2.50 6.06
>
> ### check G satisfies
> ### the first condition
> A%*%G%*%A
     [,1]     [,2]   [,3]
[1,]    3  2.0e+00  1e+00
[2,]    4  2.0e+00 -6e-14
[3,]    5  2.0e+00 -1e+00
[4,]   -1 -2.2e-14  1e+00
```

This example shows two of the arbitrarily many generalized inverses of $\boldsymbol{A}$ by generating two random versions of $\boldsymbol{Q}$ using the function `sample(.)` (type `help(sample)` for more information on this function). The seeds used for the **R** random number generator were 137 and 163 and so can be reproduced if desired.

(ii) A non-square matrix of full column rank

```
> options(digits=2)
> library(MASS)
> C<-matrix(c(1,2,3,5,1,5,6,
+ 4,5,3,1,4),4,3); C
     [,1] [,2] [,3]
[1,]    1    1    5
[2,]    2    5    3
[3,]    3    6    1
[4,]    5    4    4
> M<-ginv(C); M
       [,1]  [,2]  [,3]  [,4]
[1,] -0.134 -0.20 -0.03  0.32
[2,] -0.007  0.15  0.13 -0.14
[3,]  0.193  0.06 -0.07 -0.02
> set.seed(137)
> Q<-matrix(c(sample(1:12,
+ replace=T)),3,4);Q
     [,1] [,2] [,3] [,4]
[1,]    8   10   11    4
[2,]    5    5   10    9
[3,]   11   12    8   10
> G<-M+Q-M%*%C%*%Q%*%C%*%M
> C%*%G%*%C
     [,1] [,2] [,3]
[1,]    1    1    5
[2,]    2    5    3
[3,]    3    6    1
[4,]    5    4    4
```

```
> set.seed(163)
> Q<-matrix(c(sample(1:12,
+ replace=T)),3,4);Q
     [,1] [,2] [,3] [,4]
[1,]    4    6    2    9
[2,]    9    8    6    4
[3,]    7    1    5    9
```

```
> G<-M+Q-M%*%C%*%Q%*%C%*%M
> C%*%G%*%C
     [,1] [,2] [,3]
[1,]    1    1    5
[2,]    2    5    3
[3,]    3    6    1
[4,]    5    4    4
```

(iii) A column vector

```
> options(digits=1)
> library(MASS)
> D<-matrix(c(1,6,3),3,1)
> M<-ginv(D);M
     [,1] [,2] [,3]
[1,] 0.02  0.1 0.07
> set.seed(2015)
> Q<-matrix(c(sample(
+c(7,3,1),replace=T))
+,1,3); Q
     [,1] [,2] [,3]
[1,]    7    1    7
```

```
> G<-M+Q-M%*%D%*%Q%*%D%*%M
> G
### check G satisfies
### first condition
>D%*%G%*%D
     [,1] [,2] [,3]
[1,]    6   -3    5
     [,1]
[1,]    1
[2,]    6
[3,]    3
```

```
> set.seed(1966)
> Q<-matrix(c(sample(
+c(7,3,1),replace=T))
+,1,3); Q
     [,1] [,2] [,3]
[1,]    1    3    7
> G<-M+Q-M%*%D%*%Q%*%D%*%M
```

```
> G; D%*%G%*%D
     [,1] [,2] [,3]
[1,]  0.2   -2    4
     [,1]
[1,]    1
[2,]    6
[3,]    3
```

8.3.3.1 Solutions of linear equations

The linear equation in $\boldsymbol{x}$, $\boldsymbol{Ax} = \boldsymbol{y}$, may possess a unique solution, no solution or arbitrarily many solutions. For example, the equation $\left(\begin{smallmatrix}1 & 1\\ 2 & 2\end{smallmatrix}\right)\boldsymbol{x} = \left(\begin{smallmatrix}1\\ 1\end{smallmatrix}\right)$ has no solution since it is not possible for both $x_1 + x_2 = 1$ and $2x_1 + 2x_2 = 1$ to be true simultaneously so it is said to be ***inconsistent***. If the equation $\boldsymbol{Ax} = \boldsymbol{y}$ has a solution it is said to be ***consistent***. The equation $(1,1)\boldsymbol{x} = 1$ (where $\boldsymbol{x}$ is a 2×1 column vector has the solutions $\boldsymbol{x} = (1,q)'$ for any value of q, so there are arbitrarily many solutions.

If $\boldsymbol{Ax} = \boldsymbol{y}$ is consistent then there is a solution, $\boldsymbol{x}^\star$ say, so that $\boldsymbol{Ax}^\star = \boldsymbol{y}$. So $\boldsymbol{y} = \boldsymbol{Ax}^\star = \boldsymbol{AA}^-\boldsymbol{Ax}^\star = \boldsymbol{AA}^-\boldsymbol{y}$. Conversely, if $\boldsymbol{AA}^-\boldsymbol{y} = \boldsymbol{y}$ then let $\boldsymbol{x}^\star = \boldsymbol{A}^-\boldsymbol{y}$ so $\boldsymbol{Ax}^\star = \boldsymbol{AA}^-\boldsymbol{y} = \boldsymbol{y}$ and thus the equation is consistent. Thus a necessary and sufficient condition for the equation $\boldsymbol{Ax} = \boldsymbol{y}$ to be consistent is that $\boldsymbol{AA}^-\boldsymbol{y} = \boldsymbol{y}$. This provides a way of checking whether a system of linear equations is consistent. In practice, of course, this would be checked by using the Moore–Penrose inverse A^+ because of the ease of computation.

Clearly if A is a square non-singular matrix then $A^- = A^{-1}$ and so $AA^-y = AA^{-1}y = y$ and so the equation is consistent and $x = A^{-1}y$ is the unique solution (since A^{-1} is unique).

Further, if $Ax = y$ is consistent and if A is $m \times n$ with $m > n$ and $\rho(A) = n$, i.e., it has full column rank then $A'A$ is a non-singular $n \times n$ matrix and so possesses an inverse; consequently premultiplying both sides of the equation by $(A'A)^{-1}A'$ gives a solution as $x = (A'A)^{-1}A'y$.

Notes:

(a) This argument is only valid if $Ax = y$ is consistent because if it is not, it would depend upon a false premise, i.e., it is possible that A has full column rank but $x = (A'A)^{-1}A'y$ is **not** a solution of the equation (and indeed the equation has no solutions). This is illustrated in the first of the examples below.

(b) In the full column rank case $(A'A)^{-1}A' = A^+$; see key result (ii) in §8.3.1 on Page 125, so the solution can be written as $x = A^+y$.

(c) If $m = n$, i.e., A is square and therefore non-singular, then $(A'A)^{-1} = A^{-1}(A')^{-1}$ and so this reduces to $x = A^{-1}y$ as the (unique) solution.

(d) It will be seen that in general the solution $x = A^+y$ is unique when $m > n$ and $\rho(A) = n$, provided the equation is consistent (i.e., has any solutions at all).

Suppose $Ax = y$ is consistent (so that $AA^-y = y$), then if A^- is any generalized inverse of A we have $AA^-A = A$ so $AA^-Ax = Ax$, then if $Ax = y$ we have $A(A^-y) = y$ and so $x = A^-y$ is a solution of $Ax = y$. Conversely, suppose $Ax = y$ is consistent and has a solution $x = Gy$. Let a_j be the jth column of A and consider the equations $Ax = a_j$. Each of these has a solution $x = e_j$, the jth unit vector, i.e., a vector with j^{th} element 1 and all others 0 and so the equations are consistent. Therefore the equations $Ax = a_j$ have a solution $x = Ga_j$ for all j and so $AGa_j = a_j$ for all j and thus $AGA = A$.

Recalling (see §8.3.3) that any generalized inverse A^- can be written in the form $A^- = A^+ + Q - A^+AQAA^+$ and provided the equation is consistent, i.e., $AA^+y = y$, we can write any solution of $Ax = y$ in the form $x = (A^+ + Q - A^+AQAA^+)y = A^+y + (\mathbf{I_n} - A^+A)q$, where q is any conformable vector (i.e., $n \times 1$), writing q for Qy.

If A has full column rank then $A^+A = I_n$ (see key result (ii) in §8.3.1 on Page 125) and thus the solution above reduces to $x = A^+y$ and is unique even if A is non-square.

If A has full row rank then $AA^+ = I_m$ (see key result (i) in §8.3.1 on Page 125) and so $AA^+y = y$ for any y and so the $Ax = y$ equation is consistent for any y and has a solution $x = A^+y$. If the QR decomposition of A' is given by $A' = QR$ then $A^+ = A'(AA')^{-1} = QR(R'Q'QR)^{-1} = QRR^{-1}(R')^{-1} = Q(R')^{-1}$ as stated in §8.2.7 on Page 123.

If the equation $Ax = y$ has no solutions or has many different solutions, the question arises as to which is the best approximate solution or is the best exact solution. One method is to base the choice on the ***least squares criterion***. Consider the quantity $(\boldsymbol{Ax}-\boldsymbol{y})'(\boldsymbol{Ax}-\boldsymbol{y})$ and choose a value of $\boldsymbol{x}$ which minimises this (whether or not the equation is consistent). Let $\boldsymbol{x}^\star = \boldsymbol{A}^+\boldsymbol{y}$ then $(\boldsymbol{Ax}-\boldsymbol{y}) = \boldsymbol{A}(\boldsymbol{x}-\boldsymbol{x}^\star) - (\mathbf{I_m}-\boldsymbol{AA}^+)\boldsymbol{y}$. Noting that $(\mathbf{I_m}-\boldsymbol{AA}^+)'(\mathbf{I_m}-\boldsymbol{AA}^+) = (\mathbf{I_m}-\boldsymbol{AA}^+)$, because $\boldsymbol{AA}^+$ is symmetric by definition and $\boldsymbol{AA}^+\boldsymbol{AA}^+ = \boldsymbol{AA}^+$, and also noting that $\boldsymbol{A}'(\mathbf{I_m}-\boldsymbol{AA}^+) = \mathbf{0}$ gives $(\boldsymbol{Ax}-\boldsymbol{y})'(\boldsymbol{Ax}-\boldsymbol{y}) = (\boldsymbol{x}-\boldsymbol{A}^+\boldsymbol{y})'\boldsymbol{A}'\boldsymbol{A}(\boldsymbol{x}-\boldsymbol{A}^+\boldsymbol{y}) + \boldsymbol{y}'(\mathbf{I_m}-\boldsymbol{AA}^+)\boldsymbol{y}$. Since $(\boldsymbol{x}-\boldsymbol{A}^+\boldsymbol{y})'\boldsymbol{A}'\boldsymbol{A}(\boldsymbol{x}-\boldsymbol{A}^+\boldsymbol{y}) \geq 0$ the sum of squares $(\boldsymbol{Ax}-\boldsymbol{y})'(\boldsymbol{Ax}-\boldsymbol{y})$ is minimised when $\boldsymbol{A}(\boldsymbol{x}-\boldsymbol{A}^+\boldsymbol{y}) = \mathbf{0}$ which is true when $\boldsymbol{x} = \boldsymbol{x}^\star = \boldsymbol{A}^+\boldsymbol{y}$. So $\boldsymbol{x} = \boldsymbol{A}^+\boldsymbol{y}$ is the ***least squares solution*** of the equation $\boldsymbol{Ax} = \boldsymbol{y}$ and it may be an approximate solution or it may be an exact solution.

Summary:

- The linear equation $\boldsymbol{Ax} = \boldsymbol{y}$ may have a unique exact solution, many exact solutions or no exact solutions.
- If it has a solution (i.e., is consistent) then this is given by $\boldsymbol{x} = \boldsymbol{A}^-\boldsymbol{y}$.
- A necessary and sufficient condition for it to have any solutions is that $\boldsymbol{AA}^-\boldsymbol{y} = \boldsymbol{y}$. Essentially, this amounts to saying the solution might be $\boldsymbol{x} = \boldsymbol{A}^-\boldsymbol{y}$ (but check that it works to see whether there are any exact solutions at all).
- If $\boldsymbol{A}$ has full row rank then it has at least one solution for every value of $\boldsymbol{y}$.
- If $\boldsymbol{A}$ has full column rank and if the equation is consistent then then $\boldsymbol{x} = \boldsymbol{A}^+\boldsymbol{y}$ is the unique solution.
- Irrespective of whether the equation is consistent $\boldsymbol{x} = \boldsymbol{A}^+\boldsymbol{y}$ is the least squares solution, i.e., it minimises $(\boldsymbol{Ax}-\boldsymbol{y})'(\boldsymbol{Ax}-\boldsymbol{y})$.

Example 8.7: (*Details of the calculations are left to the exercises.*)

(i) A matrix of full column rank but equation is not consistent.

Let $\boldsymbol{A} = \begin{pmatrix} 1 & 1 \\ 2 & 2 \\ 3 & 4 \end{pmatrix}$ and $\boldsymbol{y} = \begin{pmatrix} 1 \\ 1 \\ 1 \end{pmatrix}$ then clearly the columns of $\boldsymbol{A}$ are linearly independent and so $\rho(\boldsymbol{A}) = 2$ but $\boldsymbol{AA}^+\boldsymbol{y} = \begin{pmatrix} 0.6 \\ 1.2 \\ 1.0 \end{pmatrix} \neq \boldsymbol{y}$ and so the equation $\boldsymbol{Ax} = \boldsymbol{y}$ is not consistent. This is easily seen because $\boldsymbol{Ax} = \boldsymbol{y}$ implies that $x_1 + x_2 = 1$ and $2x_1 + 2x_2 = 1$ (and $3x_1 + 4x_2 = 1$) and the first two equations cannot both be true. Note that if $\boldsymbol{y} = \begin{pmatrix} 1 \\ 2 \\ 7 \end{pmatrix}$ or $\boldsymbol{y} = \begin{pmatrix} 3 \\ 6 \\ 11 \end{pmatrix}$ then in each case $\boldsymbol{AA}^+\boldsymbol{y} = \boldsymbol{y}$ and so the equation has a solution and indeed it is unique and given by $\boldsymbol{A}^+\boldsymbol{y}$ (see exercises below).

(ii) A 3×4 matrix of rank 2.

Let $\boldsymbol{A} = \begin{pmatrix} 2 & 3 & 3 & 1 \\ 3 & 4 & 5 & 1 \\ 1 & 2 & 1 & 1 \end{pmatrix}$ and $\boldsymbol{y} = \begin{pmatrix} 14 \\ 22 \\ 6 \end{pmatrix}$

then $\boldsymbol{A}^+ = \begin{pmatrix} 0.02 & 0.05 & -0.01 \\ 0.14 & -0.16 & 0.43 \\ -0.08 & 0.30 & -0.46 \\ 0.12 & -0.21 & 0.44 \end{pmatrix}$. It is easily verified that $\boldsymbol{AA}^+\boldsymbol{y} = \boldsymbol{y}$ so the equation is consistent. Thus it has at least one solution, one of these is provided by $\boldsymbol{x} = \boldsymbol{A}^+\boldsymbol{y} = \begin{pmatrix} 1.3 \\ 1.1 \\ 2.8 \\ -0.2 \end{pmatrix}$. Other solutions are given by $\boldsymbol{x} = \boldsymbol{A}^+\boldsymbol{y} + (I_4 - \boldsymbol{A}^+\boldsymbol{A})\boldsymbol{q} = \begin{pmatrix} 1.3 \\ 1.1 \\ 2.8 \\ -0.2 \end{pmatrix} + \boldsymbol{q} - \begin{pmatrix} 0.18 & 0.24 & 0.29 & 0.06 \\ 0.24 & 0.65 & 0.06 & 0.41 \\ 0.29 & 0.06 & 0.82 & -0.24 \\ 0.06 & 0.41 & -0.24 & 0.35 \end{pmatrix} \boldsymbol{q}$, where $\boldsymbol{q}$ is any 4×1 vector. For example, taking $\boldsymbol{q} = \begin{pmatrix} 1 \\ 1 \\ 1 \\ 1 \end{pmatrix}$ gives $\boldsymbol{x} = \begin{pmatrix} 1.5 \\ 0.7 \\ 2.9 \\ 0.2 \end{pmatrix}$.

Verification that this does provide a solution and other details of the calculations are left to the exercises.

8.4 Hadamard Products

If $\boldsymbol{A}$ and $\boldsymbol{B}$ are two $m \times n$ matrices then their Hadamard product $\boldsymbol{A} \odot \boldsymbol{B}$ is the element-by-element product defined by $(\boldsymbol{A} \odot \boldsymbol{B})_{ij} = a_{ij}b_{ij}$. Hadamard products are not defined for matrices of different orders.

It is easy to see that Hadamard products follow the familiar rules of multiplicative arithmetic of scalars (commutative, associative and distributive): $\boldsymbol{A} \odot \boldsymbol{B} = \boldsymbol{B} \odot \boldsymbol{A}$, $(\boldsymbol{A} \odot \boldsymbol{B}) \odot \boldsymbol{C} = \boldsymbol{A} \odot (\boldsymbol{B} \odot \boldsymbol{C})$ and $\boldsymbol{A} \odot (\boldsymbol{B} + \boldsymbol{C}) = \boldsymbol{A} \odot \boldsymbol{B} + \boldsymbol{A} \odot \boldsymbol{C}$. Further, if λ is a scalar then clearly $\lambda(\boldsymbol{A} \odot \boldsymbol{B}) = (\lambda\boldsymbol{A}) \odot \boldsymbol{B} = \boldsymbol{A} \odot (\lambda\boldsymbol{B})$ and $(\boldsymbol{A} \odot \boldsymbol{B})' = \boldsymbol{A}' \odot \boldsymbol{B}'$.

Other properties of Hadamard products include;

- If $\boldsymbol{A}$ is a square $n \times n$ matrix $\boldsymbol{A} \odot \mathbf{I_n} = \text{diag}(\text{diag}(\boldsymbol{A}))$.
- If $\boldsymbol{u}$ and $\boldsymbol{x}$ are $m \times 1$ vectors and $\boldsymbol{v}$ and $\boldsymbol{y}$ are $n \times 1$ vectors then $(\boldsymbol{uv}') \odot (\boldsymbol{xy}') = (\boldsymbol{u} \odot \boldsymbol{x})(\boldsymbol{v} \odot \boldsymbol{y})'$.
- If $\rho(\boldsymbol{A}) = \rho(\boldsymbol{B}) = 1$ then $\rho(\boldsymbol{A} \odot \boldsymbol{B}) \leq 1$.
- $\rho(\boldsymbol{A} \odot \boldsymbol{B}) \leq \rho(\boldsymbol{A}).\rho(\boldsymbol{B})$.

- If $\boldsymbol{A}$ and $\boldsymbol{B}$ are both $m \times n$ and $\boldsymbol{x}$ is $n \times 1$ then $\text{diag}(\boldsymbol{A}\text{diag}(\text{diag}(\boldsymbol{x}))\boldsymbol{B}') = (\boldsymbol{A} \odot \boldsymbol{B})\boldsymbol{x}$.
- $\boldsymbol{x}'(\boldsymbol{A} \odot \boldsymbol{B})\boldsymbol{y} = \text{tr}(\text{diag}(\text{diag}(\boldsymbol{x}))\boldsymbol{A}\text{diag}(\text{diag}(\boldsymbol{y}))\boldsymbol{B}')$, where $\boldsymbol{x}$, $\boldsymbol{y}$, $\boldsymbol{A}$ and $\boldsymbol{B}$ are conformable so that the various products are well-defined.

Proofs of these are left as exercises below. Further key results concerning Hadamard products are given in Styan (1973). The most notable of these are:

- When $\boldsymbol{A}$ and $\boldsymbol{B}$ are both square then $\boldsymbol{\iota_n}'(\boldsymbol{A} \odot \boldsymbol{B})\boldsymbol{\iota_n} = \text{tr}(\boldsymbol{AB})$.
- When $\boldsymbol{A}$ and $\boldsymbol{B}$ are both positive [semi-]definite then so is $\boldsymbol{A} \odot \boldsymbol{B}$ (the Schur product theorem).
- If $\boldsymbol{A}$ and $\boldsymbol{B}$ are both square positive semidefinite $n \times n$ matrices and $\lambda_1 \geq \lambda_2 \geq \ldots \geq \lambda_n \geq 0$ are the eigenvalues of $\boldsymbol{A}$ and b_{min} and b_{max} are the smallest and largest diagonal elements of $\text{diag}(\boldsymbol{B})$ and μ_j is the jth largest eigenvalue of $\boldsymbol{A} \odot \boldsymbol{B}$ then $\lambda_n b_{min} \leq \mu_j \leq \lambda_1 b_{max}$.
- If $\nu_1 \geq \nu_2 \geq \ldots \geq \nu_n \geq 0$ are the eigenvalues of $\boldsymbol{B}$ then $\lambda_n \nu_n \leq \mu_j \leq \lambda_1 \nu_1$.
- $|\boldsymbol{A} \odot \boldsymbol{B}| \geq \boldsymbol{A}|.|\boldsymbol{B}|$.

For detailed proofs of these and indeed further results and applications to multivariate statistical analysis, the reader is referred to Styan (1973).

8.5 Kronecker Products and the Vec Operator

If $\boldsymbol{A}$ is a $m \times n$ matrix and $\boldsymbol{B}$ is a $p \times q$ matrix then the ***Kronecker product*** of $\boldsymbol{A}$ and $\boldsymbol{B}$, $\boldsymbol{A} \otimes \boldsymbol{B}$, is the $mp \times nq$ matrix

$$\boldsymbol{A} \otimes \boldsymbol{B} = \begin{pmatrix} a_{11}\boldsymbol{B} & a_{12}\boldsymbol{B} & \cdots & a_{1n}\boldsymbol{B} \\ a_{21}\boldsymbol{B} & a_{22}\boldsymbol{B} & \cdots & a_{2n}\boldsymbol{B} \\ \vdots & \vdots & & \vdots \\ a_{11}\boldsymbol{B} & a_{m2}\boldsymbol{B} & \cdots & a_{mn}\boldsymbol{B} \end{pmatrix}.$$

The ***vectorization*** of a matrix $\boldsymbol{A}$ is obtained by stacking the columns of $\boldsymbol{A}$ on top of one another. If the ith column is $\boldsymbol{a}_i$ then

$$\text{vec}(\boldsymbol{A}) = \begin{pmatrix} \boldsymbol{a}_1 \\ \boldsymbol{a}_2 \\ \vdots \\ \boldsymbol{a}_n \end{pmatrix} = \begin{pmatrix} a_{11} \\ \vdots \\ a_{m1} \\ a_{12} \\ \vdots \\ a_{m2} \\ \vdots \\ a_{mn} \end{pmatrix},$$

so $\text{vec}(\boldsymbol{A})$ is a $mn \times 1$ vector. A similar operator $\text{vech}(\boldsymbol{A})$ of a square $n \times n$ [symmetric] matrix $\boldsymbol{A}$ stacks just the lower triangle of the elements of $\boldsymbol{A}$ and so is a $n(n+1)/2 \times 1$ vector. The connection between Kronecker products and the vec and vech operators will become apparent below.

8.5.1 Elementary properties of Kronecker products

It is straightforward to verify from the definition that the basic properties below hold (presuming the matrices are conformable as necessary):

- $\lambda \otimes \boldsymbol{A} = \boldsymbol{A} \otimes \lambda = \lambda \boldsymbol{A}$ for any scalar λ.
- $(\lambda \boldsymbol{A}) \otimes (\mu \boldsymbol{B}) = \lambda \mu \boldsymbol{A} \otimes \boldsymbol{B}$.
- $(\boldsymbol{A} \otimes \boldsymbol{B}) \otimes \boldsymbol{C} = \boldsymbol{A} \otimes (\boldsymbol{B} \otimes \boldsymbol{C})$.
- $(\boldsymbol{A} + \boldsymbol{B}) \otimes \boldsymbol{C} = \boldsymbol{A} \otimes \boldsymbol{C} + \boldsymbol{B} \otimes \boldsymbol{C}$.
- $\boldsymbol{A} \otimes (\boldsymbol{B} + \boldsymbol{C}) = \boldsymbol{A} \otimes \boldsymbol{B} + \boldsymbol{A} \otimes \boldsymbol{C}$.
- $(\boldsymbol{A} \otimes \boldsymbol{B})' = \boldsymbol{A}' \otimes \boldsymbol{B}'$ and hence if $\boldsymbol{A}$ and $\boldsymbol{B}$ are both square symmetric matrices then so is $\boldsymbol{A} \otimes \boldsymbol{B}$.

It can be seen that in general $\boldsymbol{A} \otimes \boldsymbol{B} \neq \boldsymbol{B} \otimes \boldsymbol{A}$, for example taking

$$\begin{aligned} \boldsymbol{A} &= \begin{pmatrix} 1 & 0 \\ 0 & 0 \end{pmatrix} \text{ and } \boldsymbol{B} = \begin{pmatrix} 1 & 0 \\ 0 & 1 \end{pmatrix} \text{ gives} \\ \boldsymbol{A} \otimes \boldsymbol{B} &= \begin{pmatrix} 1 & 0 & 0 & 0 \\ 0 & 1 & 0 & 0 \\ 0 & 0 & 0 & 0 \\ 0 & 0 & 0 & 0 \end{pmatrix} \text{ and } \boldsymbol{B} \otimes \boldsymbol{A} = \begin{pmatrix} 1 & 0 & 0 & 0 \\ 0 & 0 & 0 & 0 \\ 0 & 0 & 1 & 0 \\ 0 & 0 & 0 & 0 \end{pmatrix}. \end{aligned}$$

8.5.2 Further properties of Kronecker products

Properties which are little less straightforward are

- The mixed product rule. If $\boldsymbol{A}$ is $m \times n$, $\boldsymbol{B}$ is $p \times q$, $\boldsymbol{C}$ is $n \times r$ and $\boldsymbol{D}$ is $q \times s$ then $(\boldsymbol{A} \otimes \boldsymbol{B})(\boldsymbol{C} \otimes \boldsymbol{D}) = (\boldsymbol{AC}) \otimes (\boldsymbol{BD})$ because

$$\begin{aligned} &(\boldsymbol{A} \otimes \boldsymbol{B})(\boldsymbol{C} \otimes \boldsymbol{D}) \\ &= \begin{pmatrix} a_{11}\boldsymbol{B} & a_{12}\boldsymbol{B} & \cdots & a_{1n}\boldsymbol{B} \\ a_{21}\boldsymbol{B} & a_{22}\boldsymbol{B} & \cdots & a_{2n}\boldsymbol{B} \\ \vdots & \vdots & & \vdots \\ a_{11}\boldsymbol{B} & a_{m2}\boldsymbol{B} & \cdots & a_{mn}\boldsymbol{B} \end{pmatrix} \begin{pmatrix} c_{11}\boldsymbol{D} & c_{12}\boldsymbol{D} & \cdots & c_{1n}\boldsymbol{D} \\ c_{21}\boldsymbol{D} & c_{22}\boldsymbol{D} & \cdots & c_{2n}\boldsymbol{D} \\ \vdots & \vdots & & \vdots \\ c_{11}\boldsymbol{D} & c_{m2}\boldsymbol{D} & \cdots & c_{mn}\boldsymbol{D} \end{pmatrix} \\ &= \begin{pmatrix} f_{11}\boldsymbol{BD} & f_{12}\boldsymbol{BD} & \cdots & f_{1n}\boldsymbol{BD} \\ f_{21}\boldsymbol{BD} & f_{22}\boldsymbol{BD} & \cdots & f_{2n}\boldsymbol{BD} \\ \vdots & \vdots & & \vdots \\ f_{11}\boldsymbol{ND} & f_{m2}\boldsymbol{BD} & \cdots & f_{mn}\boldsymbol{BD} \end{pmatrix}, \text{ where } f_{ij} = \sum_{k=1}^{n} a_{ik}c_{kj} = (\boldsymbol{AC})_{ij}, \\ &= (\boldsymbol{AC}) \otimes (\boldsymbol{BD}). \end{aligned}$$

- If $\boldsymbol{A}$ is $n \times n$ and $\boldsymbol{B}$ is $p \times p$ then $(\boldsymbol{A} \otimes \boldsymbol{B})^{-1} = \boldsymbol{A}^{-1} \otimes \boldsymbol{B}^{-1}$ because $(\boldsymbol{A} \otimes \boldsymbol{B})^{-1}(\boldsymbol{A}^{-1} \otimes \boldsymbol{B}^{-1}) = (\boldsymbol{A}\boldsymbol{A}^{-1}) \otimes (\boldsymbol{B}\boldsymbol{B}^{-1})$ (using the previous result) $= \mathbf{I_n} \otimes \mathbf{I_p} = \mathbf{I_{np}}$.

- If $\boldsymbol{A} = \boldsymbol{U}_A\boldsymbol{D}_A\boldsymbol{V}_A'$ and $\boldsymbol{B} = \boldsymbol{U}_B\boldsymbol{D}_B\boldsymbol{V}_B'$ are the singular value decompositions of $\boldsymbol{A}$ and $\boldsymbol{B}$ then the singular value decomposition of $\boldsymbol{A} \otimes \boldsymbol{B}$ is given by $\boldsymbol{A} \otimes \boldsymbol{B} = (\boldsymbol{U}_A \otimes \boldsymbol{U}_B)(\boldsymbol{D}_A \otimes \boldsymbol{D}_B)(\boldsymbol{V}_A \otimes \boldsymbol{V}_B)'$ which can be seen by noting that $(\boldsymbol{V}_A \otimes \boldsymbol{V}_B)' = (\boldsymbol{V}_A' \otimes \boldsymbol{V}_B')$ and applying the mixed product rule twice in succession.

- If $\boldsymbol{A}$ is $m \times n$ and $\boldsymbol{B}$ is $p \times q$ then $\rho(\boldsymbol{A} \otimes \boldsymbol{B}) = \rho(\boldsymbol{A})\rho(\boldsymbol{B})$. This follows from the previous result because $\rho(\boldsymbol{A})$ equals the number of non-zero diagonal elements of $\boldsymbol{D}_A$, (i.e., the non-zero singular values of $\boldsymbol{A}$). Similarly for $\boldsymbol{B}$ and $\rho(\boldsymbol{A} \otimes \boldsymbol{B})$ is the number of non-zero elements of $\boldsymbol{D}_A \otimes \boldsymbol{D}_B$ which is the product of the numbers of non-zero elements in $\boldsymbol{D}_A$ and in $\boldsymbol{D}_B$.

- If $\boldsymbol{A}$ and $\boldsymbol{B}$ are $m \times m$ and $n \times n$ square matrices then $\text{tr}(\boldsymbol{A} \otimes \boldsymbol{B}) = \text{tr}(\boldsymbol{A})\text{tr}(\boldsymbol{B})$ because

$$
\begin{aligned}
\text{tr}(\boldsymbol{A} \otimes \boldsymbol{B}) &= \begin{pmatrix} a_{11}\boldsymbol{B} & a_{12}\boldsymbol{B} & \cdots & a_{1m}\boldsymbol{B} \\ a_{21}\boldsymbol{B} & a_{22}\boldsymbol{B} & \cdots & a_{2m}\boldsymbol{B} \\ \vdots & \vdots & \ddots & \vdots \\ a_{m1}\boldsymbol{B} & a_{m2}\boldsymbol{B} & \cdots & a_{mm}\boldsymbol{B} \end{pmatrix} \\
&= (a_{11} + a_{22} + \ldots + a_{mm})\text{tr}(\boldsymbol{B}) = \text{tr}(\boldsymbol{A})\text{tr}(\boldsymbol{B}).
\end{aligned}
$$

- If $\boldsymbol{A}$ and $\boldsymbol{B}$ are $m \times m$ and $n \times n$ upper [lower] triangular square matrices then $\boldsymbol{A} \otimes \boldsymbol{B}$ is upper [lower] triangular because if $\boldsymbol{A}$ and $\boldsymbol{B}$ are upper triangular then $a_{ij} = 0$ if $i > j$ so

$$
\boldsymbol{A} \otimes \boldsymbol{B} = \begin{pmatrix} a_{11}\boldsymbol{B} & a_{12}\boldsymbol{B} & \cdots & a_{1m}\boldsymbol{B} \\ \boldsymbol{0} & a_{22}\boldsymbol{B} & \cdots & a_{2m}\boldsymbol{B} \\ \vdots & \vdots & \ddots & \vdots \\ \boldsymbol{0} & \boldsymbol{0} & \cdots & a_{mm}\boldsymbol{B} \end{pmatrix}
$$

  and each of the blocks $a_{ii}\boldsymbol{B}$ is upper triangular so all entries below the main diagonal are 0 and so $\boldsymbol{A} \otimes \boldsymbol{B}$ is upper triangular, similarly for lower triangular.

- If $\boldsymbol{A}$ and $\boldsymbol{B}$ are orthogonal matrices then $\boldsymbol{A} \otimes \boldsymbol{B}$ is orthogonal because we have $\boldsymbol{A}\boldsymbol{A}' = \boldsymbol{A}'\boldsymbol{A} = \mathbf{I_m}$ and $\boldsymbol{B}\boldsymbol{B}' = \boldsymbol{B}'\boldsymbol{B} = \mathbf{I_n}$ so $(\boldsymbol{A} \otimes \boldsymbol{B})(\boldsymbol{A} \otimes \boldsymbol{B})' = (\boldsymbol{A}\boldsymbol{A}') \otimes (\boldsymbol{B}\boldsymbol{B}') = \mathbf{I_m} \otimes \mathbf{I_n} = \mathbf{I_{mn}}$ by the mixed product rule. Similarly $(\boldsymbol{A} \otimes \boldsymbol{B})'(\boldsymbol{A} \otimes \boldsymbol{B}) = \mathbf{I_{mn}}$.

8.5.3 Decompositions of Kronecker products

In the same way as the singular value decomposition of $\boldsymbol{A} \otimes \boldsymbol{B}$ can be obtained as the equivalent factorization into Kronecker products of corresponding factors, as shown above by repeated application of the multiple product rule, analogous results hold for other decompositions (where suffixes A and B are used for factors in the decompositions of $\boldsymbol{A}$ and $\boldsymbol{B}$ and $\boldsymbol{A}$ and where in each case $\boldsymbol{B}$ values are assumed to have the forms required for validity of the decompositions):

- Spectral decomposition: $\boldsymbol{A} \otimes \boldsymbol{B} = (\boldsymbol{T}_A \otimes \boldsymbol{T}_B)(\Lambda_A \otimes \Lambda_B)(\boldsymbol{T}'_A \otimes \boldsymbol{T}'_B)$.
- $\boldsymbol{QR}$ decomposition: $\boldsymbol{A} \otimes \boldsymbol{B} = (\boldsymbol{Q}_A \otimes \boldsymbol{Q}_B)(\boldsymbol{R}_A \otimes \boldsymbol{R}_B)$.
- $\boldsymbol{LU}$ decomposition: $\boldsymbol{A} \otimes \boldsymbol{B} = (\boldsymbol{P}_A \otimes \boldsymbol{P}_B)(\boldsymbol{L}_A \otimes \boldsymbol{L}_B)(\boldsymbol{U}_A \otimes \boldsymbol{U}_B)$.
- Cholesky decomposition: $\boldsymbol{A} \otimes \boldsymbol{B} = (\boldsymbol{T}_A \otimes \boldsymbol{T}_B)(\boldsymbol{T}'_A \otimes \boldsymbol{T}'_B)$.
- Schur decomposition: $(\boldsymbol{Q}'_A \otimes \boldsymbol{Q}'_B)(\boldsymbol{A} \otimes \boldsymbol{B})(\boldsymbol{Q}_A \otimes \boldsymbol{Q}_B) = \boldsymbol{T}_A \otimes \boldsymbol{T}_B$.

8.5.4 Eigenanalyses and determinants of Kronecker products

Throughout this section $\boldsymbol{A}$ and $\boldsymbol{B}$ are $m \times m$ and $n \times n$ square matrices.

If $\boldsymbol{Ax} = \lambda \boldsymbol{x}$ and $\boldsymbol{By} = \mu \boldsymbol{y}$ then $(\boldsymbol{A} \otimes \boldsymbol{B})(\boldsymbol{x} \otimes \boldsymbol{y}) = \lambda\mu(\boldsymbol{x} \otimes \boldsymbol{y})$ (using the mixed product rule) and so $\boldsymbol{x} \otimes \boldsymbol{y}$ is an eigenvector of $(\boldsymbol{A} \otimes \boldsymbol{B})$ corresponding to eigenvalue $\lambda\mu$. In fact all the eigenvalues of $(\boldsymbol{A} \otimes \boldsymbol{B})$ are given by the set $\{\lambda_i\mu_j; i = 1,\ldots,m;\ j = 1,\ldots,n\}$ because, considering the Schur decomposition $(\boldsymbol{Q}'_A \otimes \boldsymbol{Q}'_B)(\boldsymbol{A} \otimes \boldsymbol{B})(\boldsymbol{Q}_A \otimes \boldsymbol{Q}_B) = \boldsymbol{T}_A \otimes \boldsymbol{T}_B$, $\text{diag}(\boldsymbol{T}_A)$ contains all the λ_i and $\text{diag}(\boldsymbol{T}_B)$ all the μ_j and the eigenvalues of $(\boldsymbol{A} \otimes \boldsymbol{B})$ are given by $\text{diag}(\boldsymbol{T}_A \otimes \boldsymbol{T}_B)$ which contains all the $\lambda_i\mu_j; i = 1,\ldots,m;\ j = 1,\ldots,n$.

However, it is not the case that all the eigenvectors of $(\boldsymbol{A} \otimes \boldsymbol{B})$ are given by $(\boldsymbol{x}_i \otimes \boldsymbol{y}_j); i = 1,\ldots,m;\ j = 1,\ldots,n$. For example if $\boldsymbol{A} = \boldsymbol{B} = \left(\begin{smallmatrix} 0 & 0 \\ 1 & 0 \end{smallmatrix}\right)$ then it is easy to verify that the eigenvalues of $\boldsymbol{A}$ and $\boldsymbol{A} \otimes \boldsymbol{A}$ are all zero but $\boldsymbol{A}$ has only one eigenvector and $\boldsymbol{A} \otimes \boldsymbol{A}$ has three eigenvectors. Demonstration of this is left as an exercise.

Noting that the determinant of a matrix is given by the product of all of its eigenvalues, $|\boldsymbol{A} \otimes \boldsymbol{B}| = |\boldsymbol{A}|^n |\boldsymbol{B}|^m$ which follows from the Schur decomposition of $\boldsymbol{A} \otimes \boldsymbol{B}$.

8.5.5 Elementary properties of the vec operator

It is straightforward to verify from the definition that the basic properties below hold (presuming the matrices are conformable as necessary):

- $\text{vec}(\lambda \boldsymbol{A}) = \lambda\,\text{vec}(\boldsymbol{A})$.
- $\text{vec}(\boldsymbol{A} + \boldsymbol{B}) = \text{vec}(\boldsymbol{A}) + \text{vec}(\boldsymbol{B})$ if $\boldsymbol{A}$ and $\boldsymbol{B}$ have the same orders.
- $\text{vec}(\boldsymbol{x}) = \text{vec}(\boldsymbol{x}')$ for all vectors $\boldsymbol{x}$.
- The trace of products rule. $(\text{vec}(\boldsymbol{A})'\text{vec}(\boldsymbol{B}) = \text{tr}(\boldsymbol{A}'\boldsymbol{B})$ because $(\text{vec}(\boldsymbol{A}))'\text{vec}(\boldsymbol{B}) = \Sigma_{ij} a_{ij} b_{ij} = \Sigma_{ij} (\boldsymbol{A}'\boldsymbol{B})_{ij} = \text{tr}(\boldsymbol{A}'\boldsymbol{B})$.

8.5.6 Connections between Kronecker products and vec operators

- $\text{vec}(\boldsymbol{x}\boldsymbol{y}') = \boldsymbol{y} \otimes \boldsymbol{x}$ because

$$\text{vec}(\boldsymbol{x}\boldsymbol{y}') = \begin{pmatrix} y_1\boldsymbol{x} \\ y_2\boldsymbol{x} \\ \vdots \\ y_n\boldsymbol{x} \end{pmatrix} = \boldsymbol{y} \otimes \boldsymbol{x}.$$

- The triple product rule. Suppose $\boldsymbol{A}$, $\boldsymbol{B}$ and $\boldsymbol{C}$ are $m \times n$, $n \times p$ and $p \times q$ matrices, then $\text{vec}(\boldsymbol{ABC}) = (\boldsymbol{C}' \otimes \boldsymbol{A})\text{vec}(\boldsymbol{B})$ because if $\boldsymbol{B} = \sum_{i-1}^{p} \boldsymbol{b})i\boldsymbol{e}_i'$ where $\boldsymbol{b}_i$ and $\boldsymbol{e}_i$ are the columns of $\boldsymbol{B}$ and $\mathbf{I_p}$ then

$$\begin{aligned}
\text{vec}(\boldsymbol{ABC}) &= \text{vec}\left(\sum_{i-1}^{p} \boldsymbol{A}\boldsymbol{b}_i\boldsymbol{e}_i'\boldsymbol{C}\right) = \sum_{i-1}^{p} \text{vec}\left((\boldsymbol{A}\boldsymbol{b}_i)(\boldsymbol{C}'\boldsymbol{e}_i)'\right) \\
&= \sum_{i-1}^{p} (\boldsymbol{C}'\boldsymbol{e}_i) \otimes (\boldsymbol{A}\boldsymbol{b}_i) \text{ (applying the result above)} \\
&= (\boldsymbol{C}' \otimes \boldsymbol{A}) \sum_{i-1}^{p} (\boldsymbol{e}_i \otimes \boldsymbol{b}_i) \text{ (applying the mixed product rule)} \\
&= (\boldsymbol{C}' \otimes \boldsymbol{A}) \sum_{i-1}^{p} \text{vec}(\boldsymbol{b}_i\boldsymbol{e}_i') = (\boldsymbol{C}' \otimes \boldsymbol{A})\text{vec}(\boldsymbol{B}).
\end{aligned}$$

- $\text{tr}(\boldsymbol{ABCD}) = \text{vec}(\boldsymbol{D}')'(\boldsymbol{C}' \otimes \boldsymbol{A})\text{vec}(\boldsymbol{B})$ because $\text{tr}(\boldsymbol{ABCD}) = \text{tr}(\boldsymbol{D}(\boldsymbol{ABC})) = \text{vec}(\boldsymbol{D}')'\text{vec}(\boldsymbol{ABC}) = \text{vec}(\boldsymbol{D}')'(\boldsymbol{C}' \otimes \boldsymbol{A})\text{vec}(\boldsymbol{B})$, noting that if $\boldsymbol{ABCD}$ is square then the product $\boldsymbol{D}(\boldsymbol{ABC})$is defined and applying results given above. Similarly, noting that $\text{tr}(\boldsymbol{ABCD}) = \text{tr}(\boldsymbol{D}'\boldsymbol{C}'\boldsymbol{B}'\boldsymbol{A}')$, we have $\text{tr}(\boldsymbol{ABCD}) = (\text{vec}(\boldsymbol{D})'(\boldsymbol{A} \otimes \boldsymbol{C}')\text{vec}(\boldsymbol{B}')$.

- Suppose $\boldsymbol{A}$ is $m \times n$ and $\boldsymbol{B}$ is $n \times p$, then $\text{vec}(\boldsymbol{A}) = (\mathbf{I_n} \otimes \boldsymbol{A})\text{vec}(\mathbf{I_n})$ because $\boldsymbol{A} = \boldsymbol{A}\mathbf{I_n}\mathbf{I_n}$ and the result follows from the triple product rule. Similarly $\text{vec}(\boldsymbol{A}) = (\mathbf{I_m} \otimes \boldsymbol{A}')\text{vec}(\mathbf{I_m})$ noting that $\boldsymbol{A} = \mathbf{I_m}\mathbf{I_m}\boldsymbol{A}$.

- $\text{vec}(\boldsymbol{AB}) = (\boldsymbol{B}' \otimes \mathbf{I_m})\text{vec}(\boldsymbol{A}) = (\boldsymbol{B}' \otimes \boldsymbol{A})\text{vec}(\mathbf{I_n}) = (\mathbf{I_p} \otimes \boldsymbol{A})\text{vec}(\boldsymbol{B})$ noting that $\boldsymbol{AB} = \mathbf{I_m}\boldsymbol{AB} = \boldsymbol{A}\mathbf{I_n}\boldsymbol{B} = \boldsymbol{AB}\mathbf{I_p}$ and applying the triple product rule.

8.5.7 Kronecker products and vec operators in R

The Kronecker product $\boldsymbol{A} \otimes \boldsymbol{B}$ is obtained in **R** using the multiplication symbol `%x%` by `A%x%B` (ordinary matrix multiplication is `A%*%B`). There is no ready made function for the vec operator in the base system of **R** nor any of the libraries supplied on first installation. The library `matrixcalc` (Novomestky, 2012) contains a function `vec(.)` (and also a function `vech(.)`) which performs the operation. This library needs to be installed using the `Packages>Install package(s)...` menu (after first choosing a CRAN mirror from which to access the package). This only needs to be done once and after that it can be loaded with `library(matrixcalc)`. This library contains various other functions which may be of interest to specialist users and a complete list can be obtained by the command `library(help=matrixcalc)`.

Alternatively, it is possible to write a simple function in **R** to perform the operation:

```
vec<-function(A) {
vec<-as.matrix(A[,1])
for (i in 2:dim(A)[2]) {vec<-rbind(vec,as.matrix(A[,i]))}
return(vec)
}
```

Note that although `A[,1]`and the `A[,i])` are columns of the matrix `A`, **R** may treat them as either row or column vectors according to context (see §2.10.1 and §2.1.2), choosing the row vector option if both possibilities are compatible. Use of the function `as.matrix(.)` forces the column vector choice.

Example 8.8: First, store the function vec(.) for use in several examples and set number of digits of output.

```
> vec<-function(A) {
+ vec<-as.matrix(A[,1])
+ for (i in 2:dim(A)[2]) {vec<-rbind(vec,as.matrix(A[,i]))}
+ return(vec)
+ }
> options(digits=2)
```

(i) Illustration of triple product rule $\text{vec}(\boldsymbol{ABC}) = (\boldsymbol{C}' \otimes \boldsymbol{A})\text{vec}(\boldsymbol{B})$:

```
> A<-matrix(c(1,0,1,0),2)
> B<-matrix(c(1,1,1,0),2)
> C<-matrix(c(1,0,1,1),2)
> A; B ;C
     [,1] [,2]
[1,]    1    1
[2,]    0    0
     [,1] [,2]
[1,]    1    1
[2,]    1    0
     [,1] [,2]
[1,]    1    1
[2,]    0    1
```

```
>
> A%x%B
     [,1] [,2] [,3] [,4]
[1,]    1    1    1    1
[2,]    1    0    1    0
[3,]    0    0    0    0
[4,]    0    0    0    0
> A%x%C
     [,1] [,2] [,3] [,4]
[1,]    1    1    1    1
[2,]    0    1    0    1
[3,]    0    0    0    0
[4,]    0    0    0    0
```

```
> vec(A%*%B%*%C)
     [,1]
[1,]    2
[2,]    0
[3,]    3
[4,]    0
```

```
> (t(C)%x%A)%*%vec(B)
     [,1]
[1,]    2
[2,]    0
[3,]    3
[4,]    0
```

(ii) Eigenanalyses of $\boldsymbol{A}$, $\boldsymbol{B}$ and $\boldsymbol{A} \otimes \boldsymbol{B}$ in example (i):

```
> eigen(A)
$values
[1] 1 0
$vectors
     [,1]  [,2]
[1,]    1 -0.71
[2,]    0  0.71
```

```
> eigen(B)
$values
[1]  1.62 -0.62
$vectors
      [,1]  [,2]
[1,] -0.85  0.53
[2,] -0.53 -0.85
```

```
> eigen(A%x%B)
$values
[1]  1.62 -0.62  0.00  0.00

$vectors
     [,1]  [,2]  [,3]  [,4]
[1,] 0.85 -0.53 -0.71  0.00
[2,] 0.53  0.85  0,00 -0.71
[3,] 0.00  0.00  0.71  0.00
[4,] 0.00  0.00  0.00  0.71
>
> eigen(A)$vec%x%eigen(B)$vec
      [,1]  [,2]  [,3]  [,4]
[1,] -0.85  0.53  0.60 -0.37
[2,] -0.53 -0.85  0.37  0.60
[3,]  0.00  0.00 -0.60  0.37
[4,]  0.00  0.00 -0.37 -0.60
```

This example shows that the eigenvalues of $\boldsymbol{A} \otimes \boldsymbol{B}$ are the products of the eigenvalues of $\boldsymbol{A}$ and $\boldsymbol{B}$ and at first sight it might appear that the eigenvectors of $\boldsymbol{A} \otimes \boldsymbol{B}$ are not given by the Kronecker products of the eigenvectors of $\boldsymbol{A}$ and $\boldsymbol{B}$. Since the last two eigenvalues of $\boldsymbol{A} \otimes \boldsymbol{B}$ are both zero there are arbitrarily many choices for the eigenvectors and it is easy to verify that $(-0.71, 0, 0.71, 0)'$ and $(0, -0.71, 0, 0.71)'$ are valid choices for eigenvectors corresponding to zero eigenvalues and also $(0.60, 0.37, -0.60, -0.37)'$ and $(-0.37, 0.60, 0.37, -0.60)'$ are equally valid choices.

8.6 Exercises

(1) If $A = \begin{pmatrix} 73 & 45 & 46 \\ 45 & 58 & 29 \\ 46 & 29 & 29 \end{pmatrix}$ find its Cholesky decomposition $A = TT'$.

(2) If $A = \begin{pmatrix} 34 & 31 & 51 \\ 31 & 53 & 32 \\ 51 & 32 & 85 \end{pmatrix}$ find its Cholesky decomposition $A = TT'$.

(3) Show that $(A^+)^+ = A$.

(4) Show that $(A')^+ = (A^+)'$.

(5) Show that $(AA^+)^+ = AA^+$.

(6) Show that $(A^+A)^+ = A^+A$.

(7) If A is symmetric (i.e., if $A' = A$) show that $AA^+ = A^+A$.

(8) Show that $A'AA^+ = A' = A^+AA'$.

(9) If $A = \begin{pmatrix} A_1 & 0 \\ 0 & A_2 \end{pmatrix}$ show $A^+ = \begin{pmatrix} A_1^+ & 0 \\ 0 & A_2^+ \end{pmatrix}$.

(10) If $x = \begin{pmatrix} 1 \\ 6 \\ 3 \end{pmatrix}$ find x^+ without using any of the **R** functions for finding Moore–Penrose inverses and check the result with `ginv(.)`.

(11) If $X = \begin{pmatrix} 4 & 12 & 8 \\ 6 & 18 & 12 \\ 5 & 15 & 10 \end{pmatrix}$ find X^+ without using any of the **R** functions for finding Moore–Penrose inverses and check the result with `ginv(.)`.

(12) If $X = \begin{pmatrix} 6 & 2 & 8 \\ 5 & 1 & 6 \\ 1 & 7 & 8 \end{pmatrix}$, find X^+ without using any of the **R** functions for finding Moore–Penrose inverses and check the result with `ginv(.)`.

(13) Let $A = \begin{pmatrix} 1 & 1 \\ 2 & 2 \\ 3 & 4 \end{pmatrix}$ and $y = \begin{pmatrix} 1 \\ 1 \\ 1 \end{pmatrix}$.

(a) Show that $AA^+y = \begin{pmatrix} 0.6 \\ 1.2 \\ 1.0 \end{pmatrix} \neq y$.

(b) What is the least squares solution to the equation $Ax = y$?

(c) If instead $y = \begin{pmatrix} 1 \\ 2 \\ 7 \end{pmatrix}$ or $y = \begin{pmatrix} 3 \\ 6 \\ 11 \end{pmatrix}$ show that in both cases the equation is consistent and find solutions. Are these solutions unique in either or both cases?

(14) Let $A = \begin{pmatrix} 2 & 3 & 3 & 1 \\ 3 & 4 & 5 & 1 \\ 1 & 2 & 1 & 1 \end{pmatrix}$ and $y = \begin{pmatrix} 14 \\ 22 \\ 6 \end{pmatrix}$.

(a) Show that the equation $Ax = y$ is consistent.

(b) Show that $x = \begin{pmatrix} 1.3 \\ 1.1 \\ 2.8 \\ -0.2 \end{pmatrix}$ and $x = \begin{pmatrix} 1.5 \\ 0.7 \\ 2.9 \\ 0.2 \end{pmatrix}$ are both solutions of the equation.

(c) Find a different solution to the equation.

(15) Show that if x is a $n \times 1$ vector then $\mathrm{tr}(xx') = \mathbf{1}_{\mathbf{n}}' x \odot x$, where $\iota_{\mathbf{n}}$ is the $n \times 1$ vector with all entries equal to one.

(16) If A is a square $n \times n$ matrix then show that $A \odot \mathbf{I_n} = \mathrm{diag}(\mathrm{diag}(A))$.

(17) If u and x are $m \times 1$ vectors and v and y are $n \times 1$ vectors then show that $(uv') \odot (xy') = (u \odot x)(v \odot y)'$.

(18) If $\rho(A) = \rho(B) = 1$ then show that $\rho(A \odot B) \leq 1$.

(19) Show that $\rho(A \odot B) \leq \rho(A).\rho(B)$.

(20) If A and B are both $m \times n$ and x is $n \times 1$ show that $\mathrm{diag}(A\mathrm{diag}(\mathrm{diag}(x))B') = (A \odot B)x$.

(21) Show that $x'(A \odot B)y = \mathrm{tr}(\mathrm{diag}(\mathrm{diag}(x))A\mathrm{diag}(\mathrm{diag}(y))B')$, where x, y, A and B are conformable so that the various products are well-defined.

(22) If A and B are $n \times n$ matrices and Λ is a $n \times n$ matrix with diagonal elements λ_i then show that $\mathrm{diag}(A\Lambda B) = (A \odot B)\mathrm{diag}(\Lambda)$.

(23) If A is $m \times n$ and B is $p \times q$ then show that $(A \otimes B)^- = A^- \otimes B^-$.

(24) If A is $m \times n$ and B is $p \times q$ then show that $(A \otimes B)^+ = A^+ \otimes B^+$.

(25) If $A = \begin{pmatrix} 0 & 0 \\ 1 & 0 \end{pmatrix}$ find the eignevectors of A and $A \otimes A$.

9

Key Applications to Statistics

9.1 Introduction

This section considers some basic results that are encountered in introductory courses on multivariate analysis and linear models using a matrix-based formulation. The purpose is to provide some details of the techniques used to establish these results, giving cross-references to those sections where the techniques are established. Substantial use is made of results from Chapter 7 and especially of §7.6 in the case of §9.2.7 on union-intersection tests and in derivation of the key results in the data analytic techniques in §9.3, §9.4 and §9.5.

Keeping this aim in mind means that only very brief descriptions are given of important topics such as principal components analysis, discriminant analysis and so on and little discussion of their purpose and interpretation is included since the information is readily available in standard specialist texts. Similarly, discussion of the evaluation of statistical tests (likelihood ratio and union-intersection) is left to other texts except that for the sake of completeness numerical solutions to some of the exercises do use **R** functions such as `mvrnorm(.)` and `pchisq(.)` without full descriptions of their use. Readers can readily find details of these in the **R** `help(.)` system.

9.2 The Multivariate Normal Distribution

The random $p \times 1$ vector $\boldsymbol{x}$ has a p-dimensional multivariate normal distribution $N_p(\mu, \Sigma)$ if the probability density function of $\boldsymbol{x}$ is

$$f_x(\boldsymbol{x}) = \frac{1}{(2\pi)^{p/2}|\Sigma|^{1/2}} \exp\{-{}^1\!/_2(\boldsymbol{x}-\mu)'\Sigma^{-1}(\boldsymbol{x}-\mu)\}.$$

We write $\boldsymbol{x} \sim N_p(\mu, \Sigma)$. Here it is taken that μ is a $p \times 1$ vector and Σ is a symmetric, non-singular positive definite $p \times p$ matrix. It is possible to extend this definition to a singular matrix Σ but that is not considered here.

9.2.1 Standardization

If $\boldsymbol{x} \sim N_p(\mu, \Sigma)$ and $\boldsymbol{y} = \Sigma^{-1/2}(\boldsymbol{x} - \mu)$ where $\Sigma^{-1/2}$ is as defined in §6.7.1.1 then

$$(\boldsymbol{x} - \mu)'\Sigma^{-1}(\boldsymbol{x} - \mu) = \boldsymbol{y}'\boldsymbol{y} = \sum_{i=1}^{p} y_i^2$$

so the density of $\boldsymbol{y}$ is $f_y(\boldsymbol{y}) = \dfrac{1}{(2\pi)^{p/2}|\Sigma|^{1/2}} \exp\{-1/2\boldsymbol{y}'\boldsymbol{y}\}J_{xy}$,

where J_{xy} is the Jacobean; see§7.4.

Since $\boldsymbol{y} = \Sigma^{-1/2}(\boldsymbol{x} - \mu)$, $\boldsymbol{x} = \Sigma^{1/2}(\boldsymbol{y} + \mu)$ and $\frac{\partial x}{\partial y} = \Sigma^{1/2}$ (see§7.4.1) so $J_{xy} = |\Sigma|^{1/2}$ and

$$f_y(\boldsymbol{y}) = \frac{1}{(2\pi)^{p/2}} \exp\{-1/2\boldsymbol{y}'\boldsymbol{y}\} = \frac{1}{(2\pi)^{p/2}} \exp\{-\sum_{i=1}^{p} y_i^2\},$$

showing that the y_i are independent univariate $N(0,1)$ variables (and thus $f_x(\boldsymbol{x}) > 0$ and integrates to 1 and so is a proper probability density function). Noting that $\mathrm{E}[\boldsymbol{y}] = 0$, $\mathrm{var}(\boldsymbol{y}) = \mathbf{I_p}$ and if $\boldsymbol{x}$ is a p-dimensional random variable and recalling (see §2.12.3) that if $\boldsymbol{A}$ is a $n \times p$ matrix, and $\boldsymbol{b}$ a $p \times 1$ vector, then $\mathrm{E}[\boldsymbol{Ax} + \boldsymbol{b}] = \boldsymbol{A}\mathrm{E}[\boldsymbol{x}] + \boldsymbol{b}$ and $\mathrm{var}(\boldsymbol{Ax} + \boldsymbol{b}) = \boldsymbol{A}\mathrm{var}(\boldsymbol{x})\boldsymbol{A}'$ gives $\mathrm{E}[x] = \mu$ and $\mathrm{var}(\boldsymbol{x}) = \Sigma^{1/2}\mathbf{I_p}\Sigma^{1/2} = \Sigma$ (since Σ is symmetric so is $\Sigma^{1/2}$).

9.2.2 Random samples

Suppose $\boldsymbol{x}_1, \boldsymbol{x}_2, \ldots, \boldsymbol{x}_n$ are independent observations of $\boldsymbol{x} \sim N_p(\mu, \Sigma)$, and $\boldsymbol{x}_i = (x_{i1}, x_{i2}, \ldots, x_{ip})'$. Recall from §2.12.3.1 that we define the sample mean vector $\bar{\boldsymbol{x}} = (\bar{x}_1, \bar{x}_2, \ldots, \bar{x}_p)' = \frac{1}{n}\boldsymbol{X}\iota_\mathbf{n}$. Let $\overline{\boldsymbol{X}} = \bar{\boldsymbol{x}}\iota_\mathbf{n}'$ (so $\overline{\boldsymbol{X}}$ is a $p \times n$ matrix with all columns equal to $\bar{\boldsymbol{x}}$). Further we define the sample variance

$$\boldsymbol{S} = \tfrac{1}{(n-1)}(\boldsymbol{X} - \overline{\boldsymbol{X}})(\boldsymbol{X} - \overline{\boldsymbol{X}})' = \tfrac{1}{(n-1)}\left\{\sum_{1}^{n} \boldsymbol{x}_i\boldsymbol{x}_i' - n\bar{\boldsymbol{x}}\bar{\boldsymbol{x}}'\right\}.$$

In §2.12.3.1 it was shown that $\mathrm{E}[\boldsymbol{S}] = \Sigma$, and thus $\bar{\boldsymbol{x}}$ and $\boldsymbol{S}$ are unbiased estimators of μ and Σ. Further, using characteristic functions of normal random variables it is possible to show that $\bar{\boldsymbol{x}} \sim N_p(\mu\,\frac{1}{n}\Sigma)$.

9.2.3 Maximum likelihood estimates

If $\boldsymbol{x}_1, \boldsymbol{x}_2, \ldots, \boldsymbol{x}_n$ are independent observations of $\boldsymbol{x} \sim N_p(\mu, \Sigma)$ then the likelihood of (μ, Σ) is

$$\mathrm{Lik}(\mu, \Sigma; \boldsymbol{X}) = \frac{1}{(2\pi)^{p/2}|\Sigma|^{n/2}} \exp\{-1/2\sum_{i=1}^{n}(\boldsymbol{x}_i - \mu)'\Sigma^{-1}(\boldsymbol{x}_i - \mu)\}$$

and the log-likelihood is

$$\begin{aligned}\ell(\mu,\Sigma;\boldsymbol{X}) &= \log_e(\mathrm{Lik}(\mu,\Sigma;\boldsymbol{X}) \\ &= -{}^1\!/_2\sum_{i=1}^{n}(\boldsymbol{x}_i-\mu)'\Sigma^{-1}(\boldsymbol{x}_i-\mu) - {}^1\!/_2 np\log(2\pi) - {}^1\!/_2 n\log(|\Sigma|) \\ &= -{}^1\!/_2\sum_{i=1}^{n}(\boldsymbol{x}_i-\bar{\boldsymbol{x}})'\Sigma^{-1}(\boldsymbol{x}_i-\bar{\boldsymbol{x}}) - {}^1\!/_2 n(\bar{\boldsymbol{x}}-\mu)'\Sigma^{-1}(\bar{\boldsymbol{x}}-\mu) \\ &\qquad - {}^1\!/_2 np\log(2\pi) - {}^1\!/_2 n\log(|\Sigma|).\end{aligned}$$

So $\frac{\partial \ell}{\partial \mu} = n\Sigma^{-1}(\bar{\boldsymbol{x}}-\mu)$ (using §7.2.3) and thus (setting the derivative equal to 0) $\hat{\mu} = \bar{\boldsymbol{x}}$.
Writing $\boldsymbol{T} = \Sigma^{-1}$ gives

$$\begin{aligned}\frac{\partial \ell}{\partial \boldsymbol{T}} &= \{n\Sigma-(n-1)\boldsymbol{S}-n(\bar{\boldsymbol{x}}-\mu)(\bar{\boldsymbol{x}}-\mu)'\} \\ &\qquad - {}^1\!/_2\mathrm{diag}\big(\mathrm{diag}((n\Sigma-(n-1)\boldsymbol{S}-n(\bar{\boldsymbol{x}}-\mu)(\bar{\boldsymbol{x}}-\mu)'))\big)\end{aligned}$$

(using §7.3.3 and §7.3.10). Setting this equal to 0 gives $\hat{\Sigma} = \frac{(n-1)}{n}\boldsymbol{S} + (\bar{\boldsymbol{x}}-\hat{\mu})(\bar{\boldsymbol{x}}-\hat{\mu})'$ and when $\hat{\mu} = \bar{\boldsymbol{x}}$ this gives the [unrestricted] maximum likelihood estimates of μ and Σ as $\hat{\mu} = \bar{\boldsymbol{x}}$ and $\hat{\Sigma} = \frac{(n-1)}{n}\boldsymbol{S}$.
More generally, whatever the maximum likelihood estimate (mle) of μ is, if $\boldsymbol{d} = \bar{\boldsymbol{x}} - \hat{\mu}$ then $\hat{\Sigma} = \frac{(n-1)}{n}\boldsymbol{S} + \boldsymbol{dd}'$. Note that this general formula applies ***only*** when Σ is an unknown symmetric matrix and other than the constraint of symmetry there are no other interdependencies between the elements of Σ so, for example, it would not be applicable if $\Sigma = \sigma^2[(1-\rho)\mathbf{I_n} + \rho\iota_\mathbf{n}\iota_\mathbf{n}']$, the equicorrelation matrix with σ and ρ unknown. This form is useful in constructing likelihood ratio tests where the null hypothesis puts some restriction on μ so that under the null hypothesis the mle of μ is not $\bar{\boldsymbol{x}}$. In these cases we can easily obtain the mle of Σ and hence the maximized likelihood under the null hypothesis.

9.2.4 The maximized log-likelihood

For the construction of likelihood ratio tests we need the actual form of the maximized likelihood under null and alternative hypotheses. Typically, the alternative hypothesis gives no restrictions on μ and Σ and so the mles under the alternative hypothesis are as given earlier (i.e., $\hat{\mu} = \bar{\boldsymbol{x}}$ and $\hat{\Sigma} = \frac{(n-1)}{n}\boldsymbol{S}$). The null hypothesis will either impose some constraint on Σ (e.g., $H_0 : \Sigma = \Sigma_0$) or some constraint on μ (e.g., $H_0 : \mu = \mu_0$ or $H_0 : \mu\mu' = 1$). In these latter cases we obtain the estimate of μ and then use the more general form for $\hat{\Sigma}$ given above. For example, under $H_0 : \mu = \mu_0$ we have $\hat{\mu} = \mu_0$ and so

$$\hat{\Sigma} = \tfrac{(n-1)}{n}\boldsymbol{S} + (\bar{\boldsymbol{x}}-\mu_0)(\bar{\boldsymbol{x}}-\mu_0)' = \tfrac{1}{n}\sum_{i}^{n}(\boldsymbol{x}_i-\mu_0)(\boldsymbol{x}_i-\mu_0)'.$$

To calculate the actual maximized likelihood in either case often requires the use of the trick in manipulating matrices discussed on Page 36 in §2.8.2 which is that $\sum_i^n \boldsymbol{y}_i'\boldsymbol{S}\boldsymbol{y}_i = \mathrm{tr}(\boldsymbol{S}\sum_i^n \boldsymbol{y}_i\boldsymbol{y}_i')$. The advantage of this is that the matrix product on the right hand side might reduce to the identity matrix whose trace is easy to calculate. Applying this result gives

$$\begin{aligned}\sum_{i=1}^{n}(\boldsymbol{x}_i-\bar{\boldsymbol{x}})'\Sigma^{-1}(\boldsymbol{x}_i-\bar{\boldsymbol{x}}) &= \operatorname{tr}\big(\Sigma^{-1}\sum_{i=1}^{n}(\boldsymbol{x}_i-\bar{\boldsymbol{x}})(\boldsymbol{x}_i-\bar{\boldsymbol{x}}')\big)\\ &= \operatorname{tr}\big(\Sigma^{-1}(n-1)\boldsymbol{S}\big)=(n-1)\operatorname{tr}\big(\Sigma^{-1}\boldsymbol{S}\big) \text{ and}\\ (\bar{\boldsymbol{x}}-\mu)'\Sigma^{-1}(\bar{\boldsymbol{x}}-\mu) &= \operatorname{tr}\big(\Sigma^{-1}(\bar{\boldsymbol{x}}-\mu)(\bar{\boldsymbol{x}}-\mu)'\big).\end{aligned}$$

Thus the log-likelihood can be written as

$$\begin{aligned}\ell(\mu,\Sigma;\boldsymbol{X}) &= -{}^1\!/_2(n-1)\operatorname{tr}\big(\Sigma^{-1}\boldsymbol{S}\big)-{}^1\!/_2 n\big(\operatorname{tr}\big(\Sigma^{-1}(\bar{\boldsymbol{x}}-\mu)(\bar{\boldsymbol{x}}-\mu)'\big)\big)\\ &\qquad -{}^1\!/_2 np\log(2\pi)-{}^1\!/_2 n\log(|\Sigma|).\end{aligned}$$

This form is often easier to work with when finding the values of the maximized log-likelihoods under null and alternative hypotheses which are needed for constructing likelihood ratio tests. With $\hat{\Sigma}=\frac{(n-1)}{n}\boldsymbol{S}+\boldsymbol{dd}'$ where $\boldsymbol{d}=\bar{\boldsymbol{x}}-\hat{\mu}$ we have

$$\max_{\mu,\Sigma}\ell(\mu,\Sigma;\boldsymbol{X}) = -{}^1\!/_2 np-{}^1\!/_2 np\log(2\pi)-{}^1\!/_2 n\log(|\tfrac{(n-1)}{n}\boldsymbol{S}+\boldsymbol{dd}'|).$$

9.2.5 Examples of likelihood ratio tests

The purpose of this section is not to present a full account of the theory of hypothesis testing but to illustrate some of the techniques and results established earlier by showing how these can be used to construct likelihood ratio test (LRT) statistics and to evaluate them numerically in cases where data have been obtained. Discussion of the evaluation of the statistical results is not given here since that is the role of more specialist courses on multivariate analysis; see for example §5.2.1 of Mardia et al. (1979).

When testing a null hypothesis H_0 against an alternative H_A the likelihood ratio test statistic is given by $\lambda=2\{\ell_{\max}(H_A)-\ell_{\max}(H_0)\}$ where $\ell_{\max}(H_0)$ and $\ell_{\max}(H_A)$ are the maximized values of the log-likelihood under H_0 and H_A respectively. General theory gives the asymptotic distribution of this statistic under various regularity conditions but that is not considered further here. Thus the procedure entails using maximum likelihood estimation to estimate any unknown parameters under the null and alternative hypotheses, substituting these into the log-likelihood and taking twice the difference. Throughout this section it will be assumed that $\boldsymbol{x}_1,\boldsymbol{x}_2,\ldots,\boldsymbol{x}_n$ are independent observations of $\boldsymbol{x}\sim N_p(\mu,\Sigma)$, and $\boldsymbol{x}_i=(x_{i1},x_{i2},\ldots,x_{ip})'$.

9.2.5.1 One-sample T^2-test

If $H_0:\mu=\mu_0$ and $H_A:\mu\neq\mu_0$ then under H_0 we have only to estimate Σ since is μ specified as μ_0 which yields

$$\ell_{\max}(H_0) = -{}^1\!/_2 np-{}^1\!/_2 np\log(2\pi)-{}^1\!/_2 n\log(|\boldsymbol{S}+\boldsymbol{dd}'|)$$

where $\boldsymbol{d}=\bar{\boldsymbol{x}}-\mu_0$. Under H_A there are no restrictions on the parameters so $\hat{\mu}=\bar{\boldsymbol{x}}$ and so

$$\ell_{\max}(H_0) = -{}^1\!/_2 np-{}^1\!/_2 np\log(2\pi)-{}^1\!/_2 n\log(|\boldsymbol{S}|),$$

giving the LRT statistic as $\lambda = \{n\log(|S+dd'|) - n\log(|S|)\} = n\log(|S+dd'|/|S|)$

$$= n\log(|S+dd'||S^{-1}|) \text{ (using §5.2.4)}$$
$$= n\log(|S^{-1}(S+dd')|) \text{ (using §4.3.2 (viii))}$$
$$= n\log(\mathbf{I_n}+S^{-1}dd')| = n\log(1+dS^{-1}d')| \text{ (using §4.6.2)}.$$

This is a monotonic function of $ndS^{-1}d = n(\bar{x}-\mu_0)'S^{-1}(\bar{x}-\mu_0)$ which is Hotelling's T^2 statistic, so demonstrating that Hotelling's T^2-test is a likelihood ratio test.

Implementation in R: Calculation of T^2 statistics in **R** can be done directly as

```
T2<-n*t(xbar-mu0)%*%solve(S)%*%(xbar-mu0)
```

The **R** library `ICSNP` (Nordhausen et al., 2012), which contains tools for various multivariate nonparametric analyses, includes a function `HotellingsT2(.)` which will perform one- and two-sample Hotelling's T^2-tests though note that the value of the test statistic produced is the scaled version which has an F-distribution.

9.2.5.2 Multisample tests: MANOVA

Suppose we have k independent samples of sizes n_i from $N_p(\mu_i,\Sigma)$ and wish to test the hypothesis $H_0 : \mu_1 = \mu_2 = \ldots = \mu_k = \mu$ against $H_A :$ at least one $\mu_i \neq \mu$. The log-likelihood is

$$\ell(\mu_1,\mu_2,\ldots,\mu_p,\Sigma;X) = -{}^1\!/_2 npk\log(2\pi) - {}^1\!/_2 nk\log(|\Sigma|)$$
$$-{}^1\!/_2\sum_{i=1}^{k}\{(n-1)\mathrm{tr}(\Sigma^{-1}S_i) + n(\mathrm{tr}(\Sigma^{-1}(\bar{x}_i-\mu_i)(\bar{x}_i-\mu_i)'))\}$$

(i.e., the sum of the k separate log-likelihoods of the individual samples).
Under H_0 we have a sample of size $n = \sum n_i$ from $N_p(\mu,\Sigma)$ so the mles are $\hat{\mu} = \bar{x}$ and $\hat{\Sigma} = \frac{(n-1)}{n}S$ and so

$$\ell_{\max}(H_0) = -{}^1\!/_2 np - {}^1\!/_2 np\log(2\pi) - {}^1\!/_2 n\log(|\tfrac{(n-1)}{n}S|), \quad \text{noting}$$

$$\sum_{i=1}^{k}\{(n-1)\mathrm{tr}(\hat{\Sigma}^{-1}S_i) + n(\mathrm{tr}(\hat{\Sigma}^{-1}(\bar{x}_i-\hat{\mu}_i)(\bar{x}_i-\hat{\mu}_i)'))\}$$
$$= \sum_{i=1}^{k}\left\{\mathrm{tr}\left(\tfrac{n}{(n-1)}S^{-1}\sum_{j=1}^{n_i}(x_{ij}-\bar{x}_i)(x_{ij}-\bar{x}_i)'\right) + n\mathrm{tr}\left(\tfrac{n}{(n-1)}S^{-1}(\bar{x}_i-\bar{x})(\bar{x}_i-\bar{x})'\right)\right\}$$

$$\text{and} \quad \sum_{i=1}^{k}\left\{\sum_{=1}^{n_i}(x_{ij}-\bar{x}_i)(x_{ij}-\bar{x}_i)' + n_i(\bar{x}_i-\bar{x})(\bar{x}_i-\bar{x})'\right\} = (n-1)S.$$

Under H_A we have $\hat{\mu}_i = \bar{x}_i$, the ith sample mean and $\hat{\Sigma} = \sum_1^k(n_i-1)S_i = \frac{(n-k)}{n}W$, where $W = \frac{1}{n-k}\sum_1^k(n_i-1)S_i$, the [pooled] ***within-groups sample variance***. So

$$\ell_{\max}(H_A) = -{}^1\!/_2 np - {}^1\!/_2 np\log(2\pi) - {}^1\!/_2 n\log(|\tfrac{(n-k)}{n}W|) \quad \text{noting}$$

$$\Sigma^{-1}(\bar{x}_i-\hat{\mu}_i)(\bar{x}_i-\hat{\mu}_i)' = \Sigma^{-1}(\bar{x}_i-\bar{x}_{ii})(\bar{x}_i-\bar{x}_{ii})' = 0.$$

Thus the LRT statistic is $\lambda = 2\{\ell_{\max}(H_A) - \ell_{\max}(H_0)\} = n\log\left|\frac{(n-1)\boldsymbol{S}}{(n-k)\boldsymbol{W}}\right|$ which is a monotonic function of $\frac{|\boldsymbol{S}|}{|\boldsymbol{W}|} = |\boldsymbol{W}^{-1}\boldsymbol{S}|$. Define the ***between-groups variance B*** by $(k-1)\boldsymbol{B} = \sum_i^k n_i(\bar{\boldsymbol{x}}_i - \bar{x})(\bar{\boldsymbol{x}}_i - \bar{x})' = (n-1)\boldsymbol{S} - (n-k)\boldsymbol{W}$ so an equivalent test statistic is $|\boldsymbol{W}^{-1}[(k-1)\boldsymbol{B} + (n-k)\boldsymbol{W}|$ or equivalently $\Lambda = |\mathbf{I_p} + \frac{k-1}{n-k}\boldsymbol{W}^{-1}\boldsymbol{B}|$. Λ is termed Wilks' Λ and has a $\Lambda(p, n-k, k-1)$ distribution.

It is in principle straightforward to calculate the various statistics using elementary **R** functions (and this is outlined in §9.4.2 below) but a ready made function `manova(.)` is available and details of its use are in the `help` system.

9.2.5.3 The hypothesis $H_0 : \mu'\mu = r_0^2$

Suppose $\boldsymbol{x} \sim N_p(\mu, \sigma^2\mathbf{I_p})$. Note that $\mu'\mu = r_0^2$ implies that μ lies on a sphere of specified radius r_0 and the alternative hypothesis is taken as $H_A = \overline{H}_0$. With $\Sigma = \sigma^2\mathbf{I_p}$ we have $\Sigma^{-1} = \sigma^{-2}\mathbf{I_p}$ and so

$$\begin{aligned}\ell(\mu, \sigma, \boldsymbol{X}) &= -\tfrac{1}{2}(n-1)\mathrm{tr}(\boldsymbol{S}\sigma^{-2}) - \tfrac{1}{2}n(\bar{\boldsymbol{x}} - \mu)'(\bar{\boldsymbol{x}} - \mu)\sigma^{-2} \\ &\qquad - \tfrac{1}{2}\log(2\pi) - \tfrac{1}{2}np\log(\sigma^2),\end{aligned}$$

noting $\log(|\sigma^2\mathbf{I_p}|) = p\log(\sigma^2)$, see §4.3.2 (ii)).

Let $\Omega = \ell(\mu, \sigma, \boldsymbol{X}) - \lambda(\mu'\mu - r_0^2)$ (where λ is a Lagrange multiplier introduced for the constrained optimization). Then $\frac{\partial\Omega}{\partial\mu} = n(\bar{\boldsymbol{x}} - \mu)\sigma^{-2} - 2\lambda\mu$ so we require $\hat{\mu} = \frac{n\bar{\boldsymbol{x}}}{n+2\lambda\sigma^2}$. Since $\hat{\mu}'\hat{\mu} = r_0^2$ we have $\hat{\mu} = \frac{\bar{\boldsymbol{x}}r_0}{\sqrt{\bar{\boldsymbol{x}}'\bar{\boldsymbol{x}}}}$ (which does not depend upon σ^2).

$$\begin{aligned}\frac{\partial\ell}{\partial\sigma} &= (n-1)\mathrm{tr}(\boldsymbol{S})\sigma^{-3} + n(\bar{\boldsymbol{x}} - \mu)'(\bar{\boldsymbol{x}} - \mu)\sigma^{-3} - np\sigma^{-1} \\ \text{so } \hat{\sigma} &= \sqrt{\frac{1}{np}\{(n-1)\mathrm{tr}(\boldsymbol{S}) + n(\bar{\boldsymbol{x}} - \hat{\mu})'(\bar{\boldsymbol{x}} - \hat{\mu})\}} \\ &= \sqrt{\frac{1}{np}\{(n-1)\mathrm{tr}(\boldsymbol{S}) + n(\sqrt{\bar{\boldsymbol{x}}'\bar{\boldsymbol{x}}} - r_0)^2\}} \\ &= \sqrt{\frac{1}{np}\left\{\sum \boldsymbol{x}_i'\boldsymbol{x}_i - 2nr_0\sqrt{\bar{\boldsymbol{x}}'\bar{\boldsymbol{x}}} + nr_0^2\right\}}.\end{aligned}$$

Thus

$\ell_{\max}(H_0) = -\tfrac{1}{2}np - \tfrac{1}{2}np\log(2\pi) - \tfrac{1}{2}n\log(\frac{1}{np}\{(n-1)\mathrm{tr}(\boldsymbol{S}) + n(\bar{\boldsymbol{x}} - \hat{\mu})'(\bar{\boldsymbol{x}} - \hat{\mu})\})$

$= -\tfrac{1}{2}np - \tfrac{1}{2}np\log(2\pi) - \tfrac{1}{2}n\log(\frac{1}{np}\left\{\sum \boldsymbol{x}_i'\boldsymbol{x}_i - 2nr_0\sqrt{\bar{\boldsymbol{x}}'\bar{\boldsymbol{x}}} + nr_0^2\right\})$.

Note that the general formula given in §9.2.4 does not apply here since that relied on differentiating with respect to Σ when Σ was a symmetric matrix and that was the only interdependency between the elements of Σ but here we have $\Sigma = \sigma^2\mathbf{I_p}$ so all diagonal elements of Σ are equal. $\ell_{\max}(H_A) = -\tfrac{1}{2}np - \tfrac{1}{2}np\log(2\pi) - \tfrac{1}{2}n\log\left(\left|\frac{(n-1)}{n}\boldsymbol{S}\right|\right)$. Thus the likelihood ratio test statistic is

$$\begin{aligned}\lambda &= 2\{\ell_{\max}(H_A) - \ell_{\max}(H_0)\} \\ &= n\log\left(\frac{1}{np}\left\{\sum \boldsymbol{x}_i'\boldsymbol{x}_i - 2nr_0\sqrt{\bar{\boldsymbol{x}}'\bar{\boldsymbol{x}}} + nr_0^2\right\} / \left|\tfrac{(n-1)}{n}\boldsymbol{S}\right|\right).\end{aligned}$$

9.2.6 Union-intersection tests

An alternative strategy for constructing hypothesis tests is based on the union-intersection principle. A test of a hypothesis expressed as an intersection of component hypotheses based on a union of one dimensional rejection regions for each one dimensional hypothesis is referred to as a ***union-intersection test*** (UIT). Although such tests do not necessarily possess the optimal properties of likelihood ratio tests (such as the asymptotic null distribution of the test statistic) they may provide additional insight into the data, in particular providing an indication of the direction of departure from a multivariate null hypothesis. A full discussion of the method is not given here; it is available in many standard texts, for example §5.2.2 of Mardia et al. (1979).

In outline, the procedure consists of projecting the data into one dimension and testing the hypothesis in that one dimension. The particular dimension chosen is that which shows the greatest deviation from the null hypothesis. Typically this leads to an optimization problem involving quadratic forms or ratios of quadratic forms in the projection vector. Often these can be solved by the techniques for constrained optimization discussed in §7.6 where the constraint might be that the projection vector is of unit length or some other scale constraint such as the denominator of a ratio of quadratic forms is unity. The selection of examples below illustrates the method.

9.2.7 Examples of union-intersection tests

Again, the purpose of this section is to illustrate some of the techniques. A full account of underlying theory is left to more specialist texts. Throughout this section it will be assumed that $\boldsymbol{x}_1, \boldsymbol{x}_2, \ldots, \boldsymbol{x}_n$ are independent observations of $\boldsymbol{x} \sim N_p(\mu, \Sigma)$, and $\boldsymbol{x}_i = (x_{i1}, x_{i2}, \ldots, x_{ip})'$ with sample mean and variance $\bar{\boldsymbol{x}}$ and $\boldsymbol{S}$. Projecting the sample into one dimension will be achieved with a projection vector β. Thus if $\boldsymbol{X}'$ is the data matrix in p-dimensions, the projected data matrix is $\boldsymbol{X}'\beta$ and individual projected observations are $\boldsymbol{x}_i'\beta$; $i = 1, \ldots, n$. The sample variance of the projected sample is $\beta'\boldsymbol{S}\beta$ .This follows directly from the definition, using algebra similar to that employed in §9.2.1 for the equivalent result for the population variance.

9.2.7.1 Two-sample T^2-test

We now test $H_0 : \mu_1 = \mu_2$ against $H_A : \mu_1 \neq \mu_2$ based on samples of size n_1 and n_2 from $N_p(\mu_i, \Sigma)$; $i = 1, 2$. Suppose the sample means and variances of the two samples are $\bar{\boldsymbol{x}}_i$, $\boldsymbol{S}_i$; $i = 1, 2$ and let $\boldsymbol{S} = \big((n_1 - 1)\boldsymbol{S}_1 + (n_2 - 1)\boldsymbol{S}_2\big)/(n-2)$ where $n = n_1 + n_2$, the pooled estimate of the common variance Σ. If the data are projected into one dimension by a projection vector β the means and variances of the project samples are $\bar{\boldsymbol{x}}_i'\beta$; $i = 1, 2$ and $\beta'\boldsymbol{S}\beta$. The appropriate test in one dimension of $H_{0\beta} : \mu_1'\beta = \mu_2'\beta$ is a two-sample Student's t-test with test statistic given by t_β where

$$t_\beta^2 = \frac{n_1 n_2 \beta'(\bar{\boldsymbol{x}}_1 - \bar{\boldsymbol{x}}_2)(\bar{\boldsymbol{x}}_1 - \bar{\boldsymbol{x}}_2)'\beta}{n\beta'\boldsymbol{S}\beta}$$

and the UIT procedure is to maximize this with respect to β.
This is essentially Example 7 (ii) on Page 109 and the maximum is given by the only non-zero eigenvalue of the rank 1 matrix $n_1n_2\boldsymbol{S}^{-1}(\bar{\boldsymbol{x}}_1-\bar{\boldsymbol{x}}_2)(\bar{\boldsymbol{x}}_1-\bar{\boldsymbol{x}}_2)'/n$ which is $n_1n_2(\bar{\boldsymbol{x}}_1-\bar{\boldsymbol{x}}_2)'\boldsymbol{S}^{-1}(\bar{\boldsymbol{x}}_1-\bar{\boldsymbol{x}}_2)/n$ (with corresponding eigenvector $\boldsymbol{S}^{-1}(\bar{\boldsymbol{x}}_1-\bar{\boldsymbol{x}}_2)$). Thus the UIT statistic is $T^2=\frac{n_1n_2}{n}(\bar{\boldsymbol{x}}_1-\bar{\boldsymbol{x}}_2)'\boldsymbol{S}^{-1}(\bar{\boldsymbol{x}}_1-\bar{\boldsymbol{x}}_2)$, i.e., a two-sample Hotelling's T^2-test.

9.2.7.2 Test of $H_0:\Sigma=\Sigma_0$ against $H_A:\Sigma\neq\Sigma_0$

If the sample variance, based on a sample of size n is $\boldsymbol{S}$ then the variance of the one-dimensional sample is $\beta'\boldsymbol{S}\beta$ and the one-dimensional component hypothesis is $H_{0\beta}:\beta'\Sigma\beta=\beta'\Sigma_0\beta$ for which the appropriate test is a [two-sided] chi-squared test based on rejecting $H_{0\beta}$ if $(n-1)\beta'\boldsymbol{S}\beta/\beta'\Sigma_0\beta$ is improbably small or large. The UIT test statistic is obtained by optimizing this with respect to β. This problem is considered in Example 7 (ii) on Page 109 where it is shown that the maximum and minimum values are given by the largest and smallest eigenvalues of $(n-1)\Sigma_0^{-1}\boldsymbol{S}$. Clearly an equivalent test can be based on the largest and smallest eigenvalues of $\Sigma_0^{-1}\boldsymbol{S}$.

In the absence of asymptotic results the test statistics would need to be evaluated by simulation techniques, generating a number K, say, of random samples of size n drawn from $N_p(\boldsymbol{0},\Sigma_0)$ and seeing in how many of the samples the smallest and largest eigenvalues of $\Sigma_0^{-1}\boldsymbol{S}_i$; $i=1,\ldots,K$ lay either above λ_1 or below λ_p where $\lambda_1\geq\lambda_2\geq\ldots\geq\lambda_p$ are the ordered eigenvalues of the observed sample value of $\Sigma_0^{-1}\boldsymbol{S}$. For a two-sided [equi-tailed] test of size α, the null hypothesis H_0 would be rejected ***either*** if fewer than $100(1-{}^1\!/_2\alpha)\%$ of the simulated largest eigenvalues of $\Sigma_0^{-1}\boldsymbol{S}_i$ are greater than λ_1 ***or*** if fewer than $100(1-{}^1\!/_2\alpha)\%$ of the simulated smallest eigenvalues of $\Sigma_0^{-1}\boldsymbol{S}_i$ are less than λ_p.

Implementation of such a simulation test in **R** is straightforward but is a little beyond the scope of this text except to note that a random sample of size n drawn from $N_p(\boldsymbol{0},\Sigma_0)$ can be obtained by using the simple function `rnorm(.)` to generate p independent univariate samples from $N(0,1)$ to form a data matrix $\boldsymbol{Y}'$ and then transforming these to a sample $\boldsymbol{X}'$ from $N_p(\boldsymbol{0},\Sigma_0)$ by $\boldsymbol{X}=\Sigma_0^{1/2}\boldsymbol{Y}$, see §9.2.1 (where $\Sigma_0^{1/2}$ is defined in §6.7.1.1). Alternatively the `MASS` library has a function `mvrnorm(.)` which will draw a random sample of specified size from a multivariate normal distribution with given mean and variance.

9.2.7.3 Multisample tests: MANOVA

The likelihood ratio test for the multivariate one-way analysis of variance was derived in §9.2.5.2 and here we consider the UIT approach. We use the notation established in that section.

Projecting the data into one dimension gives a hypothesis of equal means of k univariate Normal populations as the component hypothesis, i.e., a one-way analysis of variance, with test statistic given by the F-ratio $F_\beta=\beta'\boldsymbol{B}\beta/\beta'\boldsymbol{W}\beta$, where $\boldsymbol{B}$ and $\boldsymbol{W}$ are the between- and within-groups variances. Maximization of this ratio with respect to β was considered in Example 7 (ii) and the maximum value is the largest eigenvalue of $\boldsymbol{W}^{-1}\boldsymbol{B}$. Recall that the LRT statistic for this problem is $\Lambda = |\mathbf{I_p} + \frac{k-1}{n-k}\boldsymbol{W}^{-1}\boldsymbol{B}|$ which can be shown to be equivalent only if $k = 2$ but for $k > 2$ the two tests are different. The **R** function `manova(.)` will optionally produce the value of the UIT statistic and details are given in the `help()` system.

9.3 Principal Component Analysis

An informal account of PCA was given in §6.5 where some justification was given for the technique being regarded as a major exploratory tool of multivariate data analysis. In that section the principal components were taken as the projections of the data onto the eigenvectors of the variance matrix $\boldsymbol{S}$ but it is better to define the components differently and then show that these are given by the eigenanalysis of $\boldsymbol{S}$. If $\boldsymbol{X}'$ is a $n \times p$ data matrix then the first principal component of $\boldsymbol{X}'$ is that linear combination of the p variables $\boldsymbol{y}_1' = (\boldsymbol{X} - \overline{\boldsymbol{X}})'\boldsymbol{a}_1$ such that $\text{var}(\boldsymbol{y}_1)'$ is maximized, subject to the constraint $\boldsymbol{a}_1'\boldsymbol{a}_1 = 1$. Subsequent principal components are defined successively in a similar way, so the ith component is $\boldsymbol{y}_i' = (\boldsymbol{X} - \overline{\boldsymbol{X}})'\boldsymbol{a}_i$ such that $\text{var}(\boldsymbol{y}_i')$ is maximized, subject to the constraints $\boldsymbol{a}_i'\boldsymbol{a}_i = 1$ and $\boldsymbol{a}_i'\boldsymbol{a}_j = 0$ for all $j < i$. Note that $\boldsymbol{y}_i'$ is a $n \times 1$ vector, i.e., a set of n univariate observations which are projections of $\boldsymbol{X}'$ onto the vector $\boldsymbol{a}_i$. Some texts develop the theory of principal components with the actual data matrix $\boldsymbol{X}'$ (or else say that "without loss of generality the variable means are assumed to be zero"). Working explicitly with the mean corrected data matrix $(\boldsymbol{X} - \overline{\boldsymbol{X}})'$ rather than the actual data matrix $\boldsymbol{X}'$ emphasises this mean correction (especially important when adding in supplementary points) and conforms with the **R** functions `princomp()` and the default option of `prcomp(.)`. These functions are discussed further below.

9.3.1 Derivation

We need to maximize $\text{var}(\boldsymbol{y}_1') = \text{var}((\boldsymbol{X} - \overline{\boldsymbol{X}})'\boldsymbol{a}_1) = \boldsymbol{a}_1'\boldsymbol{S}\boldsymbol{a}_1$ subject to $\boldsymbol{a}_1'\boldsymbol{a}_1 = 1$, where $\boldsymbol{S}$ is the sample variance of $\boldsymbol{X}'$. This was considered in Example 7 (i) where the solution is given that $\boldsymbol{a}_1$ is the [normalized] eigenvector of $\boldsymbol{S}$ corresponding to its largest eigenvalue. A simple extension of the argument used in Example 7 (i) shows that the ith component is obtained by taking $\boldsymbol{a}_i$ as the eigenvector of $\boldsymbol{S}$ corresponding to its ith largest eigenvalue. Note that these eigenvectors are orthogonal (see property 6.4.2 (ii)) and so the conditions $\boldsymbol{a}_i'\boldsymbol{a}_j = 0$ for all $j < i$ are satisfied. Note further that the $\boldsymbol{a}_i$ are defined only up to a factor of ± 1 and different **R** functions may make different choices. This does not fundamentally alter the statistical properties of the transformation. A similar ambiguity of sign of eigenvectors arises in other techniques such as linear discriminant analysis and canonical correlation analysis discussed in later sections.

9.3.2 Calculations in R

Before discussing the functions provided in **R** for principal component analysis we will first indicate how to perform the calculations with more elementary **R** functions. The key step is the eigenanalysis of the covariance matrix $\boldsymbol{S}$. This can be done either with the function `eigen(.)` or with `svd(.)`, noting that $\boldsymbol{S}$ is a symmetric square matrix. The function svd(.) is known to be numerically more stable than `eigen(.)` so is the preferred choice. $\boldsymbol{S}$ can be calculated either by `S<-var(t(X))` or by the few lines of code given in §2.12.3.2. If the data matrix referred to the full set of principal components is designated as $\boldsymbol{Y}'$ then $\boldsymbol{Y}' = (\boldsymbol{X}-\overline{\boldsymbol{X}})'\boldsymbol{A}$ where $\boldsymbol{A}$ is the matrix whose columns are the eigenvectors of $\boldsymbol{S}$. To perform this in **R** we do

```
xbar<-X%*%matrix(rep(1,n),n)/n
Xbar<-xbar%*%t(matrix(rep(1,n),n))
S<-var(t(X))
Y<-t(t(X-Xbar)%*%svd(S)$v)
```

A ***score plot*** (i.e., a scatter plot of the data rotated onto principal components) is obtained by plotting the columns of $\boldsymbol{Y}'$ against each other, i.e by `plot(t(Y)[,i], t(Y)[,j])` for a plot of component i against component j.

If $\boldsymbol{Z}'$ is a $m \times p$ data matrix of m supplementary observations on the same p variables and it is required to apply the same transformation to $\boldsymbol{Z}'$ then this is achieved by $\boldsymbol{W}' = (\boldsymbol{Z}-\overline{\boldsymbol{X}}_{[m]})'\boldsymbol{A}$ where $\overline{\boldsymbol{X}}_{[m]}$ is the $p \times m$ matrix with all columns equal to $\bar{\boldsymbol{x}}$. This is achieved in **R** by

```
Xbarm<-xbar%*%t(matrix(rep(1,m),m))
W<-t(t(Z-Xbarm)%*%svd(S)$v)
```

The columns of $\boldsymbol{W}'$ can then be plotted as supplementary points on an existing score plot of the data used to determine the principal components with `points(t(W)[,i], t(W)[,j])`. Alternatively, if the analysis has been performed with `prcomp(.)` the generic function `predict(.)` can be used with `predict(object,newdata)` where `object` is an object of class `prcomp`. Type `help(predict.prcomp)` for more details.

It is sometimes preferable to perform principal component analysis using the correlation matrix $\boldsymbol{S}$ instead of the variance matrix $\boldsymbol{S}$. Calculation of $\boldsymbol{R}$ is described in §2.12.3.2.

R provides two functions for principal component analysis, `princomp(.)` and `prcomp(.)`. The first is a legacy of S-PLUS and is provided for compatibility but in general its use is deprecated. The functions will give slightly different values for the eigenvalues of the variance matrix because `princomp(.)` uses a divisor n for the variance while `prcomp(.)` uses a divisor $n-1$ (so the eigenvalues produced by princomp(.) will be $(n-1)/n \times$ those produced by prcomp(.); see Exercises 6 (7). Further, `princomp(.)` uses internally the function `eigen(.)` for the key eigenanalysis step whilst `prcomp(.)` uses the more stable `svd(.)`. A further major difference is that `princomp(.)` requires $n \geq p$ but there are no restrictions with `prcomp(.)` and it will work with $n < p$, though of course the final $p-n$ eigenvalues will be zero and so the principal components corresponding to these values will be chosen arbitrarily and so have no interpretation.

By default `prcomp(.)` subtracts variable means from the data matrix and uses the variance matrix for calculation of principal components. Both of these can be changed by additional arguments. The function returns the transformed data, the matrix of eigenvectors and the vector of eigenvalues. Full details are readily available from `help(prcomp)`.

9.3.3 Population and sample principal components

It is sometimes useful to keep in mind the distinction between population and sample principal components: given a random variable $\boldsymbol{x}$ with variance Σ its *population* principal components are given by the eigenvectors of Σ taken in the order of decreasing magnitude of the eigenvalues of Σ. Strictly the principal components are *defined* as those linear combinations of the variables which have maximal variance subject to normalizing and orthogonality constraints. Given a random sample of independent observations of $\boldsymbol{x}$ in a data matrix $\boldsymbol{X}'$ with $\text{var}(\boldsymbol{X}') = \boldsymbol{S}$ the *sample* principal components are the eigenvectors of $\boldsymbol{S}$, again conventionally taken in the order of decreasing magnitude of the eigenvalues of $\boldsymbol{S}$. It is presumed that the sample principal components are estimates of the population principal components and indeed it can be shown that asymptotically this is true; see for example Anderson (2003). In particular it is presumed that the ith sample principal component is an estimate of the ith population component but this is only true for arbitrarily large samples (i.e., asymptotically). For small samples the sampling error of the eigenvalues of $\boldsymbol{S}$ can be substantial. Anderson (2003) gives the asymptotic distribution of the ith sample eigenvalue as $N(\lambda_i, 2\lambda_i^2/n)$. If two [population] eigenvalues are close then the sample eigenvalues can '**swap**' over, so changing the order of the corresponding eigenvectors (or principal components). To illustrate this phenomenon consider the matrix

$$\Sigma = \begin{pmatrix} 1.011 & 0.212 & -0.512 & 0.289 \\ 0.212 & 1.013 & 0.187 & -0.512 \\ -0.512 & 0.187 & 1.013 & 0.212 \\ 0.289 & -0.512 & 0.212 & 1.011 \end{pmatrix}$$

which has eigenvalues 1.554, 1.500, 0.946 and 0.048 (so the first two are numerically close) and eigenvectors

$$\begin{pmatrix} 0.54 & -0.48 & -0.46 & 0.52 \\ -0.46 & -0.52 & -0.54 & -0.48 \\ -0.46 & 0.52 & -0.54 & 0.48 \\ 0.54 & 0.48 & -0.46 & -0.52 \end{pmatrix}.$$

The following transcript from an **R** session shows setting the seed for the random number generator to 137, generating a sample of size 50 from $N_4(\boldsymbol{0}, \Sigma)$, calculating the eigenanalysis of the sample covariance matrix $\boldsymbol{S}$ and showing that the first and second eigenvectors of Σ are not approximately parallel to the first two eigenvectors of $\boldsymbol{S}$ (as assessed by calculating their inner products which should be close to ± 1) but they are near parallel to the second and first eigenvectors of $\boldsymbol{S}$ (with inner products of ± 0.95), indicating that the first two eigenvalues of $\boldsymbol{S}$ have ***'swapped over'***. Indeed this is also apparent by looking at the pattern of signs of the first two eigenvectors of Σ and $\boldsymbol{S}$. In the following code the population variance Σ is denoted as `m`.

```
> options(digits=2)
> set.seed(137)
> library(MASS)
> m<-matrix(c(1.011,0.212,
+ -0.512,0.289,0.212,1.013,
+ 0.187,-0.512,-0.512,0.187,
+ 1.013,0.212,0.289,-0.512,
+ 0.212,1.011),4,4)
> eigen(m)
$values
[1] 1.554 1.500 0.946 0.048
$vectors
      [,1]  [,2]  [,3]  [,4]
[1,]  0.54 -0.48 -0.46  0.52
[2,] -0.46 -0.52 -0.54 -0.48
[3,] -0.46  0.52 -0.54  0.48
[4,]  0.54  0.48 -0.46 -0.52
> x1<-eigen(m)$vectors[,1]
> x2<-eigen(m)$vectors[,2]
> X<-t(mvrnorm(50,c(0,0,0,0),m))
> S<-var(t(X))
```

```
> eigen(S)
$values
[1] 1.771 1.114 0.888 0.047
$vectors
      [,1]  [,2] [,3]  [,4]
[1,] -0.30 -0.69 0.36  0.55
[2,] -0.66  0.22 0.56 -0.45
[3,]  0.35  0.54 0.59  0.49
[4,]  0.59 -0.43 0.45 -0.51
> y1<-eigen(S)$vectors[,1]
> y2<-eigen(S)$vectors[,2]
> t(x1)%*%y1 ; t(x2)%*%y2
     [,1]
[1,]  0.3
     [,1]
[1,] 0.29
> t(x1)%*%y2 ; t(x2)%*%y1
      [,1]
[1,] -0.95
     [,1]
[1,] 0.95
```

If the random number generator seed is changed to 163 then the inner products of corresponding eigenvectors of Σ and $\boldsymbol{S}$ are 0.93 and 0.95 and those of first and second eigenvectors -0.29 and 0.3, indicating that in this case the first two eigenvalues have not swapped places (this is left to the reader to verify). If the seed is kept as 137 but the sample size increased to 10000 then no swapping occurs.

Of course, in many practical situations the population variance is totally unknown and so it would not be possible to detect such swapping. It is not obvious from the first run that the first two population eigenvalues are close together when the sample values are 1.771 and 1.114 (see transcript above). However, a common practical situation where such swapping is a real danger is in simulations involving extraction of eigenvalues from randomly generated matrices, such as in §9.2.7.2. There the test statistic only involved eigenvalues so the bias introduced by swapping is small but if the interest is in the eigenvectors (for example obtaining an envelope of a particular principal component as a 'bootstrap' confidence band) then swapping of sample eigenvalues will result in inclusion of a sample principal component approximately orthogonal to the 'correct' direction.

These comments do not apply solely to principal components. Sample eigenvalue swapping is a danger in any technique requiring calculation of sample eigenvalues and eigenvectors, such as linear discriminant and canonical correlation analyses discussed in the next two sections and other techniques beyond the scope covered here such as biplots (Gower and Hand, 1995) and correspondence analysis (Greenacre, 2010) where bootstrap samples can be used to assess variabilities of score plots. There is some discussion of this potential pitfall in Ringrose (1996 and 2012).

9.4 Linear Discriminant Analysis

Discriminant analysis is concerned with multivariate data measured on objects divided into two or more known groups. The objective may be to classify further objects into one or other of these groups or it may be directed towards describing the differences between these groups. In both cases particular interest is in determining which linear combinations of variables best highlight the differences between the groups. As in many areas of statistics, terminology varies between authors, so these linear combinations may be referred to as linear discriminants, or more loosely as discriminant functions or canonical variates or, following Gnanadesikan (1997), as crimcoords. The term *canonical variates* is also used for a closely related (mathematically at least) set of linear combinations of variables arising in canonical correlation analysis. Here we will use the term crimcoords.

9.4.1 Derivation of crimcoords

Suppose there are n_i p-dimensional observations from Group i, $i = 1,\ldots,k$ and $\sum n_i = n$. Let the data matrix of observations from the ith group be $\boldsymbol{X}_i'$, so the observations are $\{\boldsymbol{x}_{ij}; j = 1,\ldots,n_i, i = 1,\ldots,k\}$ (so $\boldsymbol{x}_{ij}$ is a $p \times 1$ vector).
Let $\bar{\boldsymbol{x}}_i = \boldsymbol{X}_i \boldsymbol{\iota}_{\mathbf{n}}/n$ be the group i mean and

let $\boldsymbol{S}_i = \dfrac{1}{n_i - 1}(\boldsymbol{X}_i - \overline{\boldsymbol{X}_i})(\boldsymbol{X}_i - \overline{\boldsymbol{X}_i})'$ be the within-group i variance and

$\boldsymbol{W} = \dfrac{1}{n-k}\sum_{i=1}^{k}(n_i - 1)\boldsymbol{S}_i$ be the pooled within-groups variance.

Define $\boldsymbol{B} = \dfrac{1}{k-1}\sum_{i=1}^{k}(\bar{\boldsymbol{x}}_i - \bar{\boldsymbol{x}})(\bar{\boldsymbol{x}}_i - \bar{\boldsymbol{x}})'$ the between-groups variance.

Then $(n-1)\boldsymbol{S} = (\boldsymbol{X} - \overline{\boldsymbol{X}})(\boldsymbol{X} - \overline{\boldsymbol{X}})' = (n-k)\boldsymbol{W} + (k-1)\boldsymbol{B}$ which is the 'multivariate analysis of variance'.

The first crimcoord is defined as that vector $\boldsymbol{a}_1$ which maximizes the ratio $\boldsymbol{a}_1'\boldsymbol{B}\boldsymbol{a}_1/\boldsymbol{a}_1'\boldsymbol{W}\boldsymbol{a}_1$ which highlights the difference between the groups to the greatest possible extent in the sense that it maximizes the usual F-statistic for testing the hypothesis of equal group means in a one-way analysis of variance. The problem of maximizing this ratio was considered in Example 7 (ii) where it is shown that the maximum is achieved when $\boldsymbol{a}_1$ is the [right] eigenvector of $\boldsymbol{W}^{-1}\boldsymbol{B}$ corresponding to its largest eigenvector. In passing, note that the function $\boldsymbol{a}_1\boldsymbol{x}$ is referred to as ***Fisher's linear discriminant function***. The complete set of eigenvectors of $\boldsymbol{W}^{-1}\boldsymbol{B}$ corresponding to its r non-zero eigenvalues is the set of crimcoords or canonical variates. The latter term is used in particular in the context of interpreting the loadings of the variables so as to describe the nature of the difference between the groups. In general $r = \min(p, k-1)$ (unless there are collinearities between the group centroids).

In general, the crimcoords $\boldsymbol{a}_i$ are not orthogonal (though it is easy to show that $\boldsymbol{a}_i\boldsymbol{W}\boldsymbol{a}_j = 0$ if $i \neq j$ if the eigenvalues are distinct). Nevertheless, it is usual to plot the data rotated onto the non-orthogonal set of crimcoords (though drawing the axes representing the crimcoords perpendicular to each other) by constructing a scatterplot of $(\boldsymbol{X} - \overline{\boldsymbol{X}})'\boldsymbol{a}_i$ against $(\boldsymbol{X} - \overline{\boldsymbol{X}})'\boldsymbol{a}_j$ and perhaps add data measured on objects of unknown classification to such a plot as supplementary points. Again note the mean correction which is given explicitly here for conformity with the **R** function `lda(.)` in the `MASS` library. Supplementary observations need to be mean corrected by the means of the data used to determine the crimcoords.

9.4.2 Calculation in R

R provides a function `lda(.)` for performing linear discriminant analysis. It is contained in the `MASS` library so this has to be opened with `library(MASS)` before using `lda(.)`. The function may not work quite as expected because it incorporates a facility for specifying prior probabilities of classification. These are used both as weightings for the individual groups in calculation of the within-groups variance $\boldsymbol{W}$ and as prior probabilities in classifying further observations. By default the function takes the prior probabilities as proportional to the group sizes n_i so yielding the within-groups variance $\boldsymbol{W}$ as described above but then uses these weights also as prior probabilities in classifying further observations with the function `predict.lda(.)`. This can be overridden so, for example, while still using the group sizes for calculation of $\boldsymbol{W}$ (and hence the crimcoords) but using equal prior probabilities for classifying further observations by specifying `prior=c(rep(1,k)/k` in the call to `predict.lda(.)`. Alternatively the generic function `predict(.)` can be used provided its argument is of `class lda`.

We show first how to calculate the within- and between-groups variances using elementary matrix manipulations in two ways. The first involves constructing a group indicator matrix which is a little cumbersome and the second uses the advanced functions `split(.)` and `lapply(.)` whose details can be obtained from the `help()` system. We suppose the data are given in a $n \times (p-1)$ data frame `A` where the first p columns contain the observations and the final column contains a group indicator. We also suppose that group sizes n_i are known.

```
> X<-t(as.matrix(A[,1:p]) ## extract the data matrix X'
> group<-factor(A[,p+1]   ## set up the group indicator
> nobs<-c(n1,n2,...,nk)   ## set up a vector of group sizes
> n<-sum(nobs)            ## total sample size
> G<-matrix(c(rep(1,nobs[1]),rep(0,n),rep(1,nobs[2]),
+ rep(0,n),...,rep(1,nk)),n,k) ## G is the group indicator
        ## matrix G_ij=1 if and only if case i is in group j
F<-  G<-matrix(c(rep(1,nobs[1])/nobs[1],rep(0,n),
+ rep(1,nobs[2])/nobs[2],rep(0,n),...,rep(1,nk/nobs[k])),n,k)
        ## F is used in calculating group means in M
> M<-F%*%t(X)
> xbar<-t(one)%*%t(X)/n  ## overall mean vector
> W<-t(t(X)-G%*%M)%*%(t(X)-G%*%M)/(n-k) ##
> B<-t(G%*%M-one%*%xbar)%*%(G%*%M-one%*%xbar)/(k-1)
       ## W and B are within and between groups variances
```

A useful check on the calculations is to ensure that the analysis of variance is satisfied, i.e., that `(n-1)*var(t(X))=(n-k)*W+(k-1)*B` (up to rounding errors). Note that the line calculating $\boldsymbol{B}$ may appear to be different from the formula given in the previous section $\boldsymbol{B} = \frac{1}{k-1}\sum_{i=1}^{k}(\bar{\boldsymbol{x}}_i - \bar{\boldsymbol{x}})(\bar{\boldsymbol{x}}_i - \bar{\boldsymbol{x}})'$ but the **R** calculations are in terms of matrices with n_i copies of $(\bar{\boldsymbol{x}}_i - \bar{\boldsymbol{x}})$.

If the number of groups k is large (i.e., more than 5 or 6, say) then calculation of the group indicator G is cumbersome and a quicker way of doing this ***if the group sizes are equal*** is to split the dataframe into separate group dataframes with the function `split(.)` (note that the argument of `split(.)` must be of class `"dataframe"` and not of class `"matrix"`) followed by calculation of variances of each group dataframe with the function `lapply(.)` as follows:

```
Xdat<-split(A[,1:p],A[,p+1])
      ## note A is a dataframe not a matrix
      ## creates a list of dataframes, one for each group
Xdat<-lapply(Xdat,as.matrix)
      ## convert group dataframes to matrices
Xvar<-lapply(Xdat,var)
      ## find variance of each group
W<-Reduce("+",Xvar)*(n/k-1)/(n-k)
      ## assumes groups are of equal sizes
B<-((n-1)*var(X)-(W*(n-k))/k-1
```

Having obtained $\boldsymbol{W}$ and $\boldsymbol{B}$, the within- and between-groups variances, it is then easy to calculate any of the statistics used for multivariate analysis of variance referred to in §9.2.5.2and §9.2.7.3. The crimcoords can be obtained by the first $k-1$ vectors given by `eigen(solve(W)%*%B)$vectors` but note that the scaling used for the eigenvectors ensures that $\boldsymbol{y}'\boldsymbol{y} = 1$ whereas the scaling used for the crimcoords produced by the function `lda(.)` ensures that $\boldsymbol{y}'\boldsymbol{W}\boldsymbol{y} = 1$, i.e., multiplying the eigenvectors produced by `eigen(solve(W)%*%B)$vectors` by a factor $\boldsymbol{y}'\boldsymbol{y}/\boldsymbol{y}'\boldsymbol{W}\boldsymbol{y}$ will reproduce the crimcoords from `lda(.)` held in the matrix `lda(.)$scaling`.

The data transformed to crimcoords is given by $Y' = (X - \overline{X})'V$ where V is obtained either from `V=eigen(solve(W)%*%B)$vectors[,1:k-1]` or from `V=lda(.)$scaling`. The difference in scaling will not be apparent if scatterplots of the data referred to crimcoords are produced by the basic **R** command `plot(.)` but willl be noticeable if the `MASS` library routine `eqscplot(.)` is used.

9.5 Canonical Correlation Analysis

Canonical correlation analysis is concerned with investigating the relationship between two sets of variables measured on the same objects. In particular, the aim is to find which linear combination of variables of a $n \times p$ data matrix X' has the maximum correlation with which linear combination of variables of a $n \times q$ data matrix Y' amongst all such liner combinations. For example, in analyzing results of questionnaires eliciting subjects' opinions of a product, one set of variables may reflect the socio-economic aspects of the subjects and the other set may relate to their opinions on various properties of the product.

9.5.1 Derivation of canonical variates

If S_{xx}, S_{yy} and S_{xy} are the sample variances and covariance matrices respectively of X', Y' and (X', Y') and x and y are p-vectors then the correlations between $X'x$ and $Y'y$ is $x'S_{xy}y/\sqrt{x'S_{xx}xy'S_{yy}y}$. It was shown in Example 6 (iv) that this is maximized with respect to x and y by taking x to be the eigenvector of $S_{xx}^{-1}S_{xy}S_{yy}^{-1}S'_{xy}$ corresponding to its largest eigenvalue and y to be the eigenvector of $S_{yy}^{-1}S'_{xy}S_{xx}^{-1}S_{xy}$ corresponding to its largest eigenvalue. These eigenvectors are termed the first ***canonical variates*** of X' and Y'.

If the complete set of eigenpairs of $S_{xx}^{-1}S_{xy}S_{yy}^{-1}S'_{xy}$ is $(u_1, \lambda_1), \ldots, (u_r, \lambda_r)$ where $\lambda_1 \geq \lambda_2 \geq \ldots \geq \lambda_r$ and those of $S_{yy}^{-1}S'_{xy}S_{xx}^{-1}S_{xy}$ are $(v_1, \lambda_1), \ldots, (v_r, \lambda_r)$ (noting the eigenvalues are identical) where $r = \min(p, q)$ then it can be shown that the linear combinations of X' with Y' given by $(u_2, v_2) \ldots, (u_r, v_r)$ maximise the correlation between linear functions of the X' and Y' variables subject to the constraints of orthogonality with earlier ones and thus are termed the canonical variates of X' and Y'. Plots of the [mean corrected] data referred to canonical variates may provide insight into the structure of a relationship between the data sets, i.e., plots of $(X - \overline{X})'u_i$ against $(X - \overline{X})'u_j$ (typically with $j = i + 1$) or $(X - \overline{X})'u_i$ against $(Y - \overline{Y})'v_i$ and $(Y - \overline{Y})'v_i$ against $(Y - \overline{Y})'v_j$ can all be useful in informal investigation the structure of the data.

It can be shown that if the Y' variables are group indicators or a set of binary dummy variables then the canonical variates of the X' variables are precisely the linear discriminants or crimcoords between the groups. This is the reason that the latter are sometimes referred to as canonical variates — they are the linear combinations of the $\boldsymbol{X}'$ variables that 'most highly correlate' with the group structure, i.e., discriminate between them. The fact that a different scaling constraint is used in the analysis is immaterial since the result is invariant to scale.

Note that if the complete $n \times (p+q)$ data matrix is $\boldsymbol{Z}' = \begin{pmatrix} \boldsymbol{X}' & \boldsymbol{Y}' \end{pmatrix}$ then we have

$$\text{var}(\boldsymbol{Z}') = \begin{pmatrix} \boldsymbol{S}_{xx} & \boldsymbol{S}_{xy} \\ \boldsymbol{S}'_{xy} & \boldsymbol{S}_{yy} \end{pmatrix}.$$

9.5.2 Calculation in R

Suppose the data are presented in $n \times p$ and $n \times q$ data matrices $\boldsymbol{X}'$ and $\boldsymbol{Y}'$, then the calculations of the various canonical variates proceed as follows:

```
## set values of p and q
Z<-t(cbind(t(X),t(Y)))
S<-var(t(Z))
Sxx<-S[1:p,1:p]
Sxy<-S[1:p,(p+1):(p+q)]
Syy<-S[(p+1):(p+q),(p+1):(p+q)]
U<-solve(Sxx)%*%Sxy%*%solve(Syy)%*%t(Sxy)
V<-solve(Syy)%*%t(Sxy)%*%solve(Sxx)%*%Sxy
eigen(U)    ## calculate canonical variates for X variables
eigen(V)    ## calculate canonical variates for Y variables
```

The data can then be transformed to canonical variates by subtracting the variable means and multiplying by the matrix of eigenvectors

```
t(X-Xbar)%*%eigen(U)$vectors; t(Y-Ybar)%*%eigen(V)$vectors
```

Note that because the `eigen(.)` function was used for the eigenanalysis the scaling of eigenvectors here is to ensure they have unit length. The difference in scaling is only apparent if the `MASS` library function `eqscplot(.)` is used to produce scatterplots, otherwise `plot(.)` will by default adjust the scaling to fit on the page.

There are two **R** functions for performing canonical correlation analysis: `cancor(.)` in the basic `stats` library (which is always loaded in an **R** session) and `cc(.)` in the `CCA` package which must be installed and then opened before the function is used. The `cancor(.)` function uses a scaling of the eigenvectors to ensure that $\boldsymbol{x}_i'\boldsymbol{S}_{xx}^{-1}\boldsymbol{x}_i = \frac{1}{n-1} = \boldsymbol{y}_j'\boldsymbol{S}_{yy}^{-1}\boldsymbol{y}_j$ and the cc(.) function uses a scaling such that $\boldsymbol{x}_i'\boldsymbol{S}_{xx}^{-1}\boldsymbol{x}_i = 1 = \boldsymbol{y}_j'\boldsymbol{S}_{yy}^{-1}\boldsymbol{y}_j$. The `cc(.)` function in the `CCA` package has the additional advantage of providing the scores of the data on the canonical variates in `cc(t(X),t(Y))$scores$xscores` and `cc(t(X),t(Y))$scores$yscores`.

Neither `cancor(.)` nor `cc(.)` supports the generic function `predict(.)` for adding supplementary points to existing canonical variate plots so a further $m \times (p+q)$ data matrix $(\boldsymbol{A}'\ \boldsymbol{B}')$ needs to be transformed by

```
t(A-Xbarm)%*%eigen(U)$vectors; t(B-Ybarm)%*%eigen(V)$vectors,
```

where $\overline{\boldsymbol{X}}_m$ and $\overline{\boldsymbol{Y}}_m$ are $m \times p$ and $m \times q$ matrices with columns $\bar{\boldsymbol{x}}$ and $\bar{\boldsymbol{y}}$.

9.6 Classical Scaling

A rather different multivariate problem is finding a $n \times p$ data matrix $\boldsymbol{X}'$ whose inter-point distances best match a given $n \times n$ distance matrix $\boldsymbol{D}$. A distance matrix is a symmetric matrix with non-negative entries and all elements on the leading diagonal zero. A distance matrix is said to be ***Euclidean*** if there is indeed a configuration of n points in p-dimensional Euclidean space with distance matrix $\boldsymbol{D}$. The square of the inter-point distance distance between the ith and j^{th} observations in $\boldsymbol{X}'$ (regarded as points in p-dimensional Euclidean space) is $(\boldsymbol{x}_i - \boldsymbol{x}_j)'(\boldsymbol{x}_i - \boldsymbol{x}_j)$.

To produce a configuration of points with $n \times n$ distance matrix given by $\boldsymbol{D}$ first define $\boldsymbol{A}$ by $(a_{ij}) = (-{}^1\!/\!_2 d_{ij}^2)$, i.e., $\boldsymbol{A} = -{}^1\!/\!_2 \boldsymbol{D} \odot \boldsymbol{D}$ (where $\odot$ indicates the Hadamard product, §8.4) and then define $\boldsymbol{B} = \mathbf{H_n} \boldsymbol{A} \mathbf{H_n}$ where $\mathbf{H_n}$ is the $n \times n$ centering matrix (see §2.5.6.1). Then the classical scaling theorem (Mardia et al., 1979) states that if the matrix $\boldsymbol{B}$ is positive semi-definite with rank p with eigenvectors $\boldsymbol{x}_{(i)}$ scaled such that $\boldsymbol{x}'_{(i)}\boldsymbol{x}_{(i)}$ then the data matrix $\boldsymbol{X}' = (\boldsymbol{x}_{(1)}, \boldsymbol{x}_{(2)}, \ldots, \boldsymbol{x}_{(p)})$ has distance matrix given exactly by $\boldsymbol{D}$. Proof of this result is left as an exercise or it may be found in Mardia et al. (1979) or Cox (2005). Note that $\boldsymbol{B}$ is a $n \times n$ matrix so its eigenvectors $\boldsymbol{x}_{(i)}$ are $n \times 1$ vectors and thus $\boldsymbol{X}'$ is a $n \times p$ matrix.

9.6.1 Calculation in R

The function `cmdscale(.)` will produce a configuration of points with given Euclidean distance matrix. By default it produces only the first two dimensions but this can be overridden by including an option `k=p` to produce p dimensions. The following **R** code will produce a solution 'from scratch':

```
  ### set values of n and p ###
> Hn<-diag(rep(1,n))-matrix(rep(1,n*n),n,n)/n
>     ## Hn is nxn centering matrix
> A<--D*D/2  ## note Hadamard product
> B<-Hn%*%A%*%Hn
> B.eig<-eigen(B)
> X<-B.eig$vectors[,1:p] ## take first p eigenvectors
> X<-t(X%*%diag(sqrt(B.eig$values[1:p])))
>     ## X' is required data matrix.
```

If $\boldsymbol{D}$ is not Euclidean then some of the eigenvalues of the matrix $\boldsymbol{B}$ will be negative. The pragmatic solution is to take just those eigenvectors of $\boldsymbol{B}$ corresponding to positive eigenvalues but other possibilities are discussed in Mardia et al. (1979) and Cox (2005). These include using non-metric methods which are provided by function `isoMDS(.)` and `sammon(.)` in the `MASS` library.

9.7 Linear Models

9.7.1 Introduction

This section considers the central topic of linear models. First comes an account of the standard linear regression model with sections on the estimation of parameters and their properties. This is followed by discussion of variations on the standard model and a selection of particular models used in special cases. This is by no means a full account of the topic but is intended to illustrate some of the key results and techniques in the area which would be encountered early in a specialist course on the subject.

9.7.2 The standard linear regression model

The standard linear model for a set of observations $\boldsymbol{y}$ is $\boldsymbol{y} = \boldsymbol{X\beta} + \boldsymbol{\varepsilon}$ where $\boldsymbol{y}$ is a $n \times 1$ vector, $\boldsymbol{X}$ is $n \times p$ ***design matrix*** with $n \geq p$, β is a $p \times 1$ vector of unknown parameters and $\boldsymbol{\varepsilon}$ is a $n \times 1$ random vector with $\mathrm{E}[\boldsymbol{\varepsilon}] = 0$ and $\mathrm{var}(\boldsymbol{\varepsilon}) = \sigma^2\mathbf{I_n}$. It is presumed that $\boldsymbol{X}$ is a known matrix giving the values of p *independent variables* for each observation. Thus the standard model assumes that observations are independent and their values can be partitioned into a systematic [fixed] part and a random part. The contributions to $\boldsymbol{y}$ from $\boldsymbol{X\beta}$ are termed ***fixed effects***. Variants of the standard model might be to let $\mathrm{var}(\boldsymbol{\varepsilon}) = \sigma^2\boldsymbol{V}$ where $\boldsymbol{V}$ is a known non-singular $n \times n$ matrix or to extend the model to include random effects by $\boldsymbol{y} = \boldsymbol{X\beta} + \boldsymbol{Z\gamma} + \boldsymbol{\varepsilon}$ where $\boldsymbol{Z}$ is a $n \times q$ matrix of observations of q random variables and γ is a $n \times q$ of unknown parameters.

9.7.2.1 Estimation of parameters

In §8.3.3.1 it was shown that the value of β which minimizes $(\boldsymbol{y} - \boldsymbol{X\beta})'(\boldsymbol{y} - \boldsymbol{X\beta})$, i.e., the least squares solution of $\boldsymbol{X\beta} = \boldsymbol{y}$ is $\hat{\beta} = \boldsymbol{X}^+\boldsymbol{y}$ where $\boldsymbol{X}^+$ is the Moore–Penrose inverse (§8.3.1) of $\boldsymbol{X}$. If $\rho(\boldsymbol{X}) = p$ then $\boldsymbol{X}^+ = (\boldsymbol{X}'\boldsymbol{X})^{-1}\boldsymbol{X}'$ (see §8.3.1, further key result (ii)) so $\hat{\beta} = (\boldsymbol{X}'\boldsymbol{X})^{-1}\boldsymbol{X}'\boldsymbol{y}$.

An alternative derivation of this is to consider $\Omega = (\boldsymbol{y} - \boldsymbol{X\beta})'(\boldsymbol{y} - \boldsymbol{X\beta})$ (the sum of squares) and differentiate this with respect to β to obtain $-2\boldsymbol{X}'\boldsymbol{y} + 2\boldsymbol{X}'\boldsymbol{X\beta}$. Setting this equal to zero gives

$$\boldsymbol{X}'\boldsymbol{X\beta} = \boldsymbol{X}'\boldsymbol{y} \tag{9.1}$$

Equations (9.1) are known as the ***normal equations*** or as the ***least squares equations*** (in the plural because β is a vector of p unknowns). It can be shown (Guttman, 1982) that any solution of (9.1) is of the form $\boldsymbol{GX}'\boldsymbol{y}$ with $\boldsymbol{G}$ a generalized inverse of $\boldsymbol{X}'\boldsymbol{X}$. If $\boldsymbol{X}'\boldsymbol{X}$ is non-singular, i.e., if $\boldsymbol{X}$ is of full column rank and $\rho(\boldsymbol{X}) = p$, then this gives $\hat{\beta} = (\boldsymbol{X}'\boldsymbol{X})^{-1}\boldsymbol{X}'\boldsymbol{y}$. The minimized sum of squares is $\Omega_{\min} = (\boldsymbol{y} - \boldsymbol{X}\hat{\beta})'(\boldsymbol{y} - \boldsymbol{X}\hat{\beta}) = \boldsymbol{y}'\boldsymbol{y} - \hat{\beta}\boldsymbol{X}'\boldsymbol{X}\hat{\beta}$, noting $\hat{\beta}\boldsymbol{X}'\boldsymbol{X}\hat{\beta} = \boldsymbol{y}'\boldsymbol{X}\hat{\beta}$. The quantity $\boldsymbol{y}'\boldsymbol{y} - \hat{\beta}\boldsymbol{X}'\boldsymbol{X}\hat{\beta}$ is referred to as the ***residual sum of squares***.

If we make the additional assumption that $\varepsilon \sim N_n(\mathbf{0}, \sigma^2\mathbf{I_n})$ then $\boldsymbol{y} \sim N_n(\boldsymbol{X}\beta, \sigma^2\mathbf{I_n})$ and so the likelihood is

$$\text{Lik}(\beta, \sigma; \boldsymbol{y}) = \frac{1}{(2\pi)^{np/2}\sigma^n} \exp\left\{-\frac{(\boldsymbol{y}-\boldsymbol{X}\beta)'(\boldsymbol{y}-\boldsymbol{X}\beta)}{2\sigma^{2n}}\right\}$$

and so the maximum likelihood estimate of β is obtained by minimizing $(\boldsymbol{y} - \boldsymbol{X}\beta)'(\boldsymbol{y} - \boldsymbol{X}\beta)$ and thus the least squares estimator $\hat{\beta}$ is also the maximum likelihood estimator.

9.7.2.2 The Gauss–Markov theorem

$\text{E}[\hat{\beta}] = (\boldsymbol{X}'\boldsymbol{X})^{-1}\boldsymbol{X}'\text{E}[\boldsymbol{y}] = (\boldsymbol{X}'\boldsymbol{X})^{-1}\boldsymbol{X}'\boldsymbol{X}\beta = \beta$ and $\text{var}(\hat{\beta}) = \text{var}((\boldsymbol{X}'\boldsymbol{X})^{-1}\boldsymbol{X}'\boldsymbol{y}) = (\boldsymbol{X}'\boldsymbol{X})^{-1}\boldsymbol{X}'\text{var}(\boldsymbol{y})((\boldsymbol{X}'\boldsymbol{X})^{-1}\boldsymbol{X}')' = \sigma^2(\boldsymbol{X}'\boldsymbol{X})^{-1}$. Thus $\hat{\beta}$ is unbiased for β and further the ith element of $\hat{\beta}$, $\hat{\beta}_i$, is the minimum variance unbiased linear estimator of β_i as shown by the Gauss–Markov theorem: if $\theta = \boldsymbol{a}'\beta$ then the minimum variance unbiased estimator of θ which is a linear function of $\boldsymbol{y}$ is given by $\boldsymbol{a}'\hat{\beta}$. To prove this note that if $\hat{\theta} = c + \boldsymbol{b}'\boldsymbol{y}$ is an unbiased estimator for θ for any value of β (in particular $\beta = \mathbf{0}$) then we must have $c = 0$ and $\boldsymbol{a} = \boldsymbol{X}'\boldsymbol{b}$. $\text{var}(\hat{\theta}) = \sigma^2\boldsymbol{b}'\boldsymbol{b}$ and $\boldsymbol{b}'\boldsymbol{b} = (\boldsymbol{b} - \boldsymbol{X}(\boldsymbol{X}'\boldsymbol{X})^{-1}\boldsymbol{a})'(\boldsymbol{b} - \boldsymbol{X}(\boldsymbol{X}'\boldsymbol{X})^{-1}\boldsymbol{a}) + \boldsymbol{a}'(\boldsymbol{X}'\boldsymbol{X})^{-1}\boldsymbol{a}$, so $\boldsymbol{b}'\boldsymbol{b}$ is minimized when $\boldsymbol{b} = \boldsymbol{X}(\boldsymbol{X}'\boldsymbol{X})^{-1}\boldsymbol{a}$ and the minimum value is $\boldsymbol{a}'(\boldsymbol{X}'\boldsymbol{X})^{-1}\boldsymbol{a}$. Since $\text{var}(\boldsymbol{a}'\hat{\beta}) = \boldsymbol{a}'\text{var}(\hat{\beta})\boldsymbol{a} = \sigma^2\boldsymbol{a}'(\boldsymbol{X}'\boldsymbol{X})^{-1}\boldsymbol{a}$ we have established that $\boldsymbol{a}'\hat{\beta}$ is the minimum variance unbiased linear estimator of $\boldsymbol{a}'\beta$. Taking $\boldsymbol{a} = (0, 0, \ldots, 0, 1, 0, \ldots, 0)'$ (so $a_j = 1$ if $i = j$, 0 otherwise) gives the result that the ith element of $\hat{\beta}$, $\hat{\beta}_i$, is the minimum variance unbiased linear estimator of β_i.

9.7.2.3 Residuals and estimation of error variance

Having obtained an estimate $\hat{\beta}$ of β the fitted values are $\hat{\boldsymbol{y}} = \boldsymbol{X}\hat{\beta}$. The differences between the observed and fitted values are referred to as the residuals $\boldsymbol{e}$. We have $\boldsymbol{e} = \boldsymbol{y} - \boldsymbol{X}\hat{\beta} = \boldsymbol{y} - \boldsymbol{X}(\boldsymbol{X}'\boldsymbol{X})^{-1}\boldsymbol{X}'\boldsymbol{y} = (\mathbf{I_n} - \boldsymbol{X}(\boldsymbol{X}'\boldsymbol{X})^{-1}\boldsymbol{X}')\boldsymbol{y}$. The matrix $\boldsymbol{M} = \mathbf{I_n} - \boldsymbol{X}(\boldsymbol{X}'\boldsymbol{X})^{-1}\boldsymbol{X}'$ is symmetric and idempotent (§2.5.6) and $\boldsymbol{M}\boldsymbol{X} = \mathbf{0}$. The matrix $\boldsymbol{H} = \boldsymbol{X}(\boldsymbol{X}'\boldsymbol{X})^{-1}\boldsymbol{X}'$ is symmetric and idempotent and it also plays an important role and is known as the hat matrix (because for example $\boldsymbol{H}\boldsymbol{y} = \hat{\boldsymbol{y}}$). Both $\boldsymbol{H}$ and $\boldsymbol{M}$ are fundamental in calculating measures of leverage, influence and Cook's distance (concepts concerned with assessing how individual observations contribute to model estimation). This is beyond the scope of this text but is well covered in the more recent specialist texts on linear models such as Faraway (2014).

$\text{var}(\boldsymbol{e}) = \text{var}(\boldsymbol{M}\boldsymbol{y}) = \boldsymbol{M}'\text{var}(\boldsymbol{y})\boldsymbol{M} = \boldsymbol{M}\sigma^2\mathbf{I_n}\boldsymbol{M} = \sigma^2\boldsymbol{M}^2 = \sigma^2\boldsymbol{M}$.
Because $\text{E}[\boldsymbol{e}] = \mathbf{0}$ we have $\text{E}[e_i^2] = \sigma^2 m_{ii}$ where m_{ii} is the ith diagonal element of $\boldsymbol{M}$. Now $\text{E}[\boldsymbol{e}'\boldsymbol{e}] = \sum \text{E}[e_i^2] = \sigma^2 \sum m_{ii} = \sigma^2 \text{tr}(\boldsymbol{M}) = \sigma^2(n - \text{tr}(\boldsymbol{X}(\boldsymbol{X}'\boldsymbol{X})^{-1}\boldsymbol{X}')) = \sigma^2(n - \text{tr}((\boldsymbol{X}'\boldsymbol{X})^{-1}\boldsymbol{X}'\boldsymbol{X})) = \sigma^2(n - \text{tr}(\mathbf{I_p})) = (n-p)\sigma^2$ (note use of results on trace of sums and products from §2.4). Thus if $\hat{\sigma}^2 = \boldsymbol{e}'\boldsymbol{e}/(n-p)$ then $\hat{\sigma}^2$ is an unbiased estimate of σ^2.

Example 9.1. The single factor model
Suppose $\boldsymbol{y} = \boldsymbol{X}\beta + \varepsilon$ where $\boldsymbol{X}$ is the $n \times p$ matrix

$$\boldsymbol{X} = \begin{pmatrix} \iota_{\mathbf{n_1}} & \mathbf{0_{n_1}} & \cdots & \mathbf{0_{n_1}} \\ \mathbf{0_{n_2}} & \iota_{\mathbf{n_2}} & \cdots & \mathbf{0_{n_2}} \\ \vdots & \vdots & \ddots & \vdots \\ \mathbf{0_{n_p}} & \mathbf{0_{n_p}} & \cdots & \iota_{\mathbf{n_p}} \end{pmatrix} \quad \text{where } n = \sum_{i=1}^{p} n_i \text{ and } \mathbf{0_{n_i}} \text{ is a vector of } n_i \text{ 0s,}$$

and $\boldsymbol{y} = (y_{11}, y_{12}, \ldots, y_{1n_1}, y_{21}, \ldots, y_{2n_2}, \ldots, y_{pn_p})'$ then

$$\boldsymbol{X}'\boldsymbol{X} = \begin{pmatrix} n_1 & 0 & \cdots & 0 \\ 0 & n_2 & \cdots & 0 \\ \vdots & \vdots & \ddots & \vdots \\ 0 & 0 & \cdots & n_p \end{pmatrix} \quad \text{so} \quad (\boldsymbol{X}'\boldsymbol{X})^{-1} = \begin{pmatrix} 1/n_1 & 0 & \cdots & 0 \\ 0 & 1/n_2 & \cdots & 0 \\ \vdots & \vdots & \ddots & \vdots \\ 0 & 0 & \cdots & 1/n_p \end{pmatrix}.$$

Then $\hat{\beta} = (\bar{y}_1, \bar{y}_2, \ldots, \bar{y}_p)'$.

Example 9.2: The single factor model with overall mean
As an example of a case where the design matrix is not of full rank consider now the model $\boldsymbol{y} = \boldsymbol{X}\beta + \varepsilon$ where $\boldsymbol{X}$ is the $n \times p$ (where $p = k+1$) matrix

$$\boldsymbol{X} = \begin{pmatrix} \iota_{\mathbf{n_1}} & \iota_{\mathbf{n_1}} & \mathbf{0_{n_1}} & \cdots & \mathbf{0_{n_1}} \\ \iota_{\mathbf{n_2}} & \mathbf{0_{n_2}} & \iota_{\mathbf{n_2}} & \cdots & \mathbf{0_{n_2}} \\ \vdots & \vdots & \vdots & \ddots & \vdots \\ \iota_{\mathbf{n_p}} & \mathbf{0_{n_p}} & \mathbf{0_{n_k}} & \cdots & \iota_{\mathbf{n_k}} \end{pmatrix} \quad \text{i.e., the left hand column of } \boldsymbol{X} \text{ is } \iota_{\mathbf{n}}$$

and $\beta = (\mu, \mu_1, \mu_2, \ldots, \mu_k)'$. Clearly $\boldsymbol{X}$ is not of full column rank (i.e., $\rho(\boldsymbol{X}) < p$) since the left hand column is the sum of all the others so $\boldsymbol{X}'\boldsymbol{X}$ is singular. The least squares solution is still given by $\boldsymbol{X}^+\boldsymbol{y}$ but it is no longer true that $\boldsymbol{X}^+ = (\boldsymbol{X}'\boldsymbol{X})^{-1}\boldsymbol{X}'$ since this is only true in general if $\boldsymbol{X}$ is of full column rank. If all the group sizes n_i are equal to m so $n = mk$ then $\boldsymbol{X}^+ = (\boldsymbol{X}_1^+ \boldsymbol{X}_2^+ \cdots \boldsymbol{X}_k^+)$ where $\boldsymbol{X}_i^+$ is the $p \times m$ matrix with first row with all elements $1/(n+m)$, the $(i+1)^{th}$ row equal to $k/(n+m)$ and all other rows equal to $-1/(n+m)$,

$$\text{i.e., } \boldsymbol{X}_i^+ = \frac{1}{n+m} \begin{pmatrix} 1 & 1 & \cdots & 1 \\ -1 & -1 & \cdots & -1 \\ k & k & \cdots & k \\ \vdots & \vdots & \vdots & \vdots \\ -1 & -1 & \cdots & -1 \end{pmatrix} \leftarrow (i+1)^{th} \text{ row}$$

(proof of this is left as an exercise). This leads to estimates $\hat{\beta}_1 = \hat{\mu} = n\bar{y}/(n+m)$ and $\hat{\beta}_i = \hat{\mu}_i = \bar{y}_i - n\bar{y}/(n+m)$, $i = 1, \ldots, k$. $\mathrm{E}[\hat{\beta}] = \boldsymbol{X}^+\mathrm{E}[\boldsymbol{y}] = \boldsymbol{X}^+\boldsymbol{X}\beta$ and $\boldsymbol{X}^+\boldsymbol{X}$ is a $p \times p$

symmetric matrix $\mathbf{Z} = (z_{ij})$ with

$$z_{ij} = \begin{cases} +k/(k+1) & \text{if } i = j \\ +1/(k+1) & \text{if } i = 1 \text{ or if } j = 1 \ (i \neq j) \\ -1/(k+1) & \text{otherwise.} \end{cases}$$

So $\boldsymbol{X}^{+}\boldsymbol{X}\beta \neq \beta$ and thus the least squares estimate is biased.

9.7.3 Least squares estimation with constraints

An alternative way of handling such a non-full rank case is to impose constraints on the parameters. Suppose $\rho(\boldsymbol{X}) = p - q$, then impose q constraints on the parameters by $\boldsymbol{B}\beta = \delta$ where $\boldsymbol{B}$ is a $q \times p$ matrix with rows linearly independent of all the rows of $\boldsymbol{X}$ and $\rho(\boldsymbol{B}) = q$ and δ is a known $q \times 1$ vector. Constraints of this form are known as ***identifiability constraints***. We will show that the estimator

$$\tilde{\beta} = (\boldsymbol{X}'\boldsymbol{X} + \boldsymbol{B}'\boldsymbol{B})^{-1}\boldsymbol{X}'\boldsymbol{y} + (\boldsymbol{X}'\boldsymbol{X} + \boldsymbol{B}'\boldsymbol{B})^{-1}\boldsymbol{B}'\delta$$

satisfies the least squares Equations (9.1) and satisfies the constraints and is unbiased for β. To prove this first consider the matrix

$$\boldsymbol{W} = \begin{pmatrix} \boldsymbol{X} \\ \boldsymbol{B} \end{pmatrix}; \quad \boldsymbol{W}'\boldsymbol{W} = (\boldsymbol{X}' \ \ \boldsymbol{B}') \begin{pmatrix} \boldsymbol{X} \\ \boldsymbol{B} \end{pmatrix} = (\boldsymbol{X}'\boldsymbol{X} + \boldsymbol{B}'\boldsymbol{B}).$$

$\boldsymbol{W}$ is clearly of rank p because the rows of $\boldsymbol{B}$ are linearly independent of those of $\boldsymbol{X}$, so $\rho(\boldsymbol{X}'\boldsymbol{X} + \boldsymbol{B}'\boldsymbol{B}) = \rho(\boldsymbol{W}'\boldsymbol{W}) = \rho(\boldsymbol{W}) = p$ and so $(\boldsymbol{X}'\boldsymbol{X}+\boldsymbol{B}'\boldsymbol{B})$ is non-singular and thus invertible.

$$\begin{aligned} \text{Next, } \mathrm{E}[\tilde{\beta}] &= (\boldsymbol{X}'\boldsymbol{X} + \boldsymbol{B}'\boldsymbol{B})^{-1}\boldsymbol{X}'\boldsymbol{X}\beta + \gamma \ \ (\text{where } \gamma = (\boldsymbol{X}'\boldsymbol{X} + \boldsymbol{B}'\boldsymbol{B})^{-1}\boldsymbol{B}'\delta) \\ &= (\boldsymbol{X}'\boldsymbol{X} + \boldsymbol{B}'\boldsymbol{B})^{-1}(\boldsymbol{X}'\boldsymbol{X} + \boldsymbol{B}'\boldsymbol{B} - \boldsymbol{B}'\boldsymbol{B})\beta + \gamma \\ &= (\mathbf{I_p} - (\boldsymbol{X}'\boldsymbol{X} + \boldsymbol{B}'\boldsymbol{B})^{-1}\boldsymbol{B}'\boldsymbol{B})\beta + \gamma \\ &= \beta - (\boldsymbol{X}'\boldsymbol{X} + \boldsymbol{B}'\boldsymbol{B})^{-1}\boldsymbol{B}'(\boldsymbol{B}\beta) + \gamma \\ &= \beta - (\boldsymbol{X}'\boldsymbol{X} + \boldsymbol{B}'\boldsymbol{B})^{-1}\boldsymbol{B}'\delta + \gamma = \beta - \gamma + \gamma = \beta \end{aligned}$$

and thus $\tilde{\beta}$ is an unbiased estimate of β.

Since $\rho(\boldsymbol{X}) = p - q$ there are $p - q$ linearly independent rows of $\boldsymbol{X}$. Let $\boldsymbol{P}$ be the permutation matrix such that

$$\boldsymbol{P}\boldsymbol{X} = \begin{pmatrix} \boldsymbol{X}_1 \\ \boldsymbol{X}_2 \end{pmatrix} \ \text{ where } \rho(\boldsymbol{X}_1) = p - q \ \text{ and } \ \boldsymbol{X}_2 = \boldsymbol{F}\boldsymbol{X}_2 \ \text{ for some } \boldsymbol{F}.$$

The matrix $\boldsymbol{Z} = \begin{pmatrix} \boldsymbol{X}_1 \\ \boldsymbol{B} \end{pmatrix}$ is of full rank p and so is invertible.

$$\text{Let } \begin{pmatrix} \boldsymbol{X}_1 \\ \boldsymbol{B} \end{pmatrix}^{-1} = (\boldsymbol{V}_1' \ \ \boldsymbol{V}_2') \ \text{ so } \mathbf{I_p} = \begin{pmatrix} \boldsymbol{X}_1 \\ \boldsymbol{B} \end{pmatrix} (\boldsymbol{V}_1' \ \ \boldsymbol{V}_2') = \begin{pmatrix} \boldsymbol{X}_1\boldsymbol{V}_1 & \boldsymbol{X}_1\boldsymbol{V}_2 \\ \boldsymbol{B}\boldsymbol{V}_1 & \boldsymbol{B}\boldsymbol{V}_2 \end{pmatrix}.$$

So $\boldsymbol{X}_1\boldsymbol{V}_2 = \mathbf{0}$ and $\boldsymbol{B}\boldsymbol{V}_2 = \mathbf{I_q}$.

Now $PXV_2 = \begin{pmatrix} X_1V_2 \\ FX_1V_2 \end{pmatrix} = \mathbf{0}$ so $XV_2 = \mathbf{0}$. Thus$(X'X+B'B)V_2 = B'BV_2 = B'$

and so $V_2 = (X'X+B'B)^{-1}B'$ and hence we have

$$X(X'X+B'B)^{-1}B' = XV_2 = \mathbf{0} \quad \text{and} \quad B(X'X+B'B)^{-1}B' = BV_2 = \mathbf{I_q}$$

$$\text{so } B\tilde{\beta} = B(X'X+B'B)^{-1}X'y + B(X'X+B'B)^{-1}B'\delta = \mathbf{0}y + \mathbf{I_q}\delta = \delta$$

and thus $\tilde{\beta}$ satisfies the constraint that $B\beta = \delta$.
Finally, we show that $(X'X+B'B)^{-1}$ is a generalized inverse of $X'X$. This fact follows because $X'X(X'X+B'B)^{-1}X' = [(X'X+B'B) - B'B](X'X+B'B)^{-1}X' = X' - \mathbf{0} = X'$, so $X'X(X'X+B'B)^{-1}X'X = X'X$ and thus $(X'X+B'B)^{-1}$ satisfies the first of the Moore–Penrose conditions (§8.3.1) and so is a generalized inverse of $X'X$. Consequently $(X'X+B'B)^{-1}X'y$ satisfies the least squares Equations (9.1) and since $X'X(X'X+B'B)^{-1}B'\delta = 0$ we conclude that $\tilde{\beta}$ satisfies the least squares Equations (9.1). A considerably shorter proof of this and related results using linear spaces is given in Puntanen et al. (2011).

Example 9.3: The single factor model with overall mean (continued)
To illustrate this, we continue the example above with $\beta = (\mu, \mu_1, \mu_2, \ldots, \mu_k)'$ and impose the identifiability constraint that $\sum_{i=i}^{k} \mu_i = 0$, i.e., $B\beta = 0$ where $B = (0, \iota_{\mathbf{k}}')$. Then $(X'X+B'B)$ is a symmetric $p \times p$ matrix $Z = (z_{ij})$ with

$$z_{ij} = \begin{cases} n & \text{if } i = j = 1 \\ k & \text{if } i = j > 1 \\ m & \text{if } i = 1 \text{ or if } j = 1 \ (i \neq j) \\ 1 & \text{otherwise.} \end{cases}$$

and then $(X'X+B'B)^{-1}X' = (Y_1\ Y_2 \cdots Y_k)$ where Y_i is the $p \times m$ matrix with all elements on the first row equal to $1/n$, those on the $(i+1)^{th}$ row equal to m/n and on all other rows equal to $-1/n$,

$$\text{i.e., } Y_i = \frac{1}{n}\begin{pmatrix} 1 & 1 & \cdots & 1 \\ -1 & -1 & \cdots & -1 \\ m & m & \cdots & m \\ \vdots & \vdots & \vdots & \vdots \\ -1 & -1 & \cdots & -1 \end{pmatrix} \leftarrow (i+1)^{th} \text{ row}$$

(the proof of this is left as an exercise). This leads to estimates $\hat{\beta}_1 = \hat{\mu} = \bar{y}$ and $\hat{\beta}_{i+1} = \hat{\mu}_i = \bar{y}_i - \bar{y}$, $i = 1, \ldots, k$.

Calculation of this in **R** is straightforward:

```
p<-5 ; m<-3 ; n<-m*(p-1)## set values of p and m
X<-matrix(c(rep(1,n),rep(c(rep(1,m),rep(0,n)),p-2),
```

```
  rep(1,m)),n,p)             ## design matrix
B<-matrix(c(0,rep(1,p-)),1,p)  ## constraint matrix
beta<-solve(Z)%*%t(X)%*%y ## estimate of beta with data in y
```

9.7.4 Crossover trials

In a crossover clinical trial to compare two or more treatments (see, for example, Jones and Kenward, 2003) subjects are divided into groups and each subject receives each treatment but the order of treatments in different groups is different. For example in a trial to compare two treatments A and B the first group would receive treatment A and then after a period of time change to treatment B. Those in the second group receive treatment B in the first period followed by treatment A in the second period. The possible responses could include an effect due to the treatment, an effect due to the period and potentially a carryover effect of one treatment from the first period having a residual effect in the second period.

Suppose there are m subjects in Group 1 receiving the treatments in the order A $\rightarrow$ B and n in group 2 receiving them in the order B$\rightarrow$ A and that the fixed effects to be considered are μ, the overall mean, the relative treatment effects $\tau_A,\ \tau_B,\ \tau_A+\tau_B=0$, the relative period effect $\pi_1,\ \pi_2,\ \pi_1+\pi_2=0$ and relative carryover effects $\lambda_A,\ \lambda_B, \lambda_A+\lambda_B=0$ and $\alpha_i,\ i=1,2,\dots,m,m+1,m+2,\dots,m+n$, $\sum\alpha_i=0$ the fixed effects of subject i. If $\boldsymbol{y}=(y_1,y_2,\dots,y_{2(m+n)})'$ then the model can be written as $\boldsymbol{y}=\boldsymbol{X}\beta+\varepsilon$, where, $\beta=(\mu,\tau_A,\tau_B,\pi_1,\pi_2,\lambda_1,\lambda_2,\alpha_1,\alpha_2,\dots,\alpha_{(m+n)})'$, $\varepsilon\sim N_{2(m+n)}(\mathbf{0},\ \sigma^2\mathbf{I_{2(m+n)}})$ and μ

$$\boldsymbol{X}=\begin{pmatrix}
1 & 1 & 0 & 1 & 0 & 1 & 0 & 1 & 0 & 0 & \cdots & \cdots & \cdots & 0\\
1 & 1 & 0 & 1 & 0 & 1 & 0 & 0 & 1 & 0 & \cdots & \cdots & \cdots & 0\\
\vdots & \vdots & \vdots & \vdots & \vdots & \vdots & \vdots & \vdots & \ddots & \ddots & \ddots & \vdots & \vdots & \vdots\\
1 & 1 & 0 & 1 & 0 & 1 & 0 & 0 & \cdots & \ddots & 1 & 0 & \cdots & 0\\
1 & 0 & 1 & 1 & 0 & 0 & 1 & 0 & 0 & 0 & 0 & 1 & \cdots & 0\\
\vdots & \vdots & \vdots & \vdots & \vdots & \vdots & \vdots & \cdots & \cdots & \cdots & \cdots & \ddots & \ddots & 0\\
1 & 0 & 1 & 1 & 0 & 0 & 1 & \cdots & \cdots & \cdots & \cdots & \cdots & 0 & 1\\
1 & 0 & 1 & 0 & 1 & 1 & 0 & 1 & 0 & 0 & \cdots & \cdots & \cdots & 0\\
\vdots & \vdots & \vdots & \vdots & \vdots & \vdots & \ddots & \ddots & \ddots & \vdots & \vdots & \vdots & \vdots & \vdots\\
1 & 0 & 1 & 0 & 1 & 1 & 0 & 0 & \cdots & 1 & 0 & 0 & \cdots & 0\\
1 & 1 & 0 & 0 & 1 & 0 & 1 & 0 & \cdots & 0 & 1 & 0 & \cdots & 0\\
\vdots & \vdots & \vdots & \vdots & \vdots & \vdots & \vdots & \vdots & \vdots & \vdots & \ddots & \ddots & \ddots & \vdots\\
1 & 1 & 0 & 0 & 1 & 0 & 1 & 0 & \cdots & \cdots & \cdots & 0 & 1 & 0\\
1 & 1 & 0 & 0 & 1 & 0 & 1 & 0 & \cdots & \cdots & \cdots & \cdots & 0 & 1
\end{pmatrix}$$

with identifiability constraints $\boldsymbol{B}\beta = 0$ where

$$\boldsymbol{B} = \begin{pmatrix} 0 & 1 & 1 & 0 & 0 & 0 & 0 & 0 & \cdots & 0 \\ 0 & 0 & 0 & 1 & 1 & 0 & 0 & 0 & \cdots & 0 \\ 0 & 0 & 0 & 0 & 0 & 1 & 1 & 0 & \cdots & 0 \\ 0 & 0 & 0 & 0 & 0 & 0 & 0 & & \iota'_{\mathbf{2(m+n)}} & \end{pmatrix}.$$

Here the first m rows of the design matrix $\boldsymbol{X}$ relate to the subjects in group 1 in the first period, the next n rows to those in the second group in the first period and then the next m and n rows to the first and second groups respectively in the second period. Then the estimates of the parameters β are given by $\hat{\beta} = (\boldsymbol{X}'\boldsymbol{X} + \boldsymbol{B}'\boldsymbol{B})^{-1}\boldsymbol{X}'\boldsymbol{y}$ (see §9.7.3).

It is more usual to regard the patients as randomly selected from a wider population and thus that the effect of patient k is a random variable with a normal distribution $N(0, \phi^2)$. If we write $\tau = \tau_A = -\tau_B$, $\pi = \pi_1 = -\pi_2$, $\lambda = \lambda_A = -\lambda_B$ (thus eliminating the need for identifiability constraints) this leads to a mixed effects model: $\boldsymbol{y} = \boldsymbol{Z}\beta + \alpha + \varepsilon$, where $\boldsymbol{y} = (y_1, y_2, \ldots, y_{2(m+n)})'$, $\beta = (\mu, \tau, \pi, \lambda)'$, $\alpha \sim N_{2(m+n)}(\mathbf{0},\ \phi^2\mathbf{J}_2 \otimes \mathbf{I}_{(\mathbf{m+n})})$ $\varepsilon \sim N_{2(m+n)}(\mathbf{0},\ \sigma^2\mathbf{I}_{\mathbf{2(m+n)}})$ and (reordering the subjects so successive pairs of observations are the two results from subjects in period 1 and period 2)

$$\boldsymbol{Z} = \begin{pmatrix} 1 & 1 & 1 & 1 \\ 1 & -1 & -1 & 1 \\ \vdots & \vdots & \vdots & \vdots \\ 1 & 1 & 1 & 1 \\ 1 & -1 & -1 & 1 \\ 1 & -1 & 1 & -1 \\ 1 & 1 & -1 & -1 \\ \vdots & \vdots & \vdots & \vdots \\ 1 & -1 & 1 & -1 \\ 1 & 1 & -1 & -1 \end{pmatrix}.$$

Note that $\mathbf{J}_2$ is the matrix $\left(\begin{smallmatrix} 1 & 1 \\ 1 & 1 \end{smallmatrix}\right)$ and $\otimes$ indicates a Kronecker product (see §8.5) so $\mathbf{J}_2 \otimes \mathbf{I}_{(\mathbf{m+n})}$ is the $2(m+n) \times 2(m+n)$ block diagonal matrix (see §2.6.2) with 2×2 matrices $\mathbf{J_2}$ down the diagonal.

In the case $m = n$ of equal numbers of subjects in the groups, the columns of $\boldsymbol{Z}$ are orthogonal and $\boldsymbol{Z}'\boldsymbol{Z} = 4n\mathbf{I_4}$ so $(\boldsymbol{Z}'\boldsymbol{Z})^{-1}\boldsymbol{Z}' = \boldsymbol{Z}'/(4n)$ and so the least squares estimates (see §8.3.3.1 and §9.7.2.1) of the fixed effects are given by $\hat{\beta} = (\boldsymbol{Z}'\boldsymbol{Z})^{-1}\boldsymbol{Z}'\boldsymbol{y} = \boldsymbol{Z}'\boldsymbol{y}/(4n)$ giving the usual estimates of the parameters as various linear contrasts in the observations; see Jones and Kenward (2003).

9.7.4.1 Obtaining the design matrix in R

The matrix $\boldsymbol{X}$ can be calculated in two stages, first constructing the first seven columns and then including the $m+n$ columns as two identity $\mathbf{I_n}$ matrices stacked above one another.

```
# set values of m and n, the numbers of
# subjects in the two groups
k<-m+n
X<-matrix(c(rep(1,2*(m+n)), rep(1,m),rep(0,m+n),rep(1,n),
rep(0,m),rep(1,m+n),rep(0,n), rep(1,m+n),rep(0,m+n),
rep(0,m+n),rep(1,m+n), rep(1,m),rep(0,n),rep(1,m),rep(0,n),
rep(0,m), rep(1,n),rep(0,m),rep(1,n)),2*(m+n),7)
X<-cbind(X,rbind(diag(rep(1,k)),diag(rep(1,k))))
```

The constraints matrix $\boldsymbol{B}$ is constructed by

```
B<-matrix(c(0,1,1,rep(0,7+2*(m+n)),1,1,rep(0,7+2*(m+n),1,1,
   rep(0,7),rep(1,2*(m+n)),4,7,byrow=T)
```

If the data are reordered so that successive pairs of observations are the two results from subjects in periods 1 and 2 then the design matrix for the fixed effects (with constraints eliminated as above) $\boldsymbol{Z}$ can be calculated as

```
Z<-matrix(c(rep(1,2*k),rep(c(1,-1),m),rep(c(-1,1),n),
rep(c(1,-1),m+n),rep(1,2*m),rep(-1,2*n)),2k,4)
```

9.8 Exercises

(1) Suppose the random variable $\boldsymbol{x}$ has variance Σ_0, a known $p \times p$ positive definite symmetric matrix, and $\boldsymbol{X}'$ is a $n \times p$ data matrix of independent observations of $\boldsymbol{x}$ with sample variance $\boldsymbol{S}$. Find a matrix $\boldsymbol{A}$ such that the data matrix $\boldsymbol{Y}' = \boldsymbol{X}'\boldsymbol{A}$ has sample variance matrix Σ_0.

(2) Suppose $\Sigma_0 = \begin{pmatrix} 4.031 & 3.027 \\ 3.027 & 3.021 \end{pmatrix}$. Using the **R** function `mvrnorm(.)` in the `MASS` library and the previous exercise generate a sample of 47 two-dimensional observations which have a sample variance of Σ_0.

(3) With Σ_0 as given in the previous exercise generate 27 observations whose sample mean is $(20.25, 29.83)'$ and sample variance Σ_0 and a further 20 observations with sample mean $(18.95, 28.63)'$ and sample variance Σ_0. (Rao's paradox).

(4) Using the **R** function `runif(.)`, `matrix(.)`, `var(.)` and `eigen(.)` generate a random 5×5 orthogonal matrix.

(5) Suppose $\boldsymbol{x}_1, \boldsymbol{x}_2, \ldots, \boldsymbol{x}_n$ are independent observations of $N_p(\lambda\mu_0, \Sigma_0)$ where μ_0 and Σ_0 are known and λ is an unknown scalar.

(i) Show that the mle of λ is given by $\hat{\lambda} = \mu_0'\Sigma_0^{-1}\bar{\boldsymbol{x}}/\mu_0'\Sigma_0^{-1}\mu_0$.

(ii) Find the mean and variance of $\hat{\lambda}$ and hence give its distribution.

(iii) Show that the LRT statistic for testing $H_0: \ \mu = \lambda\mu_0$ for some scalar λ (where μ_0 and Σ_0 are known) is $n(\bar{\boldsymbol{x}} - \hat{\lambda}\mu_0)'\Sigma_0^{-1}(\bar{\boldsymbol{x}} - \hat{\lambda}\mu_0)$ which under H_0 follows a χ^2_p-distribution.

(iv) In a standard feeding experiment on greenfinches, four types of sunflower seeds were placed in identical quadruple compartment bird feeders in each of 27 suburban gardens. The mean weights consumed of the four types after 120 minutes were 47 g, 45 g, 39 g and 42 g. Experience from a long series of such standard experiments suggests that the standard deviations of the amounts consumed of any type of sunflower seed in a single garden can be taken to be 10 g and the pairwise correlations between the weights consumed are 0.1. Do these data suggest that greenfinches have unequal preferences for the various types of sunflower seeds?

(6) Suppose $\boldsymbol{x}_1, \boldsymbol{x}_2, \ldots, \boldsymbol{x}_n$ are independent observations of $N_p(\mu, \lambda\Sigma_0)$ where Σ_0 is a known positive definite matrix and λ is an unknown scalar and μ is not assumed to be known.

(i) Show that the mle of λ is $\hat{\lambda} = \frac{n-1}{np}\mathrm{tr}(\Sigma_0^{-1}\boldsymbol{S})$ where $\boldsymbol{S}$ is the sample variance matrix.

(ii) In a feeding experiment on sea urchins, equal amounts of three types of algæ were placed in 27 tanks each containing a single sea urchin. After 24 hours the mean weight losses over the 27 samples of the three types of algæ were 4.7 g, 3.9 g and 4.2 g with sample variance matrix $\boldsymbol{S} = \begin{pmatrix} 1.1 & 0.0 & 0.1 \\ 0.0 & 0.9 & 0.0 \\ 0.1 & 0.0 & 0.8 \end{pmatrix}$. Are these data consistent with the theory that the amounts of algæ have equal variances with pairwise correlations of 0.1?

Outline Solutions to Exercises

Chapter 1: Introduction

(1) Install **R** on your computer; see §1.4.2.

Done?

(2) Go to the CRAN home page and download (but don't print unless really desperate to do so) the manual by Venables et al. (2014); see §1.2.2.

Done? Many people find it more convenient to keep the manual in .pdf form so it can be accessed during an **R** session and use the linked contents list and/or search facility to find details of the topic of interest.

(3) Still on the CRAN home page, look at the software under Packages from the link in the menu on the left, browsing through the packages sorted by name. Find the description of the package CCA (used in §9.5) and look briefly at the reference manual.

Done? The objective is to give some idea of the range of packages available and the nature of the available documentation on them.

The aim of the next five exercises is just to provide a little familiarity with key elements of operating **R** . It will be presumed that these have all been done.

(4) Try typing direct into the **R** console window some of the examples given in §1.7.

(5) Open a new script file (§1.4.3 on P6) using the menu under File on the top left of the window.

(6) Type a few **R** commands into this script window and then highlight them with the mouse and then click on the middle icon in the top row to run them. (NB: This icon appears only when the **R** Editor window is the *active* window).

(7) Click on the **R** console window and then using the menu under File change the working directory to somewhere convenient. Usually **R** starts with the working directory (where it looks for and saves files by default) at the very top level.

(8) Making the **R** editor window the active window (by clicking the mouse when the cursor is in it) and using the File on the top left save the script file in the new working directory, using a name with extension `.R` (so that it can be opened in a a later **R** session with `File > Open script ...`, provided the working directory is the same).

Chapter 2: Vectors and Matrices

(1) Let $\boldsymbol{a} = \begin{pmatrix} 1 \\ 2 \\ 3 \end{pmatrix}$, $\boldsymbol{b} = \begin{pmatrix} 4 \\ 5 \\ 6 \end{pmatrix}$, $\boldsymbol{u} = \begin{pmatrix} 3 \\ 2 \\ 1 \end{pmatrix}$, $\boldsymbol{v} = \begin{pmatrix} 6 \\ 5 \\ 4 \end{pmatrix}$, $\boldsymbol{w} = (7, 8, 9)$.

(a) Calculate $\boldsymbol{a}+\boldsymbol{b}$, $\boldsymbol{v}-\boldsymbol{a}$, $\boldsymbol{w}'+\boldsymbol{b}$, $3\boldsymbol{u}$, $\boldsymbol{w}'-\boldsymbol{a}$, $\boldsymbol{v}/3$, $\boldsymbol{ab}'$ and $\boldsymbol{ba}'$.

(b) Repeat the calculations in (a) using **R**.

```
> a<-matrix(c(1,2,3),3,1)
> b<-matrix(c(4,5,6),3,1)
> u<-matrix(c(3,2,1),3,1)
> v<-matrix(c(6,5,4),3,1)
> w<-matrix(c(7,8,9),1,3)
> a;b;u;v;w
     [,1]
[1,]    1
[2,]    2
[3,]    3
     [,1]
[1,]    4
[2,]    5
[3,]    6
     [,1]
[1,]    3
[2,]    2
[3,]    1
     [,1]
[1,]    6
[2,]    5
[3,]    4
     [,1] [,2] [,3]
[1,]    7    8    9
```

```
> a+b
     [,1]
[1,]    5
[2,]    7
[3,]    9
> v-a
     [,1]
[1,]    5
[2,]    3
[3,]    1
> t(w)+b
     [,1]
[1,]   11
[2,]   13
[3,]   15
> 3*u
     [,1]
[1,]    9
[2,]    6
[3,]    3
> w-t(a)
     [,1] [,2] [,3]
[1,]    6    6    6
> v/3
         [,1]
[1,] 2.000000
[2,] 1.666667
[3,] 1.333333
```

```
> a%*%t(b)
     [,1] [,2] [,3]
[1,]    4    5    6
[2,]    8   10   12
[3,]   12   15   18
```

```
> b%*%t(a)
     [,1] [,2] [,3]
[1,]    4    8   12
[2,]    5   10   15
[3,]    6   12   18
```

(2) Let $\boldsymbol{x} = \begin{pmatrix} 2 \\ 2 \\ -3 \end{pmatrix}$ and $\boldsymbol{y} = \begin{pmatrix} 1 \\ -2 \\ 1 \end{pmatrix}$.

(a) Which of $\boldsymbol{a}$, $\boldsymbol{b}$, $\boldsymbol{u}$, $\boldsymbol{v}$ in Exercise (1) are orthogonal to $\boldsymbol{x}$?

(b) Which of $\boldsymbol{a}$, $\boldsymbol{b}$, $\boldsymbol{u}$, $\boldsymbol{v}$ in Exercise (1) are orthogonal to $\boldsymbol{y}$?

(c) Check the answers to (a) and (b) using **R**.

```
> x<-matrix(c(2,2,-3),3,1)
> x
     [,1]
[1,]    2
[2,]    2
[3,]   -3
> t(x)%*%a
     [,1]
[1,]   -3
> t(x)%*%b
     [,1]
[1,]    0
> t(x)%*%u
     [,1]
[1,]    7
> t(x)%*%v
     [,1]
[1,]   10
```

```
> y<-matrix(c(1,-2,1),3,1)
> y
     [,1]
[1,]    1
[2,]   -2
[3,]    1
> t(y)%*%a
     [,1]
[1,]    0
> t(y)%*%b
     [,1]
[1,]    0
> t(y)%*%u
     [,1]
[1,]    0
> t(y)%*%v
     [,1]
[1,]    0
```

```
> t(x)%*%w
Error in t(x) %*% w :
non-conformable arguments
```

```
> t(y)%*%w
Error in t(y) %*% w :
non-conformable arguments
```

Thus $\boldsymbol{x}$ is orthogonal **only** to $\boldsymbol{b}$ and $\boldsymbol{y}$ is orthogonal to $\boldsymbol{a}$, $\boldsymbol{b}$, $\boldsymbol{u}$ and $\boldsymbol{w}$. Note that a column vector cannot be orthogonal to a row vector because the transpose of one is non-conformable with the other.

(3) Let $\boldsymbol{A} = \begin{pmatrix} 1 & 2 & 3 \\ 4 & 5 & 6 \end{pmatrix}$, $\boldsymbol{B} = \begin{pmatrix} 1 & 2 \\ 3 & 4 \\ 5 & 6 \end{pmatrix}$, $\boldsymbol{U} = \begin{pmatrix} 1 & 2 \\ 3 & 4 \end{pmatrix}$, $\boldsymbol{V} = \begin{pmatrix} 5 & 6 \\ 7 & 8 \end{pmatrix}$,

$\boldsymbol{W} = \begin{pmatrix} 2 & 2 \\ 3 & 5 \end{pmatrix}$ and $\boldsymbol{Z} = \begin{pmatrix} 3 & 2 \\ 3 & 6 \end{pmatrix}$ (and use the vectors from Exercises (1) and (2)).

```
> A<- matrix(c(1,2,3,4,5,6),
+2,3,byrow=T)
> A
     [,1] [,2] [,3]
[1,]    1    2    3
[2,]    4    5    6
> B<-matrix(c(1,2,3,4,5,6),
+3,2,byrow=T)
> B
     [,1] [,2]
[1,]    1    2
[2,]    3    4
[3,]    5    6
> U<-matrix(c(1,2,3,4),
+2,2,byrow=T)
> U
     [,1] [,2]
[1,]    1    2
[2,]    3    4
```

```
> V<-matrix(c(5,6,7,8),
+2,2,byrow=T)
> V
     [,1] [,2]
[1,]    5    6
[2,]    7    8
> W<-matrix(c(2,2,3,5),
+2,2,byrow=T)
> W
     [,1] [,2]
[1,]    2    2
[2,]    3    5
> Z<- matrix(c(3,2,3,6),
+2,2,byrow=T)
> Z
     [,1] [,2]
[1,]    3    2
[2,]    3    6
```

(a) Find $\boldsymbol{AB}$, $\boldsymbol{B'A'}$, $\boldsymbol{BA}$, $\boldsymbol{a'A}$, $\boldsymbol{a'Aa}$, $\boldsymbol{V}\text{diag}(\boldsymbol{U})$, $\text{diag}(\boldsymbol{B'A'})$, $\boldsymbol{UVWZ}$, $\text{diag}(\text{diag}(\boldsymbol{UV}))$, $\text{diag}(\text{diag}(\boldsymbol{U}))\text{diag}(\text{diag}(\boldsymbol{V}))$.

```
> A%*%B
     [,1] [,2]
[1,]   22   28
[2,]   49   64
> t(B)%*%t(A)
     [,1] [,2]
[1,]   22   49
[2,]   28   64
> B%*%A
     [,1] [,2] [,3]
[1,]    9   12   15
[2,]   19   26   33
[3,]   29   40   51
> t(a)%*%A
Error in t(a) %*% A :
non-conformable arguments
> t(a)%*%A%*%a
Error in t(a) %*% A :
non-conformable arguments
```

```
> V%*%diag(U)
     [,1]
[1,]   29
[2,]   39
> diag(t(B)%*%t(A))
[1] 22 64
> U%*%V%*%W%*%Z
     [,1] [,2]
[1,]  756 1096
[2,] 1716 2488
> diag(diag(U%*%V))
     [,1] [,2]
[1,]   19    0
[2,]    0   50
> diag(diag(U))%*%
+ diag(diag(V))
     [,1] [,2]
[1,]    5    0
[2,]    0   32
```

(b) Verify that $\boldsymbol{U}$ and $\boldsymbol{V}$ do not commute but $\boldsymbol{U}$ and $\boldsymbol{W}$ commute and $\boldsymbol{U}$ and $\boldsymbol{Z}$ commute. Do $\boldsymbol{W}$ and $\boldsymbol{Z}$ commute? (Guess and verify.)

```
> U%*%V
     [,1] [,2]
[1,]   19   22
[2,]   43   50
> V%*%U
     [,1] [,2]
[1,]   23   34
[2,]   31   46
> U%*%W
     [,1] [,2]
[1,]    8   12
[2,]   18   26
> W%*%U
     [,1] [,2]
[1,]    8   12
[2,]   18   26
```

```
> U%*%Z
     [,1] [,2]
[1,]    9   14
[2,]   21   30
> Z%*%U
     [,1] [,2]
[1,]    9   14
[2,]   21   30
> W%*%Z
     [,1] [,2]
[1,]   12   16
[2,]   24   36
> Z%*%W
     [,1] [,2]
[1,]   12   16
[2,]   24   36
```

(4) Use the matrices from Exercise (3), and let $z = (2,5)'$.

(a) Calculate $z'\boldsymbol{U}z$, $z'\boldsymbol{V}z$, $\boldsymbol{x}'\boldsymbol{BAx}$ and $\boldsymbol{x}'\boldsymbol{A}'\boldsymbol{B}'\boldsymbol{x}$.

```
> t(z)%*%U%*%z
     [,1]
[1,]  154
>
> t(z)%*%V%*%z
     [,1]
[1,]  350
>
```

```
> t(x)%*%B%*%A%*%x
     [,1]
[1,]   21
>
> t(x)%*%t(A)%*%t(B)%*%x
     [,1]
[1,]   21
```

(b) Write the four results in the form $\boldsymbol{x}'\boldsymbol{Sx}$ where $\boldsymbol{S}$ is ***symmetric***.

$$\boldsymbol{U}_S = {}^1\!/_2(\boldsymbol{U}+\boldsymbol{U}') = \begin{pmatrix} 1 & 2.5 \\ 2.5 & 4 \end{pmatrix} \quad \boldsymbol{V}_S = {}^1\!/_2(\boldsymbol{V}+\boldsymbol{V}')\begin{pmatrix} 5 & 6.5 \\ 6.5 & 8 \end{pmatrix}$$

$$\boldsymbol{BA}_S = \boldsymbol{A}'\boldsymbol{B}'_S = \begin{pmatrix} 9 & 15.5 & 22 \\ 15.5 & 24 & 36.5 \\ 22 & 36.5 & 51 \end{pmatrix}.$$

Check:

```
> t(z)%*%US%*%z
     [,1]
[1,]  154
>
```

```
> t(z)%*%VS%*%z
     [,1]
[1,]  350
```

```
> ABS<-(B%*%A+t(A)
+ %*%t(B))/2
> ABS
      [,1] [,2] [,3]
[1,]   9.0 15.5 22.0
[2,]  15.5 26.0 36.5
[3,]  22.0 36.5 51.0
> t(x)%*%ABS%*%x
     [,1]
[1,]   21
```

(5) Let $\boldsymbol{A}=\begin{pmatrix} 0 & 1 \\ -1 & 0 \end{pmatrix}$, $\boldsymbol{B}=\begin{pmatrix} 0 & 1 \\ 0 & 0 \end{pmatrix}$, $\boldsymbol{C}=\begin{pmatrix} 1 & 1 \\ 1 & -1 \end{pmatrix}$, $\boldsymbol{D}=\begin{pmatrix} 1 & -1 \\ -1 & 1 \end{pmatrix}$, $\boldsymbol{E}=\begin{pmatrix} 1 & 1 \\ 1 & 1 \end{pmatrix}$ and $\boldsymbol{F}=\begin{pmatrix} 1 & 1 \\ -1 & -1 \end{pmatrix}$.

```
> A<-matrix(c(0,1,-1,0),2,2,byrow=T)
> B<-matrix(c(0,1,0,0),2,2,byrow=T)
> C<-matrix(c(1,1,1,-1),2,2,byrow=T)
> D<-matrix(c(1,-1,-1,-1),2,2,byrow=T)
> E<-matrix(c(1,1,1,1),2,2)
> F<-matrix(c(1,1,-1,-1),2,2,byrow=T)
```

then show:

(a) $\boldsymbol{A}^2 = -1$ (so $\boldsymbol{A}$ is 'like' the square root of -1).

```
> A%*%A
     [,1] [,2]
[1,]   -1    0
[2,]    0   -1
```

(b) $\boldsymbol{B}^2 = 0$ (but $\boldsymbol{B} \neq 0$), i.e., $\boldsymbol{B}$ is nilpotent; see §2.5.7.

```
> B%*%B
     [,1] [,2]
[1,]    0    0
[2,]    0    0
```

(c) $\boldsymbol{CD} = -\boldsymbol{DC}$ (but $\boldsymbol{CD} \neq 0$).

```
> C%*%D
     [,1] [,2]
[1,]    0   -2
[2,]    2    0
> D%*%C
     [,1] [,2]
[1,]    0    2
[2,]   -2    0
```

(d) $\boldsymbol{EF} = 0$ (but $\boldsymbol{E} \neq 0$ and $\boldsymbol{F} \neq 0$).

```
> E%*%F
     [,1] [,2]
[1,]    0    0
[2,]    0    0
```

Exercise (5) illustrates that some rules of scalar multiplication do not carry over to matrix multiplication. However there are some analogies:

(i) If a real square matrix $\boldsymbol{A}$ is such that $\boldsymbol{A}'\boldsymbol{A} = 0$ then we must have $\boldsymbol{A} = 0$ and the $(i, j)^{th}$ element of $\boldsymbol{A}'\boldsymbol{A}$ is $\sum_{k=1}^{n} a_{kj}^2$ so if $\boldsymbol{A}'\boldsymbol{A} = 0$ then in particular the diagonal elements of $\boldsymbol{A}'\boldsymbol{A}$ are all zero so we must have $\sum_{k=1}^{n} a_{kj}^2 = 0$ and so $a_{kj} = 0$ for all k and j and so $\boldsymbol{A} = 0$.

(ii) $\boldsymbol{AB} = 0$ if and only if $\boldsymbol{A}'\boldsymbol{AB} = 0$ since if $\boldsymbol{A}'\boldsymbol{AB} = 0$ then $\boldsymbol{B}'\boldsymbol{A}'\boldsymbol{AB} = 0$ so $(\boldsymbol{AB})'(\boldsymbol{AB}) = 0$ and the results follow from note (i) above.

(iii) $\boldsymbol{AB} = \boldsymbol{AC}$ if and only if $\boldsymbol{A}'\boldsymbol{AB} = \boldsymbol{A}'\boldsymbol{AC}$ which follows by replacing $\boldsymbol{B}$ by $\boldsymbol{B} - \boldsymbol{C}$.

(6) Show that $\text{tr}(\boldsymbol{xy}') = \boldsymbol{x}'\boldsymbol{y}$.

$(\boldsymbol{xy}')_{ij} = x_i y_j$ so $\text{tr}(\boldsymbol{xy}') = \sum_i x_i y_i = \boldsymbol{x}'\boldsymbol{y}$.

(7) Use the **R** help system to find out what the **R** functions `rep(.)` and `seq(.)` do by typing `help(rep)` and `help(seq)`.

Done?

(8) (a) Construct the sum vector ι_4 in **R** .

```
> matrix(rep(1,4),ncol=1)
     [,1]
[1,]    1
[2,]    1
[3,]    1
[4,]    1
```

(b) Construct the identity matrix $\mathbf{I_5}$.

```
> diag(rep(1,5))
     [,1] [,2] [,3] [,4] [,5]
[1,]    1    0    0    0    0
[2,]    0    1    0    0    0
[3,]    0    0    1    0    0
[4,]    0    0    0    1    0
[5,]    0    0    0    0    1
```

or `diag(5)`; see `help(diag)`.

(c) Construct the vector $\boldsymbol{e}_4$ of length 23 (i.e., the vector of length 23 with a 1 in the fourth place and zeros elsewhere).

```
> as.matrix(diag(rep(1,23))[,4])
```

Check this. Note that although this extracts the 4th column of the identity matrix $\mathbf{I_{23}}$, **R** will treat this ambiguously as a row or column vector unless it is forced to be a 23×1 matrix with the function `as.matrix(.)`.

(d) Construct the unit matrix $\mathbf{J_6}$.

`matrix(rep(1,36),6,6)` **or** `c(rep(1,6))%*%t(c(rep(1,6)))`

(e) Construct the centering matrix $\mathbf{H_3}$.

`diag(rep(1,3))-matrix(rep(1,9),3,3)/3`

To write a function to construct $\mathbf{H_n}$.

```
> Hn<- function(n)
+ {Hn<-diag(rep(1,n))-matrix(rep(1,n*n),n,n)/n
+ return(Hn)}
```

(f) Construct the vector containing all even numbers in order from 2 to 28.

```
matrix(seq(2,28,2),ncol=1)
```

(9) Suppose $\boldsymbol{A}$ is a non-singular $n \times n$ idempotent matrix. Show that $\mathbf{I_n} - \boldsymbol{A}$ is idempotent.

$(\mathbf{I_n} - \boldsymbol{A})^2 = (\mathbf{I_n} - \boldsymbol{A})(\mathbf{I_n} - \boldsymbol{A}) = (\mathbf{I_n} - \boldsymbol{A} - \boldsymbol{A} + \boldsymbol{A}^2) = (\mathbf{I_n} - \boldsymbol{A})$.

(10) Suppose $\boldsymbol{A}$ is a non-singular $n \times n$ idempotent matrix. Show that $\boldsymbol{A} = \mathbf{I_n}$.

Since $\boldsymbol{A}$ is non-singular it is invertible with inverse $\boldsymbol{A}^{-1}$ so $\mathbf{I_n} = \boldsymbol{A}\boldsymbol{A}^{-1} = \boldsymbol{A}^2\boldsymbol{A}^{-1}$ (since $\boldsymbol{A}$ is idempotent) $= \boldsymbol{A}$. So any idempotent matrix which is not the identity matrix must be singular.

(11) Suppose $\boldsymbol{A}$ and $\boldsymbol{B}$ are idempotent matrices. Show that $(\boldsymbol{A}+\boldsymbol{B})$ is idempotent if and only if $\boldsymbol{AB} = \boldsymbol{BA} = \mathbf{0}$.

$(\boldsymbol{A}+\boldsymbol{B})^2 = \boldsymbol{A}^2 + \boldsymbol{AB} + \boldsymbol{BA} + \boldsymbol{B}^2 = \boldsymbol{A} + \boldsymbol{B} + \boldsymbol{AB} + \boldsymbol{BA} = \boldsymbol{A} + \boldsymbol{B}$ if $\boldsymbol{AB} = \boldsymbol{BA} = \mathbf{0}$. $(\boldsymbol{A}+\boldsymbol{B})^2 = \boldsymbol{A}+\boldsymbol{B}$ only if $\boldsymbol{AB}+\boldsymbol{BA} = \mathbf{0}$. Premultiplying by $\boldsymbol{A}$ gives $\boldsymbol{AB}+\boldsymbol{ABA} = \mathbf{0}$ and postmultiplying by $\boldsymbol{A}$gives $\boldsymbol{ABA} + \boldsymbol{BA} = \mathbf{0}$. Subtracting these two equations gives $\boldsymbol{AB} - \boldsymbol{BA} = \mathbf{0}$ and thus $\boldsymbol{AB} = \boldsymbol{BA} = \mathbf{0}$.

(12) If $\boldsymbol{A}$ is either symmetric or skew-symmetric show that $\boldsymbol{A}^2$ is symmetric.

$(\boldsymbol{A}^2)' = ((\boldsymbol{A}')(\boldsymbol{A}'))' = \boldsymbol{A}^2$ (since $\boldsymbol{A}' = \pm\boldsymbol{A}$).

(13) Suppose $\boldsymbol{x}_1, \boldsymbol{x}_2, \ldots, \boldsymbol{x}_n$ are p-dimensional observations with sample mean and variance $\bar{\boldsymbol{x}}_n$ and $\boldsymbol{S}_n$ and nfx_n is a further observation. Show that the sample mean and variance of the augmented sample $\boldsymbol{x}_1, \boldsymbol{x}_2, \ldots, \boldsymbol{x}_n, \boldsymbol{x}_{n+1}$ are given by

$$\bar{\boldsymbol{x}}_{n+1} = \frac{n\bar{\boldsymbol{x}}_n + \boldsymbol{x}_{n+1}}{n+1} \text{ and } \boldsymbol{S}_{n+1} = \frac{1}{n}\left\{(n-1)\boldsymbol{S}_n + \frac{n}{n+1}(\boldsymbol{x}_{n+1} - \bar{\boldsymbol{x}}_n)(\boldsymbol{x}_{n+1} - \bar{\boldsymbol{x}}_n)'\right\}.$$

(These are known as the ***updating formulæ*** for mean and variance. They are appreciably more numerically stable when calculating sample variances for large quantities of data since formulæ avoid the subtraction of two similarly sized large numbers.)

$\bar{\boldsymbol{x}}_{n+1} = \frac{1}{n+1}\sum_1^{n+1}\boldsymbol{x}_i = \frac{1}{n+1}\{\sum_1^n \boldsymbol{x}_i + \boldsymbol{x}_{n+1}\} = \frac{n\bar{\boldsymbol{x}}_n + \boldsymbol{x}_{n+1}}{n+1}$

$$\begin{aligned}
\boldsymbol{S}_{n+1} &= \frac{1}{n}\sum_{i=1}^{n+1}(\boldsymbol{x}_i - \bar{\boldsymbol{x}}_{n+1})(\boldsymbol{x}_i - \bar{\boldsymbol{x}}_{n+1})' \\
&= \frac{1}{n}\sum_{i=1}^{n}(\boldsymbol{x}_i - \bar{\boldsymbol{x}}_{n+1})(\boldsymbol{x}_i - \bar{\boldsymbol{x}}_{n+1})' + \frac{1}{n}(\boldsymbol{x}_{n+1} - \bar{\boldsymbol{x}}_{n+1})(\boldsymbol{x}_{n+1} - \bar{\boldsymbol{x}}_{n+1})'.
\end{aligned}$$

Now

$$\begin{aligned}\sum_{i=1}^{n}[\boldsymbol{x}_i-\bar{\boldsymbol{x}}_{n+1}][\boldsymbol{x}_i-\bar{\boldsymbol{x}}_{n+1}]' &= \sum_{i=1}^{n}[\boldsymbol{x}_i-\frac{n\bar{\boldsymbol{x}}_n+\boldsymbol{x}_{n+1}}{n+1}][\boldsymbol{x}_i-\frac{n\bar{\boldsymbol{x}}_n+\boldsymbol{x}_{n+1}}{n+1}]' \\ &= \sum_{i=1}^{n}[\boldsymbol{x}_i-\bar{\boldsymbol{x}}_n+\frac{\bar{\boldsymbol{x}}_n-\boldsymbol{x}_{n+1}}{n+1}][\boldsymbol{x}_i-\bar{\boldsymbol{x}}_n+\frac{\bar{\boldsymbol{x}}_n-\boldsymbol{x}_{n+1}}{n+1}]' \\ &= (n-1)\boldsymbol{S}_n+\frac{n}{(n+1)^2}(\bar{\boldsymbol{x}}_n-\boldsymbol{x}_{n+1})(\bar{\boldsymbol{x}}_n-\boldsymbol{x}_{n+1})',\end{aligned}$$

noting that the cross-product terms vanish. Further

$$\begin{aligned}(\boldsymbol{x}_{n+1}-\bar{\boldsymbol{x}}_{n+1})(\boldsymbol{x}_{n+1}-\bar{\boldsymbol{x}}_{n+1})' &= \left(\boldsymbol{x}_{n+1}-\frac{n\bar{\boldsymbol{x}}_n+\boldsymbol{x}_{n+1}}{n+1}\right)\left(\boldsymbol{x}_{n+1}-\frac{n\bar{\boldsymbol{x}}_n+\boldsymbol{x}_{n+1}}{n+1}\right)' \\ &= \frac{n^2}{(n+1)^2}(\boldsymbol{x}_{n+1}-\bar{\boldsymbol{x}}_n)(\boldsymbol{x}_{n+1}-\bar{\boldsymbol{x}}_n)'\end{aligned}$$

and so

$$\boldsymbol{S}_{n+1}=\frac{1}{n}\left\{(n-1)\boldsymbol{S}_n+\frac{n}{n+1}(\boldsymbol{x}_{n+1}-\bar{\boldsymbol{x}}_n)(\boldsymbol{x}_{n+1}-\bar{\boldsymbol{x}}_n)'\right\}.$$

Chapter 3: Rank of Matrices

(1) Let $\boldsymbol{X}_1=\begin{pmatrix}1.3 & 9.1\\ 1.2 & 8.4\end{pmatrix}$, $\boldsymbol{X}_2=\begin{pmatrix}1.2 & 9.1\\ 1.3 & 8.4\end{pmatrix}$, $\boldsymbol{X}_3=\begin{pmatrix}1 & 2 & 3\\ 2 & 1 & 9\end{pmatrix}$

$\boldsymbol{X}_4=\begin{pmatrix}1 & 2\\ 3 & 9\\ 2 & 1\end{pmatrix}$, $\boldsymbol{X}_5=\begin{pmatrix}1 & 2 & 9\\ 2 & 1 & 3\\ 9 & 3 & 0\end{pmatrix}$ and $\boldsymbol{X}_6=\begin{pmatrix}6 & 2 & 8\\ 5 & 1 & 6\\ 1 & 7 & 8\end{pmatrix}$.

(a) What is the rank of each of $\boldsymbol{X}_1,\ldots,\boldsymbol{X}_6$?

(i) $\boldsymbol{X}_1=\begin{pmatrix}1.3 & 9.1\\ 1.2 & 8.4\end{pmatrix}$, $\boldsymbol{X}_1\neq 0$ so $\rho(\boldsymbol{X}_1)=1$ or 2. If $\rho(\boldsymbol{X}_1)=1$ then the second column of $\boldsymbol{X}_1$ is a multiple of the first column; 9.1/1.3=7, 8.4/7=1.2 so the second column is a multiple of 7 of the first and $\rho(\boldsymbol{X}_1)=1$. (Note that similarly the second row is a multiple 0.923 of the first.)

(ii) $\boldsymbol{X}_2=\begin{pmatrix}1.2 & 9.1\\ 1.3 & 8.4\end{pmatrix}$, $\boldsymbol{X}_2\neq 0$ so $\rho(\boldsymbol{X}_2)=1$ or 2. $9.1/1.2=7.583$ and $7.583\times 1.2=9.858\neq 8.4$ and so the second column is not a multiple of the first and so $\rho(\boldsymbol{X}_2)=2$.

(iii) $\boldsymbol{X}_3=\begin{pmatrix}1 & 2 & 3\\ 2 & 1 & 9\end{pmatrix}$. $\boldsymbol{X}_3$ is a 2×3 matrix $\rho(\boldsymbol{X}_3)=1$ or 2, since $\rho(\boldsymbol{X}_3)\leq \min(2,3)$. If $\rho(\boldsymbol{X}_3)=1$ then the second row is a multiple of the first. It is easy to see that this is not so and so $\rho(\boldsymbol{X}_3) = 2$.

(iv) $\boldsymbol{X}_4 = \begin{pmatrix} 1 & 2 \\ 3 & 9 \\ 2 & 1 \end{pmatrix}$. $\boldsymbol{X}_4$ is a 3×2 matrix $\rho(\boldsymbol{X}_4) = 1$ or 2, since $\rho(\boldsymbol{X}_4) \leq \min(3,2)$. If $\rho(\boldsymbol{X}_4) = 1$ then the second column is a multiple of the first. It is easy to see that this is not so and so $\rho(\boldsymbol{X}_4) = 2$.

(v) $\boldsymbol{X}_5 = \begin{pmatrix} 1 & 2 & 9 \\ 2 & 1 & 3 \\ 9 & 3 & 0 \end{pmatrix}$. $\boldsymbol{X}_5$ is a 3×3 matrix $\rho(\boldsymbol{X}_5) = 1$, 2 or 3. $\rho(\boldsymbol{X}_5) \neq 1$ since, for example, the second column is not a multiple of the first. If $\rho(\boldsymbol{X}_5) = 2$ then there are constants a_1, a_2 and a_3 (not all zero) such that $a_1 + 2a_2 + 9a_3 = 0, 2a_1 + a_2 + 3a_3 = 0$ and $9a_1 + 3a_2 = 0$. The third implies $a_2 = -3a_1$; substituting this into the second and third gives $-5a_1 + 9a_3 = 0$ and $-a_1 + 3a_3 = 0$. Subtracting three times this last from the preceding one shows that $a_1 = 0$ and hence $a_2 = a_3 = 0$. Thus $\rho(\boldsymbol{X}_5) = 3$.

(vi) $\boldsymbol{X}_6 = \begin{pmatrix} 6 & 2 & 8 \\ 5 & 1 & 6 \\ 1 & 7 & 8 \end{pmatrix}$. Clearly $\rho(\boldsymbol{X}_6) = 2$ or 3.

If $\rho(\boldsymbol{X}_6) = 2$ then there are constants a_1, $a_2$and a_3 (not all zero) such that $6a_1 + 2a_2 + 8a_3 = 0$, $5a_1 + a_2 + 6a_3 = 0$ and $a_1 + 7a_2 + 8a_3 = 0$. Subtracting the third from the first gives $5a_1 - 5a_2 = 0$, so $a_1 = a_2$. Substituting this in any of the three original equations gives $a_1 = -a_3$. So taking $a_1 = a_2 = 1$ and $a_3 = -1$ satisfies all three equations and so we have found suitable constants a_1, a_2 and a_3 (not all zero); thus $\rho(\boldsymbol{X}_6) = 2$.

(b) Find constants a_1, a_2, a_3 such that $a_1\boldsymbol{c}_{31} + a_2\boldsymbol{c}_{32} + a_3\boldsymbol{c}_{33} = 0$ where $\boldsymbol{c}_{3j}, j = 1,2,3$ are the three columns of $\boldsymbol{X}_3$.

$\boldsymbol{X}_3 = \begin{pmatrix} 1 & 2 & 3 \\ 2 & 1 & 9 \end{pmatrix}$ so we require constants a_1, $a_2$and a_3 (not all zero) such that $a_1 + 2a_2 + 3a_3 = 0$ and $2a_1 + a_2 + 9a_3 = 0$. Subtracting three times the first from the second gives $-a_1 - 5a_2 = 0$, so we could take $a_2 = 1$, $a_1 = -5$ and putting these in either equation gives $a_3 = 1$. So we have $(a_1, a_2, a_3) = (-5, 1, 1)$.

(c) Find constants a_1, a_2, a_3 such that $a_1\boldsymbol{r}_{41} + a_2\boldsymbol{r}_{42} + a_3\boldsymbol{r}_{43} = 0$ where $\boldsymbol{r}_{4j}, j = 1,2,3$ are the three rows of $\boldsymbol{X}_4$.

$\boldsymbol{X}_4 = \begin{pmatrix} 1 & 2 \\ 3 & 9 \\ 2 & 1 \end{pmatrix}$, so we require constants a_1, $a_2$and a_3 (not all zero) such that $a_1 + 3a_2 + 2a_3 = 0$ and $2a_1 + 9a_2 + a_3 = 0$. These equations are much the same as in (b) but with a_2 and a_3 interchanged so it is easy to see that we can take $(a_1, a_2, a_3) = (-5, 1, 1)$.

(2) Let $X_7 = \begin{pmatrix} 4 & 5 & 6 \\ 8 & 10 & 12 \\ 12 & 15 & 18 \end{pmatrix}$ and $X_8 = \begin{pmatrix} 4 & 12 & 8 \\ 6 & 18 & 12 \\ 5 & 15 & 10 \end{pmatrix}$.

(a) Show that X_7 and X_8 are both of rank 1.

(i) It is easy to see that the second and third rows of X_7 are multiples by 2 and 3 of the first row (or that the second and third columns are multiples by 1.25 and 1.5 of the first). So $\rho(X_7) = 1$.

(ii) It is easy to see that the second and third columns of X_8 are multiples by 3 and 2 of the first (or that the second and third rows are multiples by 1.5 and 1.25 of the first). So $\rho(X_8) = 1$.

(b) Find vectors $\boldsymbol{a}$ and $\boldsymbol{b}$ such that $X_7 = \boldsymbol{ab}'$.

Since $\rho(X_7) = 1$ there must exist vectors $\boldsymbol{a}$ and $\boldsymbol{b}$ such that $X_7 = \boldsymbol{ab}'$ (see §3.2.1). Noting the solution to (a) (i) it is sensible to take the first row of X_7 and the sequence of multiples of it identified. So, first try $\boldsymbol{a} = (4, 5, 6)'$ and $\boldsymbol{b} = (1, 2, 3)'$ which gives $\boldsymbol{ab}' = X_7$.

(c) Find vectors $\boldsymbol{u}$ and $\boldsymbol{v}$ such that $X_8 = \boldsymbol{uv}'$.

Since $\rho(X_8) = 1$ there must exist vectors $\boldsymbol{u}$ and $\boldsymbol{v}$ such that $X_8 = \boldsymbol{uv}'$. Noting the discussion in the solution to (a) (ii) and trying $u = (4, 6, 5)'$ and $\boldsymbol{v} = (1, 3, 2)'$ would give $\boldsymbol{uv}' = X_8'$ so instead $\boldsymbol{u} = (1, 3, 2)'$ and $\boldsymbol{v} = (4, 6, 5)'$.

(3) Let $X_9 = X_3X_4$ and $X_{10} = X_4X_3$.

(a) Evaluate X_9 and X_{10} in **R** .

```
> x3<-matrix(c(1,2,3,2,1,9), 2,3, byrow=T)
> x3
     [,1] [,2] [,3]
[1,]    1    2    3
[2,]    2    1    9
> x4<-matrix(c(1,2,3,9,2,1),3,2,byrow=T)
> x4
     [,1] [,2]
[1,]    1    2
[2,]    3    9
[3,]    2    1
> x9<-x3%*%x4
> x9
     [,1] [,2]
```

```
[1,]   13   23
[2,]   23   22
> x10<-x4%*%x3
> x10
     [,1] [,2] [,3]
[1,]    5    4   21
[2,]   21   15   90
[3,]    4    5   15
```

(b) What is the rank of X_9?

$\rho(X_9) = 2$ since clearly the second column is not a multiple of the first.

(c) What is the rank of X_{10}?

$\rho(x_{10}) = \rho(X_4X_3) \leq \min(\rho(X_4), \rho(X_3)) = \min(2,2) = 2$. $\rho(X_{10}) \neq 1$ since clearly the second column is not a multiple of the first so $\rho(X_{10}) = 2$.

(4) If x is a $n \times 1$ vector show that $\rho(xx' - x'x\mathbf{I_n}) < n$.

$x'(xx' - x'x\mathbf{I_n}) = x'xx' - x'(x'x)\mathbf{I_n} = (x'x)x' - (x'x)x'\mathbf{I_n} = (x'x)x' - (x'x)x' = 0$ so the columns of $(xx' - x'x\mathbf{I_n})$ are linearly dependent and thus the matrix cannot be of full rank.

(5) If $\rho(X) < n$ and $Xx = \lambda x$ show that:

(a) $\rho(X + \lambda xx') < n$
$X + \lambda xx' = X + Xxx' = X(\mathbf{I_n} + xx')$
so $\rho(X + x\mathbf{I_n}) \leq \min(\rho(X), \rho(\mathbf{I_n} + xx')) < n$.

(b) $\rho(X + xy') < n$ for any $n \times 1$ vector y.

$X + xy' = X + \lambda x(y'/\lambda) = X + Xx(y'/\lambda) = X(\mathbf{I_n} + x(y'/\lambda))$
so $\rho(X + xy') \leq \min(\rho(X), \rho((\mathbf{I_n} + x(y'/\lambda)) < n$.

(6) If A is $m \times n$, B is $n \times m$ and $\rho(AB) = m$ show that $\rho(A) = \rho(B) = m$ and thus BA is singular unless both A and B are square.

$\rho(AB) \leq \min(\rho(A), \rho(B))$ and $\rho(A) \leq \min(m,n)$ so $m \leq n$ and thus $\rho(A) = m$. Similarly $\rho(B) \leq \min(m,n)$ so $\rho(B) = m$. BA is $n \times n$ so $\rho(BA) \leq n$ but also $\rho(BA) \leq \min(\rho(A), \rho(B)) = m$ and $m \leq n$ so $\rho(BA) = n$ only if $m = n$, i.e., unless both A and B are square.

(7) If A is $m \times n$ with $m \geq n$ and $\rho(A) = n$ show that $\rho(AB) = \rho(B)$ for any conformable matrix B.

Suppose B is $n \times p$. $\rho(AB) \leq \min(\rho(A), \rho(B)) = \min(n, r)$ where $\rho(B) = r$. Suppose (without losing any generality) that the first n rows of A are linearly independent and the first r columns of B are linearly independent. Then the first r columns of AB are linearly independent and so $\rho(AB) \geq r$ but $\rho(AB) \leq r$ so $\rho(AB) = r$.

(8) If $\boldsymbol{X} = \boldsymbol{AB}$ where $\boldsymbol{A}$ is $m \times n$, $\boldsymbol{B}$ is $n \times m$ with $\rho(\boldsymbol{A}) = \rho(\boldsymbol{B}) = n$ show that $\rho(\boldsymbol{X}) = \rho(\boldsymbol{X}^2)$ if and only if $\rho(\boldsymbol{BA}) = n$.

Because $\rho(\boldsymbol{A}) = n$ we have $\rho(\boldsymbol{X}) = \rho(\boldsymbol{AB}) = n$ (by the previous exercise). So $\rho(\boldsymbol{X}^2) = \rho(\boldsymbol{A}(\boldsymbol{BA})\boldsymbol{B}) \leq \min(n, \rho(\boldsymbol{BA}))$ so if $n = \rho(\boldsymbol{X}^2)$ then $\rho(\boldsymbol{BA}) \geq n$ but $\rho(\boldsymbol{BA}) \leq n$ so $\rho(\boldsymbol{BA}) = n$. Suppose now that $\rho(\boldsymbol{BA}) = n$ then $\rho((\boldsymbol{BA})^2) = \rho(\boldsymbol{BA})$ by the previous exercise and applying it again gives $\rho((\boldsymbol{BA})^3) = \rho((\boldsymbol{BA})^2) = n$ but $\rho((\boldsymbol{BA})^3) = \rho(\boldsymbol{BX}^2\boldsymbol{A}) \leq \min(n, \rho(\boldsymbol{X}^2))$ so $\rho(\boldsymbol{X}^2) \geq n$ and $\rho(\boldsymbol{X}^2) \leq n$ so $\rho(\boldsymbol{X}^2) = n$.

(9) Suppose $\boldsymbol{A}$ is $m \times n$ and let $\boldsymbol{B}$ and $\boldsymbol{C}$ be $m \times m$ and $n \times n$ non-singular matrices.

(i) Show that $\rho(\boldsymbol{BAC}) = \rho(\boldsymbol{A})$.

$\rho(\boldsymbol{A}) = \rho(\boldsymbol{B}^{-1}\boldsymbol{BACC}^{-1}) \leq \rho(\boldsymbol{BAC}) \leq \rho(\boldsymbol{A})$ so $\rho(\boldsymbol{BAC}) = \rho(\boldsymbol{A})$.

(ii) Deduce $\rho(\boldsymbol{BA}) = \rho(\boldsymbol{AC}) = \rho(\boldsymbol{A})$.

Put $\boldsymbol{C} = \mathbf{I_n}$ and then $\boldsymbol{B} = \mathbf{I_m}$.

(10) Show that $\rho\begin{pmatrix} \boldsymbol{A} & \boldsymbol{0} \\ \boldsymbol{0} & \boldsymbol{D} \end{pmatrix} = \rho(\boldsymbol{A}) + \rho(\boldsymbol{D})$.

Consider the matrices $\boldsymbol{A}_0 = \begin{pmatrix} \boldsymbol{A} \\ \boldsymbol{0} \end{pmatrix}$ and $\boldsymbol{D}_0 = \begin{pmatrix} \boldsymbol{0} \\ \boldsymbol{D} \end{pmatrix}$ and let $\boldsymbol{a}_0$ and $\boldsymbol{d}_0$ be non-zero columns of $\boldsymbol{A}_0$ and $\boldsymbol{D}_0$. Then $\alpha_1\boldsymbol{a}_0 + \alpha_2\boldsymbol{d}_0 = \alpha_1\begin{pmatrix} \boldsymbol{a} \\ \boldsymbol{0} \end{pmatrix} + \alpha_2\begin{pmatrix} \boldsymbol{0} \\ \boldsymbol{d} \end{pmatrix} = \begin{pmatrix} \alpha_1\boldsymbol{a} \\ \alpha_2\boldsymbol{d} \end{pmatrix}$ so if $\alpha_1\boldsymbol{a}_0 + \alpha_2\boldsymbol{d}_0 = 0$ then $\alpha_1 = \alpha_2 = 0$ and thus the non-zero columns of $\boldsymbol{A}_0$ and $\boldsymbol{D}_0$ are linearly independent and the results follows.

NB: Some parts of the following exercises require use of properties (§4.3.2).

(11) Suppose $\boldsymbol{A}$ is $m \times m$ and non-singular, $\boldsymbol{D}$ is $n \times n$.

Let $\boldsymbol{Z} = \begin{pmatrix} \boldsymbol{A} & \boldsymbol{B} \\ \boldsymbol{0} & \boldsymbol{D} \end{pmatrix}$ and $\boldsymbol{W} = \begin{pmatrix} \mathbf{I_m} & -\boldsymbol{A}^{-1}\boldsymbol{B} \\ \boldsymbol{0} & \mathbf{I_n} \end{pmatrix}$.

(i) Show that $\boldsymbol{W}$ is non-singular.

$\boldsymbol{W}$ is a triangular matrix with all diagonal elements 1 and so has determinant 1 (see §4.3.2 property (vii)) and so is non-singular by property (v) of §4.3.2.

(ii) Show that $\boldsymbol{ZW} = \begin{pmatrix} \boldsymbol{A} & \boldsymbol{0} \\ \boldsymbol{0} & \boldsymbol{D} \end{pmatrix}$.

$$\boldsymbol{ZW} = \begin{pmatrix} \boldsymbol{A}\mathbf{I_m} + \boldsymbol{B0} & -\boldsymbol{AA}^{-1}\boldsymbol{B} + \boldsymbol{B}\mathbf{I_n} \\ \boldsymbol{0A} + \boldsymbol{D0} & -\boldsymbol{0A}^{-1}\boldsymbol{B} + \boldsymbol{D}\mathbf{I_n} \end{pmatrix} = \begin{pmatrix} \boldsymbol{A} & \boldsymbol{0} \\ \boldsymbol{0} & \boldsymbol{D} \end{pmatrix}.$$

(iii) Show that $\rho(\boldsymbol{Z}) = \rho(\boldsymbol{A}) + \rho(\boldsymbol{D})$.

$\rho(\boldsymbol{Z}) = \rho(\boldsymbol{ZW})$ ($\boldsymbol{W}$ is non-singular, Exercise (9)) $= \rho\begin{pmatrix} \boldsymbol{A} & \boldsymbol{0} \\ \boldsymbol{0} & \boldsymbol{D} \end{pmatrix} = \rho(\boldsymbol{A}) + \rho(\boldsymbol{D})$ (using Exercise (10) above).

(12) Suppose $\boldsymbol{Z} = \begin{pmatrix} \boldsymbol{A} & \boldsymbol{B} \\ \boldsymbol{0} & \boldsymbol{D} \end{pmatrix}$ with $\boldsymbol{B}$ non-singular.

(i) Show that $\begin{pmatrix} \mathbf{I_m} & \mathbf{0} \\ -\boldsymbol{DB}^{-1} & \mathbf{I_n} \end{pmatrix}\begin{pmatrix} \boldsymbol{A} & \boldsymbol{B} \\ \mathbf{0} & \boldsymbol{D} \end{pmatrix}\begin{pmatrix} \mathbf{0} & \mathbf{I_p} \\ \mathbf{I_m} & -\boldsymbol{B}^{-1}\boldsymbol{A} \end{pmatrix} = \begin{pmatrix} \boldsymbol{B} & \mathbf{0} \\ \mathbf{0} & -\boldsymbol{DB}^{-1}\boldsymbol{A} \end{pmatrix}$.

$\begin{pmatrix} \mathbf{I_m} & \mathbf{0} \\ -\boldsymbol{DB}^{-1} & \mathbf{I_n} \end{pmatrix}\begin{pmatrix} \boldsymbol{A} & \boldsymbol{B} \\ \mathbf{0} & \boldsymbol{D} \end{pmatrix} = \begin{pmatrix} \boldsymbol{A}+\mathbf{0} & \boldsymbol{B}+\mathbf{0} \\ -\boldsymbol{DB}^{-1}\boldsymbol{A}+\mathbf{0} & -\boldsymbol{DB}^{-1}\boldsymbol{B}+\boldsymbol{D} \end{pmatrix}$ and $\begin{pmatrix} \boldsymbol{A} & \boldsymbol{B} \\ -\boldsymbol{DB}^{-1}\boldsymbol{A} & \mathbf{0} \end{pmatrix}\begin{pmatrix} \boldsymbol{B} & \mathbf{0} \\ \mathbf{0} & -\boldsymbol{DB}^{-1}\boldsymbol{A} \end{pmatrix} = \begin{pmatrix} \boldsymbol{B} & \mathbf{0} \\ \mathbf{0} & -\boldsymbol{DB}^{-1}\boldsymbol{A} \end{pmatrix}$.

(ii) Show that the first and third matrices in part (i) are non-singular.

Noting that these matrices are triangular with all diagonal elements 1 this follows by properties (v) and (vii) of §4.3.2.

(iii) Show that $\rho(Z) = \rho(\boldsymbol{B}) + \rho(\boldsymbol{DB}^{-1}\boldsymbol{A})$.

This follows using the results of part (i) and Exercises (9) and (10).

(13) Suppose $\boldsymbol{Z} = \begin{pmatrix} \boldsymbol{A} & \boldsymbol{B} \\ \mathbf{0} & \boldsymbol{D} \end{pmatrix}$ with $\boldsymbol{A}$ $m \times n$, $\boldsymbol{B}$ $m \times q$ and $\boldsymbol{D}$ $p \times q$ with neither $\boldsymbol{A}$, $\boldsymbol{B}$ nor $\boldsymbol{D}$ necessarily non-singular. Show that $\rho(\boldsymbol{Z}) \geq \rho(\boldsymbol{A}) + \rho(\boldsymbol{D})$.

Suppose $\rho(\boldsymbol{A}) = r$ and $\rho(\boldsymbol{D}) = s$. Consider the $n+q$ columns of $\boldsymbol{Z}$

$$\begin{pmatrix} \boldsymbol{a}_1 \\ \boldsymbol{0} \end{pmatrix}, \begin{pmatrix} \boldsymbol{a}_2 \\ \boldsymbol{0} \end{pmatrix}, \ldots, \begin{pmatrix} \boldsymbol{a}_r \\ \boldsymbol{0} \end{pmatrix}, \ldots, \begin{pmatrix} \boldsymbol{a}_n \\ \boldsymbol{0} \end{pmatrix}$$

$$\begin{pmatrix} \boldsymbol{b}_1 \\ \boldsymbol{d}_1 \end{pmatrix}, \begin{pmatrix} \boldsymbol{b}_2 \\ \boldsymbol{d}_2 \end{pmatrix}, \ldots, \begin{pmatrix} \boldsymbol{b}_s \\ \boldsymbol{d}_s \end{pmatrix}, \ldots, \begin{pmatrix} \boldsymbol{b}_q \\ \boldsymbol{d}_q \end{pmatrix}$$

where $\boldsymbol{a}_1, \boldsymbol{a}_2, \ldots, \boldsymbol{a}_r$ and $\boldsymbol{d}_1, \boldsymbol{d}_2, \ldots, \boldsymbol{d}_s$ are linearly independent columns of $\boldsymbol{A}$ and $\boldsymbol{D}$ and $\boldsymbol{b}_1, \boldsymbol{b}_2, \ldots, \boldsymbol{b}_s$ are the columns of $\boldsymbol{B}$ corresponding to those of $\boldsymbol{D}$. Suppose these columns are linearly dependent, then there are α_i; $i = 1, \ldots, r$ and β_j; $j = 1, \ldots, s$ such that $\sum \alpha_i \boldsymbol{a}_i + \sum \beta_j \boldsymbol{b}_j = 0$ and $\sum \beta_j \boldsymbol{d}_j = 0$. Since the $\boldsymbol{d}_j$ are linearly independent we must have $\beta_j = 0$ and thus $\sum \alpha_i \boldsymbol{a}_i = 0$ and so $\alpha_I = 0$. Thus there are at least $r+s$ linearly independent columns in $\boldsymbol{Z}$, i.e., $\rho(\boldsymbol{Z}) \geq \rho(\boldsymbol{A}) + \rho(\boldsymbol{D})$.

(14) Suppose $\boldsymbol{A}$, $\boldsymbol{B}$ and $\boldsymbol{C}$ are matrices such that the product $\boldsymbol{ABC}$ is defined.

(i) Show that

$$\begin{pmatrix} \mathbf{I_m} & -\boldsymbol{A} \\ \mathbf{0} & \mathbf{I_n} \end{pmatrix}\begin{pmatrix} \mathbf{0} & \boldsymbol{AB} \\ \boldsymbol{BC} & \boldsymbol{B} \end{pmatrix}\begin{pmatrix} \mathbf{I_q} & \mathbf{0} \\ -\boldsymbol{C} & \mathbf{I_p} \end{pmatrix} = \begin{pmatrix} -\boldsymbol{ABC} & \mathbf{0} \\ \mathbf{0} & \boldsymbol{B} \end{pmatrix}.$$

The product of the first two matrices is $\begin{pmatrix} -\boldsymbol{ABC} & \mathbf{0} \\ \boldsymbol{BC} & \boldsymbol{AB}+\boldsymbol{B} \end{pmatrix}$ and then the result follows.

(ii) Show that the first and third matrices in part (i) are non-singular.

Noting that these matrices are triangular with all diagonal elements 1 this follows by properties (v) and (vii) of §4.3.2.

(iii) Show that $\rho\begin{pmatrix} \mathbf{0} & \boldsymbol{AB} \\ \boldsymbol{BC} & \boldsymbol{B} \end{pmatrix} = \rho(\boldsymbol{ABC}) + \rho(\boldsymbol{B})$.

This follows using the results of part (i) and Exercises (9) and (10).

(15) Suppose $\boldsymbol{A}$, $\boldsymbol{B}$ and $\boldsymbol{C}$ are matrices such that the product $\boldsymbol{ABC}$ is defined. Show that $\rho(\boldsymbol{ABC}) \geq \rho(\boldsymbol{AB}) + \rho(\boldsymbol{BC}) - \rho(\boldsymbol{B})$. (This is known as the ***Frobenius inequality***)

From the previous exercise $\rho\begin{pmatrix} \mathbf{0} & \boldsymbol{AB} \\ \boldsymbol{BC} & \boldsymbol{B} \end{pmatrix} = \rho(\boldsymbol{ABC}) + \rho(\boldsymbol{B})$.

Also $\rho\begin{pmatrix} \mathbf{0} & \boldsymbol{AB} \\ \boldsymbol{BC} & \boldsymbol{B} \end{pmatrix} = \rho\begin{pmatrix} \boldsymbol{AB} & \mathbf{0} \\ \boldsymbol{B} & \boldsymbol{BC} \end{pmatrix} \geq \rho(\boldsymbol{AB}) + \rho(\boldsymbol{BC})$ (interchanging columns and using Exercise (13)). Thus $\rho(\boldsymbol{ABC}) \geq \rho(\boldsymbol{AB}) + \rho(\boldsymbol{BC}) - \rho(\boldsymbol{B})$.

(16) Suppose $\boldsymbol{A}$ is $m \times n$ and $\boldsymbol{B}$ is $n \times p$ show $\rho(\boldsymbol{AB}) \geq \rho(\boldsymbol{A}) + \rho(\boldsymbol{B}) - n$. (This is known as ***Sylvester's inequality***.)

$\rho(\boldsymbol{A}\mathbf{I_n}\boldsymbol{B}) \geq \rho(\boldsymbol{A}\mathbf{I_n}) + \rho(\mathbf{I_n}\boldsymbol{B}) - \rho(\mathbf{I_n})$ (by the Frobenius inequality) and so $\rho(\boldsymbol{AB}) \geq \rho(\boldsymbol{A}) + \rho(\boldsymbol{B}) - n$.

(17) If $\boldsymbol{A}$ is $m \times n$ and $\boldsymbol{B}$ is $n \times p$ and $\boldsymbol{AB} = \mathbf{0}$ show that $\rho(\boldsymbol{A}) \leq n - \rho(\boldsymbol{B})$.

$\rho(\boldsymbol{A}) + \rho(\boldsymbol{B}) - n \leq \rho(\boldsymbol{AB}) = 0$ (Sylvester's inequality) and so $\rho(\boldsymbol{A}) \leq n - \rho(\boldsymbol{B})$.

(18) If $\rho(\boldsymbol{A}^k) = \rho(\boldsymbol{A}^{k+1})$ show that $\rho(\boldsymbol{A}^{k+1}) = \rho(\boldsymbol{A}^{k+2})$.

$\rho(\boldsymbol{A}^{k+2}) = \rho(\boldsymbol{A}\boldsymbol{A}^{k+1}) \leq \min(\rho(\boldsymbol{A}), \rho(\boldsymbol{A}^{k+1}))$. Also, by the Frobenius inequality, $\rho(\boldsymbol{A}^{k+2}) = \rho(\boldsymbol{A}\boldsymbol{A}^k\boldsymbol{A}) \geq \rho(\boldsymbol{A}^{k+1}) + \rho(\boldsymbol{A}^{k+1}) - \rho(\boldsymbol{A}^k) = 2\rho(\boldsymbol{A}^{k+1}) - \rho(\boldsymbol{A}^{k+1}) = \rho(\boldsymbol{A}^{k+1})$. So $\rho(\boldsymbol{A}^{k+1}) = \rho(\boldsymbol{A}^{k+2})$.

(19) If $\boldsymbol{A}$ is $n \times n$ show that there is a k, $0 < k \leq n$, such that $\rho(\boldsymbol{A}) > \rho(\boldsymbol{A}^2) > \ldots > \rho(\boldsymbol{A}^k) = \rho(\boldsymbol{A}^{k+1}) = \ldots$.

$\rho(\boldsymbol{A}^{r+1}) \leq \min(\boldsymbol{A}, \rho(\boldsymbol{A}^r))$. If $\rho(\boldsymbol{A}^r) = \rho(\boldsymbol{A}^{r+1})$ then $\rho(\boldsymbol{A}^{r+1}) = \rho(\boldsymbol{A}^{r+2}) = \ldots$ by the previous exercise. If $\rho(\boldsymbol{A}^{r+1}) < \rho(\boldsymbol{A}^r)$ then consider $\rho(\boldsymbol{A}^{r+2})$, since $n \geq \rho(\boldsymbol{A}^{r+1}) > 0$ there must be a k, $0 < k \leq n$ such that $\rho(\boldsymbol{A}^k) = \rho(\boldsymbol{A}^{k+1})$.

(20) If $\boldsymbol{A}$ is $n \times n$ show that $\rho(\boldsymbol{A}^{k+1}) - 2\rho(\boldsymbol{A}^k) + \rho(\boldsymbol{A}^{k-1}) \geq 0$.

$\rho(\boldsymbol{A}^{k+1}) = \rho(\boldsymbol{A}\boldsymbol{A}^{k-1}\boldsymbol{A}) \geq 2\rho(\boldsymbol{A}^k) - \rho(\boldsymbol{A}^{k-1})$ by the Frobenius inequality so $\rho(\boldsymbol{A}^{k+1}) - 2\rho(\boldsymbol{A}^k) + \rho(\boldsymbol{A}^{k-1}) \geq 0$.

Chapter 4: Determinants

(1) Find the determinants of $\boldsymbol{A} = \begin{pmatrix} 2 & 3 \\ 2 & 4 \end{pmatrix}$ and $\boldsymbol{B} = \begin{pmatrix} -4 & 3 \\ 2 & -2 \end{pmatrix}$.

```
> A<-matrix(c(2,2,3,4)
+,2,2)
> A;det(A)
     [,1] [,2]
[1,]    2    3
[2,]    2    4
[1] 2
```

```
> B<-matrix(c(-4,-2,3,-2)
+,2,2)
> B;det(B)
     [,1] [,2]
[1,]   -4    3
[2,]   -2   -2
[1] 14
```

(2) Find the determinant of $\boldsymbol{X} = \begin{pmatrix} 4 & 5 & 6 \\ 8 & 10 & 12 \\ 12 & 15 & 18 \end{pmatrix}$.

```
> X<-matrix(c(4,8,12,5,10,15,6,12,18),3,3)
> X;det(X)
     [,1] [,2] [,3]
[1,]    4    5    6
[2,]    8   10   12
[3,]   12   15   18
[1] 0
```

It is easy to see that rows 2 and 3 are multiples (by 2 and 3) of the first row and so $\boldsymbol{X}$ has rank 1 and therefore $|\boldsymbol{X}| = 0$. The **R** calculations confirm this.

(3) Find the determinant of $\boldsymbol{X} = \begin{pmatrix} 1 & 2 & 9 \\ 2 & 1 & 3 \\ 9 & 3 & 0 \end{pmatrix}$.

```
> X<-matrix(c(1,2,9,2,1,3,9,3,0),3,3)
> X; det(X)
     [,1] [,2] [,3]
[1,]    1    2    9
[2,]    2    1    3
[3,]    9    3    0
[1] 18
```

(4) Find the determinants of $\boldsymbol{X} = \begin{pmatrix} 2 & 3 & 3 \\ 3 & 3 & 3 \\ 3 & 3 & 2 \end{pmatrix}$ and $\boldsymbol{Y} = \begin{pmatrix} 3 & 2 & 3 \\ 2 & 6 & 6 \\ 3 & 6 & 11 \end{pmatrix}$.

```
> X<-matrix(c(2,3,3,3,
+ 3,3,3,3,2),3,3)
> X; det(X)
     [,1] [,2] [,3]
[1,]    2    3    3
[2,]    3    3    3
[3,]    3    3    2
[1] 3
```

```
> Y<-matrix(c(3,2,3,2,
+ +6,6,3,6,11),3,3)
> Y ; det(Y)
     [,1] [,2] [,3]
[1,]    3    2    3
[2,]    2    6    6
[3,]    3    6   11
[1] 64
```

(5) Find the determinants of $\boldsymbol{S} = \begin{pmatrix} 1+\alpha & 1 & \beta \\ 1 & 1+\alpha & \beta \\ \beta & \beta & \alpha+\beta^2 \end{pmatrix}$.

$$|\boldsymbol{S}| = \begin{vmatrix} 1+\alpha & 1 & \beta \\ 1 & 1+\alpha & \beta \\ \beta & \beta & \alpha+\beta^2 \end{vmatrix} = \begin{vmatrix} \alpha & -\alpha & 0 \\ 1 & 1+\alpha & \beta \\ \beta & \beta & \alpha+\beta^2 \end{vmatrix}$$

(subtracting the second row from the first)

$$= \begin{vmatrix} \alpha & 0 & 0 \\ 1 & 2+\alpha & \beta \\ \beta & 2\beta & \alpha+\beta^2 \end{vmatrix} \text{ (adding the first column to the second)}$$

$$= \alpha\left((2+\alpha)(\alpha+\beta^2) - 2\beta^2\right) = \alpha^2(2+\alpha+\beta^2).$$

(6) Find the determinants of $\boldsymbol{S} = \begin{pmatrix} 2 & 1 & 3 \\ 1 & 2 & 3 \\ 3 & 3 & 10 \end{pmatrix}$.

$$\begin{aligned} |\boldsymbol{S}| &= \begin{vmatrix} 2 & 1 & 3 \\ 1 & 2 & 3 \\ 3 & 3 & 10 \end{vmatrix} = \begin{vmatrix} 1 & -1 & 0 \\ 1 & 2 & 3 \\ 3 & 3 & 10 \end{vmatrix} \\ &= \begin{vmatrix} 1 & 0 & 0 \\ 1 & 3 & 3 \\ 3 & 6 & 10 \end{vmatrix} = 1.(30-18) = 12 \end{aligned}$$

(first subtracting the second row from the first and then adding the first column to the second).
Check in **R**:

```
> det(matrix(c(2,1,3,1,2,3,3,3,10),3,3))
[1] 12
```

Note the nested commands `det(matrix(c(..)))`.

(7) (Equicorrelation matrix) If $X = \sigma^2 \begin{pmatrix} 1 & \rho & \rho & \cdots & \rho \\ \rho & 1 & \rho & \cdots & \rho \\ \vdots & \vdots & \ddots & \vdots & \vdots \\ \rho & \cdots & \cdots & \ddots & \rho \\ \rho & \rho & \cdots & \rho & 1 \end{pmatrix}$ show that $X = \sigma^2[(1-\rho)\mathbf{I_n} + \rho \iota_n \iota_n']$ and hence that $|X| = \sigma^{2n}[1+(n-1)\rho](1-\rho)^{(n-1)}$.

Note that $(1-\rho)\mathbf{I_n} + \rho\iota_n\iota_n' = (1-\rho)\mathbf{I_n} + \rho\mathbf{J_n}$ (where $\mathbf{J_n}$ is the $n \times n$ matrix with all elements 1). The diagonal elements of this are $(1-\rho)+\rho = 1$ and the off-diagonal elements are $0+\rho = \rho$, thus $X = \sigma^2[(1-\rho)\mathbf{I_n} + \rho\iota_n\iota_n']$. We have

$$\begin{aligned} |X| &= \left|\sigma^2[(1-\rho)\mathbf{I_n} + \rho\iota_n\iota_n']\right| = \sigma^{2n}(1-\rho)^n \left|\mathbf{I_n} + \frac{\rho}{1-\rho}\iota_n\iota_n'\right| \\ &= \sigma^{2n}(1-\rho)^n \left|\mathbf{I_1} + \frac{\rho}{1-\rho}\iota_n'\iota_n\right| \quad \text{(noting special case (i) in §4.6.2)} \\ &= \sigma^{2n}(1-\rho)^n\left(1+\frac{\rho}{1-\rho}n\right) = \sigma^{2n}[1+(n-1)\rho](1-\rho)^{(n-1)}. \end{aligned}$$

(8) If $X = \begin{pmatrix} 1+\alpha & 1 & \beta \\ 1 & 1+\alpha & \beta \\ \beta & \beta & \alpha+\beta^2 \end{pmatrix}$ show that $|X| = \alpha^2(2+\alpha+\beta^2)$ by showing that $X = \alpha\mathbf{I_3} + xx'$ for a suitable choice of x.

If $x = (1,1,\beta)$ then $x = \begin{pmatrix} 1 & 1 & \beta \\ 1 & 1 & \beta \\ \beta & \beta & \beta^2 \end{pmatrix}$ and so $X = \alpha\mathbf{I_3} + xx'$.

Thus $|X| = |\alpha\mathbf{I_3} + xx'| = \alpha^3\,|\mathbf{I_1} + x'x/\alpha|$ and since $x'x = \sum_i x_i^2 = 2+\beta^2$ we have $|X| = \alpha^2(2+\alpha+\beta^2)$.

(9) Use the results of the previous exercise to evaluate the determinant of

$$S = \begin{pmatrix} 2 & 1 & 3 \\ 1 & 2 & 3 \\ 3 & 3 & 10 \end{pmatrix}.$$

Inspection shows that taking $\alpha = 1$ and $\beta = 3$ in the previous exercise gives the matrix $S = \mathbf{I_3} + (1,1,3)(1,1,3)'$ and so $|S| = 1^2(2+1+3^2) = 12$, in agreement with Exercise (6).

(10) To show $\begin{vmatrix} \boldsymbol{A} & \boldsymbol{B} \\ \boldsymbol{B} & \boldsymbol{A} \end{vmatrix} = |\boldsymbol{A}+\boldsymbol{B}|\,|\boldsymbol{A}-\boldsymbol{B}|$ for $n \times n$ matrices $\boldsymbol{A}$ and $\boldsymbol{B}$:

(a) Show $\begin{pmatrix} \mathbf{I_n} & \mathbf{I_n} \\ \mathbf{0} & \mathbf{I_n} \end{pmatrix} \begin{pmatrix} \boldsymbol{A} & \boldsymbol{B} \\ \boldsymbol{B} & \boldsymbol{A} \end{pmatrix} = \begin{pmatrix} \boldsymbol{A}+\boldsymbol{B} & \boldsymbol{B}+\boldsymbol{A} \\ \boldsymbol{B} & \boldsymbol{A} \end{pmatrix}$.

This follows directly from the rule for multiplying partitioned matrices; see §2.6.2.

(b) Show $\begin{pmatrix} \boldsymbol{A}+\boldsymbol{B} & \mathbf{0} \\ \mathbf{0} & \mathbf{I_n} \end{pmatrix} \begin{pmatrix} \mathbf{I_n} & \mathbf{I_n} \\ \boldsymbol{B} & \boldsymbol{A} \end{pmatrix} = \begin{pmatrix} \boldsymbol{A}+\boldsymbol{B} & \boldsymbol{B}+\boldsymbol{A} \\ \boldsymbol{B} & \boldsymbol{A} \end{pmatrix}$.

This follows directly from the rule for multiplying partitioned matrices; see §2.6.2.

(c) Show $\begin{pmatrix} \mathbf{I_n} & \mathbf{I_n} \\ \boldsymbol{B} & \boldsymbol{A} \end{pmatrix} = \begin{pmatrix} \mathbf{I_n} & \mathbf{0} \\ \boldsymbol{B} & \mathbf{I_n} \end{pmatrix} \begin{pmatrix} \mathbf{I_n} & \mathbf{I_n} \\ \mathbf{0} & \boldsymbol{A}-\boldsymbol{B} \end{pmatrix}$.

This follows directly from the rule for multiplying partitioned matrices; see §2.6.2.

(d) Show $\begin{vmatrix} \boldsymbol{A} & \boldsymbol{B} \\ \boldsymbol{B} & \boldsymbol{A} \end{vmatrix} = |\boldsymbol{A}+\boldsymbol{B}|\,|\boldsymbol{A}-\boldsymbol{B}|$.

Thus

$$\begin{pmatrix} \mathbf{I_n} & \mathbf{I_n} \\ \mathbf{0} & \mathbf{I_n} \end{pmatrix} \begin{pmatrix} \boldsymbol{A} & \boldsymbol{B} \\ \boldsymbol{B} & \boldsymbol{A} \end{pmatrix} = \begin{pmatrix} \boldsymbol{A}+\boldsymbol{B} & \mathbf{0} \\ \mathbf{0} & \mathbf{I_n} \end{pmatrix} \begin{pmatrix} \mathbf{I_n} & \mathbf{0} \\ \boldsymbol{B} & \mathbf{I_n} \end{pmatrix} \begin{pmatrix} \mathbf{I_n} & \mathbf{I_n} \\ \mathbf{0} & \boldsymbol{A}-\boldsymbol{B} \end{pmatrix}.$$

Noting $\begin{vmatrix} \mathbf{I_n} & \mathbf{I_n} \\ \mathbf{0} & \mathbf{I_n} \end{vmatrix} = 1 = \begin{vmatrix} \mathbf{I_n} & \mathbf{0} \\ \boldsymbol{B} & \mathbf{I_n} \end{vmatrix}$

and $\begin{vmatrix} \boldsymbol{A}+\boldsymbol{B} & \mathbf{0} \\ \mathbf{0} & \mathbf{I_n} \end{vmatrix} = |\boldsymbol{A}+\boldsymbol{B}|$ and $\begin{vmatrix} \mathbf{I_n} & \mathbf{I_n} \\ \mathbf{0} & \boldsymbol{A}-\boldsymbol{B} \end{vmatrix} = |\boldsymbol{A}-\boldsymbol{B}|$

we have $\begin{vmatrix} \boldsymbol{A} & \boldsymbol{B} \\ \boldsymbol{B} & \boldsymbol{A} \end{vmatrix} = |\boldsymbol{A}+\boldsymbol{B}|\,|\boldsymbol{A}-\boldsymbol{B}|$.

(11) Let $\boldsymbol{A} = \begin{pmatrix} 2 & 3 \\ 3 & 3 \\ 3 & 3 \end{pmatrix}$ and $\boldsymbol{B} = \begin{pmatrix} 3 \\ 3 \\ 2 \end{pmatrix}$ and $\boldsymbol{X} = \begin{pmatrix} \boldsymbol{A} & \boldsymbol{B} \end{pmatrix}$.

```
> A<-matrix(c(2,3,3,3,3,3),3)
> B<-matrix(C(3,3,2),3)
> X<-cbind(A,B)
> X
     [,1] [,2] [,3]
[1,]    2    3    3
[2,]    3    3    3
[3,]    3    3    2
```

```
> A;B
     [,1] [,2]
[1,]    2    3
[2,]    3    3
[3,]    3    3
     [,1]
[1,]    3
[2,]    3
[3,]    2
```

(a) Find $\begin{vmatrix} A'A & A'B \\ B'A & B'B \end{vmatrix}$.

```
> C<-cbind(t(A)%*%A,t(A)%*%B)
> D<-cbind(t(B)%*%A,t(B)%*%B)
> det(rbind(C,D))
[1] 9
```

(b) Find $|AA'+BB'|$.

```
> det(A%*%t(A)+B%*%t(B))
[1] 9
```

(c) Find $|X|$.

```
> det(X)
[1] 3
```

(12) If A and B are $n\times m$ and $n\times(n-m)$ matrices and $X=\begin{pmatrix} A & B \end{pmatrix}$ prove that $|X|^2=|AA'+BB'|=\begin{vmatrix} A'A & A'B \\ B'A & B'B \end{vmatrix}$.

$|X|^2=|X||X'|=|XX'|=\left|\begin{pmatrix} A & B \end{pmatrix}\begin{pmatrix} A & B \end{pmatrix}'\right|$

$=\left|\begin{pmatrix} A & B \end{pmatrix}\begin{pmatrix} A' \\ B' \end{pmatrix}\right|=|AA'+BB'|.$

$|X|^2=|X'||X|=|X'X|=\left|\begin{pmatrix} A' \\ B' \end{pmatrix}\begin{pmatrix} A & B \end{pmatrix}\right|=\begin{vmatrix} A'A & A'B \\ B'A & B'B \end{vmatrix}.$

(13) Show that

$$\begin{vmatrix} 1+\lambda_1 & \lambda_2 & \cdots & \lambda_n \\ \lambda_1 & 1+\lambda_2 & \cdots & \lambda_n \\ \vdots & \vdots & \ddots & \vdots \\ \lambda_1 & \lambda_2 & \cdots & 1+\lambda_n \end{vmatrix}=1+\lambda_1+\lambda_2+\cdots+\lambda_n.$$

$$\begin{vmatrix} 1+\lambda_1 & \lambda_2 & \cdots & \lambda_n \\ \lambda_1 & 1+\lambda_2 & \cdots & \lambda_n \\ \vdots & \vdots & \ddots & \vdots \\ \lambda_1 & \lambda_2 & \cdots & 1+\lambda_n \end{vmatrix} = |\mathbf{I_n}+\iota_{\mathbf{n}}\lambda'| \text{ where } \lambda=(\lambda_1,\lambda_2,\ldots,\lambda_n)'$$

$$= |1+\lambda'\iota_{\mathbf{n}}|=1+\lambda_1+\lambda_2+\cdots+\lambda_n.$$

(14) By considering the matrix

$$\begin{pmatrix} E & 0 \\ 0 & \mathbf{I_n} \end{pmatrix} \text{ show that } \begin{vmatrix} EA & EB \\ C & D \end{vmatrix}=|E|\begin{vmatrix} A & B \\ C & D \end{vmatrix}.$$

$$\begin{pmatrix} E & 0 \\ 0 & \mathbf{I_n} \end{pmatrix}\begin{pmatrix} A & B \\ C & D \end{pmatrix}=\begin{pmatrix} EA & EB \\ C & D \end{pmatrix} \text{ and } \begin{vmatrix} E & 0 \\ 0 & \mathbf{I_n} \end{vmatrix}=|E|$$

and the result follows.

(15) By considering the matrix

$$\begin{pmatrix} \mathbf{I_m} & 0 \\ E & \mathbf{I_n} \end{pmatrix} \text{ show that } \begin{vmatrix} A & B \\ C+EA & D+EB \end{vmatrix}=\begin{vmatrix} A & B \\ C & D \end{vmatrix}.$$

$$\begin{pmatrix} \mathbf{I_m} & \mathbf{0} \\ \boldsymbol{E} & \mathbf{I_n} \end{pmatrix}\begin{pmatrix} \boldsymbol{A} & \boldsymbol{B} \\ \boldsymbol{C} & \boldsymbol{D} \end{pmatrix} = \begin{pmatrix} \boldsymbol{A} & \boldsymbol{B} \\ \boldsymbol{C}+\boldsymbol{EA} & \boldsymbol{D}+\boldsymbol{EB} \end{pmatrix} \text{ and } \begin{vmatrix} \mathbf{I_m} & \mathbf{0} \\ \boldsymbol{E} & \mathbf{I_n} \end{vmatrix} = 1$$

and the result follows.

(16) Suppose

$$\boldsymbol{Z} = \begin{pmatrix} \boldsymbol{A} & \boldsymbol{B} \\ \boldsymbol{C} & \boldsymbol{D} \end{pmatrix} \text{ and } \boldsymbol{Z}^{-1} = \begin{pmatrix} \boldsymbol{P} & \boldsymbol{Q} \\ \boldsymbol{R} & \boldsymbol{S} \end{pmatrix}.$$

(i) Express $\boldsymbol{ZZ}^{-1}$ in terms of the eight sub-blocks of $\boldsymbol{Z}$ and $\boldsymbol{Z}^{-1}$.

$$\boldsymbol{ZZ}^{-1} = \begin{pmatrix} \boldsymbol{AP}+\boldsymbol{BR} & \boldsymbol{AQ}+\boldsymbol{BS} \\ \boldsymbol{CP}+\boldsymbol{DR} & \boldsymbol{CQ}+\boldsymbol{DS} \end{pmatrix}$$

(ii) Show that

$$\begin{pmatrix} \boldsymbol{A} & \boldsymbol{B} \\ \boldsymbol{C} & \boldsymbol{D} \end{pmatrix}\begin{pmatrix} \mathbf{I_m} & \boldsymbol{Q} \\ \mathbf{0} & \boldsymbol{S} \end{pmatrix} = \begin{pmatrix} \boldsymbol{A} & \mathbf{0} \\ \boldsymbol{C} & \mathbf{I_n} \end{pmatrix}.$$

$$\begin{pmatrix} \boldsymbol{A} & \boldsymbol{B} \\ \boldsymbol{C} & \boldsymbol{D} \end{pmatrix}\begin{pmatrix} \mathbf{I_m} & \boldsymbol{Q} \\ \mathbf{0} & \boldsymbol{S} \end{pmatrix} = \begin{pmatrix} \mathbf{I_m} & \boldsymbol{AQ}+\boldsymbol{BS} \\ \mathbf{0} & \boldsymbol{CQ}+\boldsymbol{DS} \end{pmatrix} = \begin{pmatrix} \boldsymbol{A} & \mathbf{0} \\ \boldsymbol{C} & \mathbf{I_n} \end{pmatrix},$$

noting $\boldsymbol{AQ}+\boldsymbol{BS} = \mathbf{0}$ and $\boldsymbol{CQ}+\boldsymbol{DS} = \mathbf{I_n}$ because $\boldsymbol{ZZ}^{-1} = \begin{pmatrix} \mathbf{I_m} & \mathbf{0} \\ \mathbf{0} & \mathbf{I_n} \end{pmatrix}$.

(iii) Show that

$$\begin{pmatrix} \boldsymbol{A} & \boldsymbol{B} \\ \boldsymbol{C} & \boldsymbol{D} \end{pmatrix}\begin{pmatrix} \boldsymbol{P} & \mathbf{0} \\ \boldsymbol{R} & \mathbf{I_n} \end{pmatrix} = \begin{pmatrix} \boldsymbol{A} & \mathbf{0} \\ \boldsymbol{C} & \mathbf{I_n} \end{pmatrix}.$$

Similarly $\boldsymbol{AP}+\boldsymbol{BR} = \mathbf{I_m}$ and $\boldsymbol{CP}+\boldsymbol{DR} = \mathbf{0}$ and the result follows.

(iv) Show that

$$|\boldsymbol{Z}| = \frac{|\boldsymbol{A}|}{|\boldsymbol{S}|} = \frac{|\boldsymbol{D}|}{|\boldsymbol{P}|}.$$

Note that

$$\begin{vmatrix} \boldsymbol{A} & \mathbf{0} \\ \boldsymbol{C} & \mathbf{I_n} \end{vmatrix} = |\boldsymbol{A}| \text{ and } \begin{vmatrix} \mathbf{I_m} & \boldsymbol{Q} \\ \mathbf{0} & \boldsymbol{S} \end{vmatrix} = |\boldsymbol{S}| \text{ (etc.) gives } |\boldsymbol{Z}| = \frac{|\boldsymbol{A}|}{|\boldsymbol{S}|} = \frac{|\boldsymbol{D}|}{|\boldsymbol{P}|}.$$

(17) If $|\boldsymbol{A}| \neq 0$ and $\boldsymbol{A}$ and $\boldsymbol{C}$ commute show that $\begin{vmatrix} \boldsymbol{A} & \boldsymbol{B} \\ \boldsymbol{C} & \boldsymbol{D} \end{vmatrix} = |\boldsymbol{AD}-\boldsymbol{CB}|$.

$\begin{vmatrix} \boldsymbol{A} & \boldsymbol{B} \\ \boldsymbol{C} & \boldsymbol{D} \end{vmatrix} = |\boldsymbol{A}||\boldsymbol{D}-\boldsymbol{CA}^{-1}\boldsymbol{B}|$ (see §4.6)

$= |\boldsymbol{AD}-\boldsymbol{ACA}^{-1}\boldsymbol{B}| = |\boldsymbol{AD}-\boldsymbol{CAA}^{-1}\boldsymbol{B}| = |\boldsymbol{AD}-\boldsymbol{CB}|$.

Chapter 5: Inverses

(1) Find the inverses, both 'by hand' and with **R** of

(i) $A_1 = \begin{pmatrix} 12 & 7 \\ -4 & 6 \end{pmatrix}$.

$|A_1| = 12 \times 6 - 7 \times (-4) = 100;\ A_1^{-1} = \frac{1}{100}\begin{pmatrix} 6 & -7 \\ 4 & 12 \end{pmatrix}$.

```
> A1<-matrix(c(12,-4,7,6),2,2)
> A1;det(A1);solve(A1)
```

```
     [,1] [,2]
[1,]   12    7
[2,]   -4    6
[1] 100
```

```
     [,1]  [,2]
[1,] 0.06 -0.07
[2,] 0.04  0.12
```

(ii) $A_2 = \begin{pmatrix} 0 & 5 \\ -4 & 1 \end{pmatrix}$.

$|A_2| = 0 \times 1 - 5 \times (-4) = 20;\ A_2^{-1} = \frac{1}{20}\begin{pmatrix} 1 & -5 \\ 4 & 0 \end{pmatrix}$.

```
> A2<-matrix(c(0,-4,5,1),2,2)
> A2;det(A2);solve(A2)
```

```
     [,1] [,2]
[1,]    0    5
[2,]   -4    1
[1] 20
```

```
     [,1]  [,2]
[1,] 0.05 -0.25
[2,] 0.20  0.00
```

(iii) $A_3 = \begin{pmatrix} 1 & 5 \\ -4 & 0 \end{pmatrix}$.

$|A_3| = 1 \times 0 - 5 \times (-4) = 20;\ A_3^{-1} = \frac{1}{20}\begin{pmatrix} 0 & -5 \\ 4 & 1 \end{pmatrix}$.

```
> A3<-matrix(c(1,-4,5,0),2,2)
> A3;det(A3);solve(A3)
```

```
     [,1] [,2]
[1,]    1    5
[2,]   -4    0
[1] 20
```

```
     [,1]  [,2]
[1,]  0.0 -0.25
[2,]  0.2  0.05
```

(iv) $A_4 = \begin{pmatrix} 0 & 5 \\ -4 & 0 \end{pmatrix}$.

$|A_4| = 0 \times 0 - 5 \times (-4) = 20;\ A_4^{-1} = \frac{1}{20}\begin{pmatrix} 0 & -5 \\ 4 & 0 \end{pmatrix}$.

```
> A4<-matrix(c(0,-4,5,0),2,2)
>  A4;det(A4);solve(A4)
```

```
     [,1] [,2]
[1,]    0    5
[2,]   -4    0
[1] 20
```

```
     [,1]  [,2]
[1,]  0.0 -0.25
[2,]  0.2  0.00
```

(v) $A_5 = \begin{pmatrix} 0 & 0 & 0 & 5 \\ 0 & 0 & 4 & 0 \\ 0 & 7 & 0 & 0 \\ 3 & 0 & 0 & 0 \end{pmatrix}$.

$|A_5| = 3 \times 7 \times 4 \times 5 = 420$; $A_5^{-1} = \begin{pmatrix} 0 & 0 & 0 & 3^{-1} \\ 0 & 0 & 7^{-1} & 0 \\ 0 & 4^{-1} & 0 & 0 \\ 5^{-1} & 0 & 0 & 0 \end{pmatrix}$.

```
> A5<-matrix(c(0,0,0,3,0,0,7,0,0,4,0,0,5,0,0,0),4,4)
>  A5;det(A5);solve(A5)
```

```
     [,1] [,2] [,3] [,4]
[1,]    0    0    0    5
[2,]    0    0    4    0
[3,]    0    7    0    0
[4,]    3    0    0    0
[1] 420
```

```
     [,1] [,2] [,3] [,4]
[1,]  0.0 0.00 0.00 0.33
[2,]  0.0 0.00 0.14 0.00
[3,]  0.0 0.25 0.00 0.00
[4,]  0.2 0.00 0.00 0.00
>
```

(2) Suppose $\boldsymbol{AB} = \boldsymbol{BA}$ and that $\boldsymbol{A}$ is a non-singular $n \times n$ matrix. Show that $\boldsymbol{A}^{-1}\boldsymbol{B} = \boldsymbol{BA}^{-1}$.

$\boldsymbol{AB} = \boldsymbol{BA}$ so $\boldsymbol{ABA}^{-1} = \boldsymbol{BAA}^{-1} = \boldsymbol{B}\mathbf{I_n} = \boldsymbol{B}$ (postmultiplying by $\boldsymbol{A}^{-1}$), so $\boldsymbol{A}^{-1}\boldsymbol{ABA}^{-1} = \boldsymbol{A}^{-1}\boldsymbol{B}$ (premultiplying by $\boldsymbol{A}^{-1}$), so $\boldsymbol{BA}^{-1} = \boldsymbol{A}^{-1}\boldsymbol{B}$.

(3) Suppose $\boldsymbol{A}$ is an $n \times n$ orthogonal matrix and $\boldsymbol{B}$ is $n \times n$. Show that $\boldsymbol{AB}$ is orthogonal if and only if $\boldsymbol{B}$ is orthogonal.

$\boldsymbol{A}$ is orthogonal so $\boldsymbol{AA}' = \boldsymbol{A}'\boldsymbol{A} = \mathbf{I_n}$. Suppose $\boldsymbol{B}$ is orthogonal, then $\boldsymbol{BB}' = \boldsymbol{B}'\boldsymbol{B}$ so $\boldsymbol{AB}(\boldsymbol{AB})' = \boldsymbol{ABB}'\boldsymbol{A}' = \boldsymbol{A}\mathbf{I_n}\boldsymbol{A}' = \boldsymbol{AA}' = \mathbf{I_n}$. Similarly $(\boldsymbol{AB})'\boldsymbol{AB} = \mathbf{I_n}$ and thus $\boldsymbol{AB}$ is orthogonal.

Suppose now $\boldsymbol{AB}$ is orthogonal then $\boldsymbol{AB}(\boldsymbol{AB})' = \mathbf{I_n}$, i.e., $\boldsymbol{ABB}'\boldsymbol{A}' = \mathbf{I_n}$ so $\boldsymbol{A}'\boldsymbol{ABB}'\boldsymbol{A} = \boldsymbol{A}'\boldsymbol{A}$ (premultiplying by $\boldsymbol{A}'$ and postmultiplying by $\boldsymbol{A}$) so $\boldsymbol{BB}' = \mathbf{I_n}$. Similarly $(\boldsymbol{AB})'\boldsymbol{AB} = \mathbf{I_n}$ so $\boldsymbol{B}'\boldsymbol{B} = \mathbf{I_n}$ and thus $\boldsymbol{B}$ is orthogonal.

(4) Suppose $\boldsymbol{X}$ and $\boldsymbol{Y}$ are non-singular $n \times n$ matrices and all other matrices stated in this exercise are also non-singular.

(a) Show that $(\mathbf{I_n}+X)^{-1}=X^{-1}(\mathbf{I_n}+X^{-1})^{-1}$.

$(\mathbf{I_n}+X)X^{-1}(\mathbf{I_n}+X^{-1})^{-1}=X^{-1}(\mathbf{I_n}+X^{-1})^{-1}+XX^{-1}(\mathbf{I_n}+X^{-1})^{-1}$
$=(\mathbf{I_n}+X^{-1})(X^{-1}+\mathbf{I_n})^{-1}=\mathbf{I_n}$.

(b) Show that $(X+YY')^{-1}Y=X^{-1}Y(\mathbf{I_n}+Y'X^{-1}Y)^{-1}$.

$\left((X+YY')\right)\left(X^{-1}Y(\mathbf{I_n}+Y'X^{-1}Y)^{-1}Y^{-1}\right)$
$=(XX^{-1}Y+YY'X^{-1}Y)(\mathbf{I_n}+Y'X^{-1}Y)^{-1}Y^{-1}$
$=Y(\mathbf{I_n}+Y'X^{-1}Y)(\mathbf{I_n}+Y'X^{-1}Y)^{-1}Y^{-1}$
$=Y\mathbf{I_n}Y^{-1}=\mathbf{I_n}$.

(c) Show that $(X+Y)^{-1}=X^{-1}(X^{-1}+Y^{-1})^{-1}Y^{-1}=Y^{-1}(X^{-1}+Y^{-1})^{-1}X^{-1}$.

$(X+Y)X^{-1}(X^{-1}+Y^{-1})^{-1}Y^{-1}=(\mathbf{I_n}+YX^{-1})(X^{-1}+Y^{-1})^{-1}Y^{-1}$
$=Y(Y^{-1}+X^{-1})(X^{-1}+Y^{-1})^{-1}Y^{-1}=Y\mathbf{I_n}Y^{-1}=\mathbf{I_n}$
So $(X+Y)^{-1}=X^{-1}(X^{-1}+Y^{-1})^{-1}Y^{-1}$ and also
$(X+Y)^{-1}=Y^{-1}(X^{-1}+Y^{-1})^{-1}X^{-1}$ by symmetry.

(d) Show that $X^{-1}+Y^{-1}=X^{-1}(X+Y)Y^{-1}$.

$X^{-1}(X+Y)Y^{-1}=(\mathbf{I_n}+X^{-1}Y)Y^{-1}=Y^{-1}+X^{-1}\mathbf{I_n}=X^{-1}+Y^{-1}$.

(e) Show that $X-X(X+Y)^{-1}X=Y-Y(X+Y)^{-1}Y$.

$X-X(X+Y)^{-1}X=X(X+Y)^{-1}((X+Y)-X)$
$=(X+Y-Y)(X+Y)^{-1}Y=Y-Y(X+Y)^{-1}Y$.

(f) Show that if $(X+Y)^{-1}=X^{-1}+Y^{-1}$ then $XY^{-1}X=YX^{-1}Y$.

If $(X+Y)^{-1}=X^{-1}+Y^{-1}$ then $\mathbf{I_n}=(X+Y)(X^{-1}+Y^{-1})=2\mathbf{I_n}+XY^{-1}+YX^{-1}$, so $X+Y+XY^{-1}X=0$.
Similarly $X+Y+YX^{-1}Y=0$ so $XY^{-1}X=YX^{-1}Y$.

(g) Show that $(\mathbf{I_n}+XY)^{-1}=\mathbf{I_n}-X(\mathbf{I_n}+YX)^{-1}Y$.

$(\mathbf{I_n}+XY)(\mathbf{I_n}-X(\mathbf{I_n}+YX)^{-1}Y)$
$=\mathbf{I_n}+XY-X(\mathbf{I_n}+YX)^{-1}Y-XYX(\mathbf{I_n}+YX)^{-1}Y$
$=\mathbf{I_n}+XY-X((\mathbf{I_n}+YX)^{-1}+YX(\mathbf{I_n}+YX)^{-1})Y$
$=\mathbf{I_n}+XY-X(\mathbf{I_n}+YX)^{-1}(\mathbf{I_n}+YX)Y=\mathbf{I_n}+XY-X\mathbf{I_n}Y=\mathbf{I_n}$.

(h) Show that $(\mathbf{I_n}+XY)^{-1}X=X(\mathbf{I_n}+YX)^{-1}$.

$(\mathbf{I_n}+XY)^{-1}X=(\mathbf{I_n}-X(\mathbf{I_n}+YX)^{-1}Y)X=X(\mathbf{I_n}-(\mathbf{I_n}+YX)^{-1}YX)$
$=X(\mathbf{I_n}+(\mathbf{I_n}+YX)^{-1}-(\mathbf{I_n}+YX)^{-1}-(\mathbf{I_n}+YX)^{-1}YX)$
$=X(\mathbf{I_n}+(\mathbf{I_n}+YX)^{-1}-(\mathbf{I_n}+YX)^{-1}(\mathbf{I_n}+YX))$
$=X(\mathbf{I_n}+YX)^{-1}$.

(5) Show that $(A+BCB')^{-1}=A^{-1}-A^{-1}B[C^{-1}+B'A^{-1}B]^{-1}B'A^{-1}$ where A and C are non-singular $m\times m$ and $n\times n$ matrices and B is $m\times n$.

Putting $D=[C^{-1}+B'A^{-1}B]$ gives

$$\begin{aligned}(A+BCB')(A^{-1}&-A^{-1}B[C^{-1}+B'A^{-1}B]^{-1}B'A^{-1})\\ &=\mathbf{I_n}-BD^{-1}B'A^{-1}+BCB'A^{-1}-BCB'A^{-1}BD^{-1}B'A^{-1}\\ &=\mathbf{I_n}-BC\{-C^{-1}D^{-1}+\mathbf{I_n}-B'A^{-1}BD^{-1}\}B'A^{-1}\\ &=\mathbf{I_n}-BC\{\mathbf{I_n}-DD^{-1}\}B'A^{-1}=\mathbf{I_n}.\end{aligned}$$

(6) Show that $\begin{pmatrix} A & B \\ 0 & D \end{pmatrix}^{-1} = \begin{pmatrix} A^{-1} & X \\ 0 & D^{-1} \end{pmatrix}$ for a suitable X.

$$\begin{pmatrix} A & B \\ 0 & D \end{pmatrix}\begin{pmatrix} A^{-1} & X \\ 0 & D^{-1} \end{pmatrix} = \begin{pmatrix} AA^{-1} & AX+BD^{-1} \\ 0 & DD^{-1} \end{pmatrix} = \mathbf{I_{m+n}}$$

if $X = -A^{-1}BD^{-1}$.

(7) Let $A = \begin{pmatrix} \mathbf{I_n} & \lambda\iota_\mathbf{n} \\ \lambda\iota_\mathbf{n}' & 1 \end{pmatrix}$.

(i) For what values of λ is A non-singular?

Using the general result in §4.6 gives

$$\begin{vmatrix} \mathbf{I_n} & \lambda\iota_\mathbf{n} \\ \lambda\iota_\mathbf{n}' & 1 \end{vmatrix} = |\mathbf{I_n}||1-\lambda\iota_\mathbf{n}'\mathbf{I_n}^{-1}\lambda\iota_\mathbf{n}| = 1-n\lambda^2$$

So the matrix is non-singular for all values of $\lambda \neq \pm n^{-1/2}$.

(ii) Find A^{-1} when it exists.

Using result (xii) of §5.5.1 gives

$$A^{-1} = \begin{pmatrix} \mathbf{I_n} + \frac{\lambda^2}{1-n\lambda^2}\mathbf{J_n} & -\frac{\lambda}{1-n\lambda^2}\iota_\mathbf{n} \\ -\frac{\lambda}{1-n\lambda^2}\iota_\mathbf{n}' & \frac{1}{1-n\lambda^2} \end{pmatrix}$$

and it is easy to verify that $AA^{-1} = A^{-1}A = \mathbf{I_n}$.

(8) Suppose A is skew-symmetric and non-singular.

(i) Show that A^{-1} is skew-symmetric.
$(A^{-1})' = (A')^{-1} = -A^{-1}$ and so A^{-1} is skew-symmetric.

(ii) Show that $(\mathbf{I_n}-A)(\mathbf{I_n}+A)^{-1}$ is orthogonal.

$$\begin{aligned}(\mathbf{I_n}-A)(\mathbf{I_n}+A)^{-1}&((\mathbf{I_n}-A)(\mathbf{I_n}+A)^{-1})' \\ &= (\mathbf{I_n}-A)(\mathbf{I_n}+A)^{-1}\left((\mathbf{I_n}+A)'\right)^{-1}(\mathbf{I_n}-A)' \\ &= (\mathbf{I_n}-A)(\mathbf{I_n}+A)^{-1}\left((\mathbf{I_n}-A)\right)^{-1}(\mathbf{I_n}+A) \\ &= (\mathbf{I_n}-A)(\mathbf{I_n}+A)^{-1}(\mathbf{I_n}+A)(\mathbf{I_n}-A)^{-1} = \mathbf{I_n}\end{aligned}$$

(noting that $(\mathbf{I_n}-A)$ and $(\mathbf{I_n}+A)$ commute)

and thus $(\mathbf{I_n}-A)(\mathbf{I_n}+A)^{-1}$ is orthogonal.

(9) Suppose that $AX = \mathbf{0}$ and A is idempotent. Let $B = (X-A)^{-1}$. Prove that

(i) $XB = \mathbf{I_n} - A$.
$XB = X(X-A)^{-1} = (A+X-A)(X-A)^{-1} = A(X-A)^{-1} + \mathbf{I_n}$
$= AB + \mathbf{I_n}$.
Now $-A = AX - A = AX - A^2 = A(X-A) = AB^{-1}$ so $AB = -A$ and thus
$XB = \mathbf{I_n} - A$.

(ii) $XBX = X$.

$$XBX = (\mathbf{I_n} - A)X = X - AX = X.$$

(iii) $XBA = \mathbf{0}$.

$$XBA = (\mathbf{I_n} - A)A = A - A^2 = A - A = \mathbf{0}.$$

(10) Suppose $A = \mathbf{I_n} - 2xx'$ with $x'x = 1$. Find A^{-1}.

Suppose $A^{-1} = \lambda\mathbf{I_n}yy'$ then we require
$\mathbf{I_n} = (\mathbf{I_n} - 2xx')(\lambda\mathbf{I_n} - yy') = \lambda\mathbf{I_n} - 2\lambda xx' - yy' + 2xx'yy'$.
This is satisfied by taking $\lambda = 1$ and $y = x$. So $A^{-1} = A$.

(11) Show that $\begin{pmatrix} \mathbf{0} & A \\ \lambda & x' \end{pmatrix}^{-1} = \frac{1}{\lambda}\begin{pmatrix} -x'A^{-1} & 1 \\ \lambda A^{-1} & \mathbf{0} \end{pmatrix}$.

$$\begin{pmatrix} \mathbf{0} & A \\ \lambda & x' \end{pmatrix}\begin{pmatrix} -x'A^{-1} & 1 \\ \lambda A^{-1} & \mathbf{0} \end{pmatrix} = \begin{pmatrix} \mathbf{0} + \lambda\mathbf{I_n} & \mathbf{0} + \mathbf{0} \\ -\lambda x'A^{-1} + \lambda x'A^{-1} & \lambda + 0 \end{pmatrix} = \lambda\mathbf{I_n}.$$

Chapter 6: Eigenanalysis

(1) Let $X = \begin{pmatrix} 0 & 0 & 6 \\ 1/2 & 0 & 0 \\ 0 & 1/3 & 0 \end{pmatrix}$.

(i) Show that $|X| = 1$.

$$|X| = 6\begin{vmatrix} 1/2 & 0 \\ 0 & 1/3 \end{vmatrix} = 6 \times 1/2 \times 1/3 = 1.$$

(ii) Show that $\lambda = 1$ is an eigenvalue of X.

If $\lambda = 1$ then $|X - \lambda\mathbf{I_3}| = \begin{vmatrix} -1 & 0 & 6 \\ 1/2 & -1 & 0 \\ 0 & 1/3 & -1 \end{vmatrix} = 0$ (adding twice the second row to the first and then expanding by the first column). So 1 is an eigenvalue of X.

(iii) Show that $X^3 = \mathbf{I_n}$ but $X^2 \neq \mathbf{I_n}$.

Either by hand or in **R**:

```
> X<-matrix(c(0,1/2,0,0,0,1/3,6,0,0),3,3)
```

```
> X%*%X
     [,1] [,2] [,3]
[1,] 0.00    2    0
[2,] 0.00    0    3
[3,] 0.17    0    0
```

```
> X%*%X%*%X
     [,1] [,2] [,3]
[1,]    1    0    0
[2,]    0    1    0
[3,]    0    0    1
```

(2) Find the eigenvalues of the $n \times n$ matrix $X = \begin{pmatrix} 1 & \rho & \rho & \cdots & \rho \\ \rho & 1 & \rho & \cdots & \rho \\ \vdots & \vdots & \ddots & \vdots & \vdots \\ \rho & \cdots & \cdots & \ddots & \rho \\ \rho & \rho & \cdots & \rho & 1 \end{pmatrix}$ and show that one eigenvector is proportional to $\iota_\mathbf{n}$.

$X = (1-\rho)\mathbf{I_n} + \rho\iota_\mathbf{n}\iota_\mathbf{n}'$. The eigenvalues are the roots of $|X - \lambda\mathbf{I_n}| = |(1-\rho-\lambda)\mathbf{I_n} + \rho\iota_\mathbf{n}\iota_\mathbf{n}'| = (1-\rho-\lambda)^n|\mathbf{I_1} + \rho\iota_\mathbf{n}'\iota_\mathbf{n}/(1-\rho-\lambda)| = (1-\rho-\lambda)^{n-1}(1-\rho-\lambda+n\rho)$ and so the eigenvalues are $(1-\rho)$ (with multiplicity $(n-1)$ and $(1+(n-1)\rho)$.

$X\iota_\mathbf{n} = (1+(n-1)\rho, 1+(n-1)\rho, \ldots, 1+(n-1)\rho)' = (1+(n-1)\rho)\iota_\mathbf{n}$ so $\iota_\mathbf{n}$ is an eigenvector of X with eigenvalue $1+(n-1)\rho$. Since any scalar multiple of an eigenvector is also an eigenvector with the same eigenvalue it follows that the eigenvector of X corresponding to the eigenvalue $1+(n-1)\rho$ is proportional to $\iota_\mathbf{n}$.

(3) If X has eigenvalues 2λ, $\lambda+\alpha$ and $\lambda+3\alpha$ and $|X| = 80$ and $\text{tr}(X) = 16$ find the eigenvalues of X.

Since the trace and determinant of a matrix are equal to the sum and product of its eigenvalues we have $4\lambda + 4\alpha = 16$, so $\lambda + \alpha = 4$, and $2\lambda(\lambda+\alpha)(\lambda+3\alpha) = 80$ so $2\lambda(4)(4+2\alpha) = 80$ so $(4-\alpha)(2+\alpha) = 5$, giving $\alpha^2 - 2\alpha - 3 = 0$ so $\alpha = -1$ and $\lambda = 5$ or $\alpha = 3$ and $\lambda = 1$, giving the eigenvalues as 2, 5 and 10 in either case.

(4) If λ is an eigenvalue of X show that $\lambda + \alpha$ is an eigenvalue of $X + \alpha\mathbf{I_n}$.

Since λ is an eigenvalue of X with eigenvector x (say) we have $Xx = \lambda x$ so $(X + \alpha\mathbf{I_n})x = \lambda x + \alpha x = (\lambda+\alpha)x$ so $\lambda + \alpha$ is an eigenvalue of $X + \alpha\mathbf{I_n}$ with the same eigenvector x.

(5) If x is a $n \times 1$ vector find the spectral decomposition of xx'.

$xx'x = (x'x)x$ so x is an eigenvector of xx' with eigenvalue $x'x$ and this is the only non-zero eigenvalue since $xx'$is symmetric and of rank 1. A normalized eigenvector of xx' is $x/(x'x)^{1/2}$ and thus $\left(x/(x'x)^{1/2}\right)\left(\text{diag}(x'x)\right)\left(x/(x'x)^{1/2}\right)'$ is the spectral decomposition of xx'.

(6) If x is a $n \times 1$ vector and S is a symmetric $n \times n$ matrix find the spectral decomposition of xSx'.

$\left(Sx/(x'Sx)^{1/2}\right)\left(\text{diag}(x'Sx)\right)\left(Sx/(x'Sx)^{1/2}\right)'$ is the spectral decomposition of xSx'.

(7) If $T = kS$ where k is a scalar show that T and S have identical eigenvectors and the eigenvalues of T are obtained by multiplying those of S by k.

If (x, λ) is an eigenpair of S then $Sx = \lambda x$ so $T = kS = (k\lambda)x$ so $(x, k\lambda)$ is an eigenpair of T.

(8) Find the square root of the matrix

$$S = \begin{pmatrix} 2 & 1 & 1 \\ 1 & 2 & 1 \\ 1 & 1 & 3 \end{pmatrix}.$$

```
> S<-matrix(c(2,1,1,1,2,
+ 1,1,1,3),3,3)
> Z<-svd(S)
> SQ<-Z$u%*%sqrt(diag(Z$d))
+ %*%t(Z$v)
>
```

```
#### Check:
> SQ%*%SQ
     [,1] [,2] [,3]
[1,]    2    1    1
[2,]    1    2    1
[3,]    1    1    3
```

To write a function to calculate matrix square roots:

```
> SQR<-function(S){
+ Z<-svd(S)
+ SQR<-Z$u%*%sqrt(diag(Z$d))
+ %*%t(Z$v)
+ return(SQR)
+ }
>
```

```
#### Check:
> W<-SQR(S)
> W%*%W
     [,1] [,2] [,3]
[1,]    2    1    1
[2,]    1    2    1
[3,]    1    1    3
```

(9) If $\boldsymbol{X}$ is 2×2 show that $|\boldsymbol{X}+\mathbf{I_n}| = |\boldsymbol{X}|+1$ if and only if $\mathrm{tr}(\boldsymbol{X}) = 0$.

Suppose the eigenvalues of $\boldsymbol{X}$ are λ_1 and λ_2, then $|\boldsymbol{X}| = \lambda_1\lambda_2$. Also the eigenvalues of $|\boldsymbol{X}+\mathbf{I_n}|$ are λ_1+1 and λ_2+1 (see the previous exercise) so $|\boldsymbol{X}+\mathbf{I_n}| = (\lambda_1+1)(\lambda_2+1) = \lambda_1\lambda_2 + (\lambda_1+\lambda_2)+1 = |\boldsymbol{X}| + \mathrm{tr}(\boldsymbol{X}) + 1$ and the result follows.

(10) Suppose $\boldsymbol{X}$ is $n \times n$ matrix with distinct eigenvalues.

Let $\boldsymbol{X} = \boldsymbol{T}^{-1}\Lambda\boldsymbol{T}$ where Λ is a diagonal matrix with the eigenvalues λ_i, $i = 1,\ldots,n$ of $\boldsymbol{X}$. Then $\exp(\boldsymbol{X}) = \boldsymbol{T}^{-1}\mathrm{diag}(\exp(\lambda_i))\boldsymbol{T}$.

(a) Show that $\exp(\alpha\boldsymbol{X})\exp(\beta\boldsymbol{X}) = \exp((\alpha+\beta)\boldsymbol{X})$.

The eigenvalues of $\alpha\boldsymbol{X}$ are $\alpha\lambda_i$, $i = 1,\ldots,n$ so

$$\begin{aligned}\exp(\alpha\boldsymbol{X})\exp(\beta\boldsymbol{X}) &= \boldsymbol{T}^{-1}\mathrm{diag}(\exp(\alpha\lambda_i))\boldsymbol{T}\boldsymbol{T}^{-1}\mathrm{diag}(\exp(\beta\lambda_i))\boldsymbol{T} \\ &= \boldsymbol{T}^{-1}\mathrm{diag}((\exp(\alpha+\beta)\lambda_i))\boldsymbol{T} = \exp((\alpha+\beta)\boldsymbol{X}).\end{aligned}$$

(b) Show that $\boldsymbol{X}\exp(\boldsymbol{X}) = \exp(\boldsymbol{X})\boldsymbol{X}$.

$\boldsymbol{X}\exp(\boldsymbol{X}) = \boldsymbol{X}(\sum_{r=1}^{\infty}\frac{\boldsymbol{X}^r}{r!}) = \sum_{r=1}^{\infty}\frac{\boldsymbol{X}^{r+1}}{r!} = (\sum_{r=1}^{\infty}\frac{\boldsymbol{X}^r}{r!})\boldsymbol{X} = \exp(\boldsymbol{X})\boldsymbol{X}$.

(c) Show that $(\exp(\boldsymbol{X}))^r = \exp(r\boldsymbol{X})$ where r is an integer and $r > 0$.

$$\begin{aligned}(\exp(\boldsymbol{X}))^r &= (\boldsymbol{T}^{-1}\text{diag}(\exp(\lambda_i))\boldsymbol{T})^r \\ &= \boldsymbol{T}^{-1}(\text{diag}(\exp(\lambda_i)))^r\boldsymbol{T} \\ &= \boldsymbol{T}^{-1}\text{diag}(\exp(r\lambda_i))\boldsymbol{T} = \exp(r\boldsymbol{X})\end{aligned}$$

noting that $r\lambda_i$ are the eigenvalues of $r\boldsymbol{X}$.

(d) Show that $(\exp(\boldsymbol{X}))^{-1} = \exp(-\boldsymbol{X})$.
This follows by an argument similar to that in the previous exercise.

(e) Show that $(\exp(\boldsymbol{X}))' = \exp(\boldsymbol{X}')$.
$(\exp(\boldsymbol{X}))' = (\sum_{r=1}^{\infty}\frac{\boldsymbol{X}^r}{r!})' = \sum_{r=1}^{\infty}\frac{(\boldsymbol{X}')^r}{r!} = \exp(\boldsymbol{X}')$.

(f) Show that $|\exp(\boldsymbol{X})| = \exp(\text{tr}(\boldsymbol{X}))$.
$|\exp(\boldsymbol{X})| = |\boldsymbol{T}|^{-1}|\text{diag}(\exp(\lambda_i))||\boldsymbol{T}| = \prod_i \exp(\lambda_i) = \exp\sum_i \lambda_i = \exp(\text{tr}(\boldsymbol{X}))$.

(11) If $\boldsymbol{X}$ and $\boldsymbol{Y}$ are $n \times n$ matrices, each with distinct eigenvalues and $\boldsymbol{XY} = \boldsymbol{YX}$, show that $\exp(\boldsymbol{X}+\boldsymbol{Y}) = \exp(\boldsymbol{X})\exp(\boldsymbol{Y}) = \exp(\boldsymbol{Y})\exp\boldsymbol{X})$.

$$\begin{aligned}\exp(\boldsymbol{X})\exp(\boldsymbol{Y}) &= \sum_{j=0}^{\infty}\sum_{k=0}^{\infty}\frac{\boldsymbol{X}^j\boldsymbol{Y}^k}{j!k!} = \sum_{j=0}^{\infty}\sum_{k=0}^{j}\frac{\boldsymbol{X}^{j-k}\boldsymbol{Y}^k}{(j-k)!k!} \\ &= \sum_{j=0}^{\infty}\frac{1}{j!}\sum_{k=0}^{j}\binom{j}{k}\boldsymbol{X}^{j-k}\boldsymbol{Y}^k = \sum_{j=0}^{\infty}\frac{(\boldsymbol{X}+\boldsymbol{Y})^j}{j!} = \exp(\boldsymbol{X}+\boldsymbol{Y})\end{aligned}$$

where the rearrangement of the infinite summations relies on the absolute convergence of the exponential series and the binomial expansion of $(\boldsymbol{X}+\boldsymbol{Y})^j$ relies on the commutativity of $\boldsymbol{X}$ and $\boldsymbol{Y}$ and their powers. Clearly $\exp(\boldsymbol{X})\exp(\boldsymbol{Y}) = \exp(\boldsymbol{Y})\exp\boldsymbol{X})$.

(12) Suppose $\boldsymbol{A}$ is idempotent and symmetric,

(i) Show that the eigenvalues of $\boldsymbol{A}$ are 1 or 0.
Suppose $(\lambda, \boldsymbol{x})$ is an eigenpair of $\boldsymbol{A}$ then $\boldsymbol{Ax} = \lambda\boldsymbol{x}$ so $\boldsymbol{A}^2\boldsymbol{x} = \boldsymbol{Ax} = \lambda\boldsymbol{Ax} = \lambda^2\boldsymbol{x}$ which is true only if $\lambda = 1$ or 0. Thus all eigenvalues of $\boldsymbol{A}$ are 0 or 1.

(ii) Show that $\text{tr}(\boldsymbol{A}) = \rho(\boldsymbol{A})$.
Since $\rho(\boldsymbol{A})$ = number of non-zero eigenvalues (noting $\boldsymbol{A}$ is symmetric (see §6.4.7) and since $\text{tr}(\boldsymbol{A}) = \sum\lambda_i$ (see property 6.4.4 (v)) and all eigenvalues are 0 or 1 we have $\text{tr}(\boldsymbol{A}) = \rho(\boldsymbol{A})$. $(\mathbf{I_n} - \boldsymbol{A})(\mathbf{I_n} - \boldsymbol{A}) = \mathbf{I_n} - 2\boldsymbol{A} + \boldsymbol{A}^2 = \mathbf{I_n} - \boldsymbol{A}$

(iii) Show that $\rho(\mathbf{I_n} - \boldsymbol{A}) = n - \rho(\boldsymbol{A})$.
Since $(\mathbf{I_n} - \boldsymbol{A})$ is idempotent (see Exercise (9)) $\rho(\mathbf{I_n} - \boldsymbol{A}) = \text{tr}(\mathbf{I_n} - \boldsymbol{A}) = \text{tr}(\mathbf{I_n}) - \text{tr}(\boldsymbol{A}) = n - \rho(\boldsymbol{A})$.

(13) Where is the fallacy in the 'deceptively obvious proof' of the Cayley-Hamilton theorem "$p_A(\lambda) = |\boldsymbol{A} - \lambda\mathbf{I_n}|$ so $p_A(\boldsymbol{A}) = |\boldsymbol{A} - \boldsymbol{A}\mathbf{I_n}| = |\boldsymbol{A} - \boldsymbol{A}| = 0$"?

$p_A(\lambda) = |\boldsymbol{A} - \lambda \mathbf{I_n}|$ is a scalar equation involving the scalar λ. It is invalid to substitute a $n \times n$ matrix for a scalar and further the Cayley-Hamilton theorem is that $p_A(\boldsymbol{A}) = \boldsymbol{0}$ – the zero matrix not the scalar zero. For a full proof see Banerjee and Roy (2014) or Abadir and Magnus (2005); it involves considering the Schur decomposition (see §8.2.4) of $\boldsymbol{A}$.

(14) If $\boldsymbol{A}$ is nilpotent show that all eigenvalues of $\boldsymbol{A}$ are 0.

Suppose $(\lambda, \boldsymbol{x})$ is an eigenpair of $\boldsymbol{A}$ then $\boldsymbol{Ax} = \lambda \boldsymbol{x}$ so $\boldsymbol{Ax} = \lambda \boldsymbol{x}$ so $\boldsymbol{A}^2\boldsymbol{x} = \lambda \boldsymbol{Ax}$, i.e., $\lambda^2 \boldsymbol{x} = \boldsymbol{0}$ so $\lambda = 0$.

(15) If all eigenvalues of $\boldsymbol{A}$ are 0 show that $\boldsymbol{A}$ is nilpotent.

If all the eigenvalues of $\boldsymbol{A}$ are 0 then its characteristic polynomial is $p_A(\lambda) = \lambda^n$ and thus, by the Cayley-Hamilton theorem (see Page 84), $\boldsymbol{A}^n = 0$ so $\boldsymbol{A}$ is nilpotent.

(16) Suppose $\boldsymbol{A}$ is a real $n \times n$ skew-symmetric matrix.

(i) Show that the only real eigenvalue of $\boldsymbol{A}$ is 0.

Suppose $(\lambda, \boldsymbol{x})$ is an eigenpair of $\boldsymbol{A}$ then $\boldsymbol{Ax} = \lambda \boldsymbol{x}$ then $\boldsymbol{A}'\boldsymbol{Ax} = \lambda \boldsymbol{A}'\boldsymbol{x} = -\lambda^2 \boldsymbol{x}$, i.e., $(-\lambda^2, \boldsymbol{x})$ is an eigenpair of the positive semi-definite symmetric matrix $\boldsymbol{A}'\boldsymbol{A}$, but all eigenvalues of a real positive semi-definite symmetric matrix are non-negative (see §6.4.6 (ii)), i.e., $-\lambda^2 \geq 0$ and so either $\lambda = 0$ or λ is imaginary.

(ii) Show that if n is odd then $\boldsymbol{A}$ has at least one eigenvalue of 0.

Since $\boldsymbol{A}$ is real $|\boldsymbol{A}|$ is real but $|\boldsymbol{A}| = \prod_i \lambda_i$ and the product of an odd number of imaginary values is imaginary, there must be an even number of imaginary values and thus at least one must be 0.

Chapter 7: Vector and Matrix Calculus

(1) If $\boldsymbol{x} = (x_1, x_2, \ldots, x_n)'$ and $f(\boldsymbol{x}) = \boldsymbol{x}$ find $\frac{\partial f}{\partial \boldsymbol{x}}$.
$\left(\frac{\partial f}{\partial \boldsymbol{x}}\right)_{ij} = \frac{\partial x_i}{\partial x_j} = \mathbf{I_n}$ since $\frac{\partial x_i}{\partial x_j} = 1$ if $i = j$ and 0 otherwise.

(2) If $\boldsymbol{A}$, $\boldsymbol{X}$ and $\boldsymbol{B}$ are $m \times n$, $n \times p$ and $p \times m$ matrices (where $n \neq p$) find $\frac{\partial \text{tr}(\boldsymbol{AXB})}{\partial \boldsymbol{X}}$.

Note that because $n \neq p$ $\boldsymbol{X}$ is not symmetric. $\text{tr}(\boldsymbol{AXB}) = \text{tr}(\boldsymbol{X}(\boldsymbol{BA}))$ so $\frac{\partial \text{tr}(\boldsymbol{AXB})}{\partial \boldsymbol{X}} = (\boldsymbol{BA})' = \boldsymbol{A}'\boldsymbol{B}'$ (using §7.3.4).

(3) If $\boldsymbol{A}$, $\boldsymbol{X}$ and $\boldsymbol{B}$ are $n \times n$ matrices find $\frac{\partial \text{tr}(\boldsymbol{AXB})}{\partial \boldsymbol{X}}$.

$\text{tr}(\boldsymbol{AXB}) = \text{tr}(\boldsymbol{XBA})$ so $\frac{\partial \text{tr}(\boldsymbol{AXB})}{\partial \boldsymbol{X}} = \boldsymbol{BA} + \boldsymbol{A}'\boldsymbol{B}' - \text{diag}(\text{diag}(\boldsymbol{BA}))$ or $\boldsymbol{A}'\boldsymbol{B}'$ according as $\boldsymbol{X}$ is symmetric or not (using §7.3.4 and §7.3.5).

(4) If $\boldsymbol{A}$, $\boldsymbol{X}$ and $\boldsymbol{B}$ are $n \times n$ matrices (where $n \neq p$) find $\frac{\partial \text{tr}(\boldsymbol{AX'B})}{\partial \boldsymbol{X}}$.

$\text{tr}(\boldsymbol{AX'B}) = \text{tr}((\boldsymbol{AX'B})') = \text{tr}(\boldsymbol{B'XA'})$ so $\frac{\partial \text{tr}(\boldsymbol{AX'B})}{\partial \boldsymbol{X}} = \frac{\partial \text{tr}(\boldsymbol{B'XA'})}{\partial \boldsymbol{X}}$
$= \boldsymbol{BA} + \boldsymbol{A'B'} - \text{diag}(\text{diag}(\boldsymbol{A'B'}))$ or $\boldsymbol{BA}$ according as $\boldsymbol{X}$ is symmetric or not.
(Note that $\text{diag}(\text{diag}(\boldsymbol{A'B'})) = \text{diag}(\text{diag}(\boldsymbol{BA}))$, verifying that
$\frac{\partial \text{tr}(\boldsymbol{AX'B})}{\partial \boldsymbol{X}} = \frac{\partial \text{tr}(\boldsymbol{AXB})}{\partial \boldsymbol{X}}$ when $\boldsymbol{X}$ is symmetric.)

(5) Find $\frac{\partial \text{tr}(\boldsymbol{X}^2)}{\partial \boldsymbol{X}}$.

$$\text{tr}(\boldsymbol{X}^2) = \sum_{i=1}^{n}\sum_{k=1}^{n} x_{ik}x_{ki} \text{ so } \frac{\partial \text{tr}(\boldsymbol{X}^2)}{\partial x_{rs}} = 1.x_{sr} + x_{sr}.1 \text{ and thus } \frac{\partial \text{tr}(\boldsymbol{X}^2)}{\partial \boldsymbol{X}} = 2\boldsymbol{X}'.$$

(6) If $\boldsymbol{X}$ is a $n \times n$ matrix with elements x_{ij} and is not symmetric (i.e., $x_{ij} \neq x_{ji}$) show that $\frac{\partial \boldsymbol{X}^{-1}}{\partial x_{ij}} = -\boldsymbol{X}^{-1}\Delta_{[ij]}\boldsymbol{X}^{-1}$ where $\Delta_{[ij]}$ is a $n \times n$ matrix with $\Delta_{rs} = \delta_{ri}\delta_{sj}$ (where δ is Kronecker's delta function, i.e., the elements of Δ are all zero except in the $(i, j)^{th}$ place where there is a 1.

The 'trick' here is to consider the identity $\boldsymbol{XX}^{-1} = \mathbf{I_n}$ and differentiate both sides: $\frac{\partial \boldsymbol{XX}^{-1}}{\partial x_{ij}} = \frac{\partial \mathbf{I_n}}{\partial x_{ij}} = 0$ so $\boldsymbol{X}\frac{\partial \boldsymbol{X}^{-1}}{\partial x_{ij}} + \frac{\partial \boldsymbol{X}}{\partial x_{ij}}\boldsymbol{X}^{-1} = 0$ and since clearly $\frac{\partial \boldsymbol{X}}{\partial x_{ij}} = \Delta_{[ij]}$ we have $\boldsymbol{X}\frac{\partial \boldsymbol{X}^{-1}}{\partial x_{ij}} + \Delta_{[ij]}\boldsymbol{X}^{-1} = 0$ so $\frac{\partial \boldsymbol{X}^{-1}}{\partial x_{ij}} = -\boldsymbol{X}^{-1}\Delta_{[ij]}\boldsymbol{X}^{-1}$.

(7) If $\boldsymbol{X}$ is a $n \times n$ matrix with elements x_{ij} and is symmetric (i.e., $x_{ij} = x_{ji}$) show that $\frac{\partial \boldsymbol{X}^{-1}}{\partial x_{ij}} = -\boldsymbol{X}^{-1}(\Delta_{[ij]} + \Delta_{[ji]})\boldsymbol{X}^{-1}$.
If $\boldsymbol{X}$ is symmetric then $\frac{\partial \boldsymbol{X}}{\partial x_{ij}} = \Delta_{[ij]} + \Delta_{[ji]}$ and so as above we have
$\frac{\partial \boldsymbol{X}^{-1}}{\partial x_{ij}} = -\boldsymbol{X}^{-1}(\Delta_{[ij]} + \Delta_{[ji]})\boldsymbol{X}^{-1}$.

(8) Find the maximum value of $\boldsymbol{x'Xaa'X'x}/\boldsymbol{x'XX'x}$.

Since this is invariant with respect to scalar multiplication of $\boldsymbol{x}$ we can impose the constraint $\boldsymbol{x'XX'x} = 1$ and introduce a Lagrange multiplier λ and define $\Omega = \boldsymbol{x'Xaa'X'x} - \lambda(\boldsymbol{x'XX'x} - 1)$. Differentiating with respect to $\boldsymbol{x}$ and setting the derivative equal to zero shows $\boldsymbol{Xaa'X'x} = \lambda\boldsymbol{XX'x}$, i.e., $(\boldsymbol{XX'})^{-1}\boldsymbol{Xaa'X'x} = \lambda\boldsymbol{x}$. So we require $\boldsymbol{x}$ as the eigenvector of $(\boldsymbol{XX'})^{-1}\boldsymbol{Xaa'X'}$ corresponding to its only non-zero eigenvalues (noting that it is of rank 1). This eigenvector is $(\boldsymbol{XX'})^{-1}\boldsymbol{Xa}$ with eigenvalue $\boldsymbol{a'X'}(\boldsymbol{XX'})^{-1}\boldsymbol{Xa}$ and the maximum value is $\lambda = \boldsymbol{a'X'}(\boldsymbol{XX'})^{-1}\boldsymbol{Xa}$.

(9) Find the maximum value of $\boldsymbol{x'A'By}$ with respect to $\boldsymbol{x}$ and $\boldsymbol{y}$ subject to the constraints $\boldsymbol{x'x} = \boldsymbol{y'y} = 1$ where $\boldsymbol{A}$ and $\boldsymbol{B}$ are $n \times p$ and $n \times m$ matrices.

This is essentially a special case of §7.6 (iv) with $\boldsymbol{S} = \mathbf{I_p}$ and $\boldsymbol{T} = \mathbf{I_m}$ and $\boldsymbol{S} = \boldsymbol{A'B}$. This formulation is that given as the starting point of partial least squares regression (see, for example, Cox, 2005) where $\boldsymbol{A}$ is a matrix of observations of independent variables and $\boldsymbol{B}$ are observations of the responses. The maximum value is given as the square root of the largest eigenvalue of $\boldsymbol{A'BB'A}$ (or equivalently the largest singular value of $\boldsymbol{A'B}$).

(10) Show that $\frac{\partial \boldsymbol{AB}}{\partial x} = \boldsymbol{A}\frac{\partial \boldsymbol{B}}{\partial x} + \frac{\partial \boldsymbol{A}}{\partial x}\boldsymbol{B}$.

$(\boldsymbol{AB})_{ij} = \sum_k a_{ik}b_{kj}$ so

$$\frac{\partial(\boldsymbol{AB})_{ij}}{\partial x_{rs}} = \frac{\partial \sum_k a_{ik}b_{kj}}{\partial x_{rs}} = \sum_k a_{ik}\frac{\partial b_{kj}}{\partial x_{rs}} + \sum_k \frac{\partial a_{ik}}{\partial x_{rs}} b_{kj} = \boldsymbol{A}\frac{\partial \boldsymbol{B}}{\partial x} + \frac{\partial \boldsymbol{A}}{\partial x}\boldsymbol{B}.$$

(11) Show that $\frac{\partial \boldsymbol{A}\odot\boldsymbol{B}}{\partial x}. = \boldsymbol{A}\odot\frac{\partial \boldsymbol{B}}{\partial x} + \frac{\partial \boldsymbol{A}}{\partial x}\odot\boldsymbol{B}$ where $\odot$ indicates the Hadamard product (see §8.4).

$(\boldsymbol{A}\odot\boldsymbol{B})_{ij} = a_{ij}b_{ij}$ so

$$\frac{\partial(\boldsymbol{A}\odot\boldsymbol{B})_{ij}}{\partial x_{rs}} = \frac{\partial a_{ij}b_{ij}}{\partial x_{rs}} = a_{ij}\frac{\partial b_{ij}}{\partial x_{rs}} + \frac{\partial a_{ij}}{\partial x_{rs}} b_{ij} = \boldsymbol{A}\odot\frac{\partial \boldsymbol{B}}{\partial x_{rs}} + \frac{\partial \boldsymbol{A}}{\partial x_{rs}}\odot\boldsymbol{B}.$$

(12) Show that $\frac{\partial \boldsymbol{A}\otimes\boldsymbol{B}}{\partial x} = \boldsymbol{A}\otimes\frac{\partial \boldsymbol{B}}{\partial x} + \frac{\partial \boldsymbol{A}}{\partial x}\otimes\boldsymbol{B}$ where $\otimes$ indicates the Kronecker product (see §8.5).

$(\boldsymbol{A}\otimes\boldsymbol{B})_{ij} = a_{ij}\boldsymbol{B}$ so

$$\frac{\partial(\boldsymbol{A}\otimes\boldsymbol{B})_{ij}}{\partial x_{rs}} = \frac{\partial a_{ij}\boldsymbol{B}}{\partial x_{rs}} = a_{ij}\frac{\partial \boldsymbol{B}}{\partial x_{rs}} + \frac{\partial a_{ij}}{\partial x_{rs}}\boldsymbol{B} = \boldsymbol{A}\otimes\frac{\partial \boldsymbol{B}}{\partial x_{rs}} + \frac{\partial \boldsymbol{A}}{\partial x_{rs}}\otimes\boldsymbol{B}.$$

Chapter 8: Further Topics

(1) If $\boldsymbol{A} = \begin{pmatrix} 73 & 45 & 46 \\ 45 & 58 & 29 \\ 46 & 29 & 29 \end{pmatrix}$ find its Cholesky decomposition $\boldsymbol{A} = \boldsymbol{TT}'$.

```
> A<-matrix(c(73,45,46,
+ 45,58,29,46,29,29),3,3)
A
     [,1] [,2] [,3]
[1,]   73   45   46
[2,]   45   58   29
[3,]   46   29   29
> t(A)-A
     [,1] [,2] [,3]
[1,]    0    0    0
[2,]    0    0    0
[3,]    0    0    0
> eigen(A)$values
[1]  1.4e+02  2.2e+01 -2.3e-16
```

So $\boldsymbol{A}$ is symmetric and positive semidefinite, noting one zero eigenvalue.

```
> U<-chol(A,pivot=TRUE)
Warning message:
In chol.default(A,
pivot = TRUE) :
the matrix is either
 rank-deficient
 or indefinite
> t(U)
     [,1] [,2]     [,3]
[1,]  8.5 0.00  0.0e+00
[2,]  5.3 5.50  0.0e+00
[3,]  5.4 0.12 -7.1e-15
attr(,"pivot")
[1] 1 2 3
attr(,"rank")
[1] 2
```

Since $\boldsymbol{A}$ is not positive definite we may not obtain an exact decomposition with $\boldsymbol{A} = \boldsymbol{TT}'$ (taking $\boldsymbol{T}$=t(U)).

```
> t(U)%*%U
     [,1] [,2] [,3]
[1,]   73   45   46
[2,]   45   58   29
[3,]   46   29   29
```

```
> A
     [,1] [,2] [,3]
[1,]   73   45   46
[2,]   45   58   29
[3,]   46   29   29
```

So in this case we do obtain the Cholesky decomposition $\boldsymbol{A} = \boldsymbol{TT}'$ (taking $\boldsymbol{T}$=t(U)) without further steps but there is no guarantee this would be the case as the following exercise illustrates. So it is advisable always to use the extra steps given in the next exercise.

(2) If $\boldsymbol{A} = \begin{pmatrix} 34 & 31 & 51 \\ 31 & 53 & 32 \\ 51 & 32 & 85 \end{pmatrix}$ find its Cholesky decomposition $\boldsymbol{A} = \boldsymbol{TT}'$.

```
> A<-matrix(c(34,31,51,31,
+ 53,32,51,32,85),3,3)
> A
     [,1] [,2] [,3]
[1,]   34   31   51
[2,]   31   53   32
[3,]   51   32   85
> t(A)-A
     [,1] [,2] [,3]
[1,]    0    0    0
[2,]    0    0    0
[3,]    0    0    0
> eigen(A)$values
[1]  1.4e+02  3.3e+01 -1.8e-15
```

So $\boldsymbol{A}$ is symmetric and positive semi-definite, noting one zero eigenvalue.

```
> U<-chol(A,pivot=TRUE)
Warning message:
In chol.default(A,
pivot = TRUE) : the
 matrix is either
 rank-deficient
 or indefinite
> t(U)
     [,1] [,2] [,3]
[1,]  9.2  0.0    0
[2,]  3.5  6.4    0
[3,]  5.5  1.8    0
attr(,"pivot")
[1] 3 2 1
attr(,"rank")
[1] 2
```

Since $\boldsymbol{A}$ is not positive definite we may not obtain an exact decomposition with $\boldsymbol{A} = \boldsymbol{TT}'$ (taking $\boldsymbol{T}$=t(U)).

```
> t(U)%*%U
     [,1] [,2] [,3]
[1,]   85   32   51
[2,]   32   53   31
[3,]   51   31   34
```

```
> A
     [,1] [,2] [,3]
[1,]   34   31   51
[2,]   31   53   32
[3,]   51   32   85
```

This is the case in which `t(U)%*%U` gives $\boldsymbol{A}$ with permuted rows and columns, so we need extra steps:

```
> key <- attr(U, "pivot")
> V<-(U[, order(key)])
> T<-t(V)
> T
     [,1] [,2] [,3]
[1,]  5.5  1.8    0
[2,]  3.5  6.4    0
[3,]  9.2  0.0    0
```

Check $\boldsymbol{TT}' = \boldsymbol{A}$.

```
> T%*%t(T)
     [,1] [,2] [,3]
[1,]   34   31   51
[2,]   31   53   32
[3,]   51   32   85
> A
     [,1] [,2] [,3]
[1,]   34   31   51
[2,]   31   53   32
[3,]   51   32   85
```

(3) Show that $(\boldsymbol{A}^+)^+ = \boldsymbol{A}$.

We need to show that $\boldsymbol{A}$ satisfies the four Moore–Penrose conditions to be the MP-inverse of $\boldsymbol{A}^+$, i.e., that $\boldsymbol{A}^+\boldsymbol{A}\boldsymbol{A}^+ = \boldsymbol{A}^+$, $\boldsymbol{A}\boldsymbol{A}^+\boldsymbol{A} = \boldsymbol{A}$, $(\boldsymbol{A}\boldsymbol{A}^+)' = \boldsymbol{A}\boldsymbol{A}^+$ and $(\boldsymbol{A}^+\boldsymbol{A})' = \boldsymbol{A}^+\boldsymbol{A}$. But these are just the four conditions that $\boldsymbol{A}^+$ satisfies to be the MP-inverse of $\boldsymbol{A}$ in a different order ((i) and (ii) are interchanged and (iii) and (iv) are interchanged).

(4) Show that $(\boldsymbol{A}')^+ = (\boldsymbol{A}^+)'$.

We need to show that $(\boldsymbol{A}^+)'$ satisfies the four Moore–Penrose conditions to be the MP-inverse of $\boldsymbol{A}'$, i.e., that $\boldsymbol{A}'(\boldsymbol{A}^+)'\boldsymbol{A}' = \boldsymbol{A}'$, $(\boldsymbol{A}^+)'\boldsymbol{A}'(\boldsymbol{A}^+)' = (\boldsymbol{A}^+)'$, $(\boldsymbol{A}'(\boldsymbol{A}^+)')' = \boldsymbol{A}'(\boldsymbol{A}^+)'$ and $((\boldsymbol{A}^+)'\boldsymbol{A}')' = (\boldsymbol{A}^+)'\boldsymbol{A}'$. These follow immediately from the four conditions for $\boldsymbol{A}^+$ to be the MP-inverse of $\boldsymbol{A}$ by transposing each of the four equations and reordering them (interchanging (i) and (ii) and interchanging (iii) and (iv)).

(5) Show that $(\boldsymbol{A}\boldsymbol{A}^+)^+ = \boldsymbol{A}\boldsymbol{A}^+$.

$\boldsymbol{A}\boldsymbol{A}^+\boldsymbol{A}\boldsymbol{A}^+\boldsymbol{A}\boldsymbol{A}^+ = (\boldsymbol{A}\boldsymbol{A}^+\boldsymbol{A})(\boldsymbol{A}^+\boldsymbol{A}\boldsymbol{A}^+) = \boldsymbol{A}\boldsymbol{A}^+$ because $\boldsymbol{A}^+$ is the MP-inverse of $\boldsymbol{A}$ and so satisfies MP conditions (i) and (ii). Thus $\boldsymbol{A}\boldsymbol{A}^+$ satisfies MP conditions (i) and (ii) as MP-inverse of $\boldsymbol{A}\boldsymbol{A}^+$. $(\boldsymbol{A}\boldsymbol{A}^+\boldsymbol{A}\boldsymbol{A}^+)' = (\boldsymbol{A}\boldsymbol{A}^+)'(\boldsymbol{A}\boldsymbol{A}^+)' = \boldsymbol{A}\boldsymbol{A}^+\boldsymbol{A}\boldsymbol{A}^+$ (because $\boldsymbol{A}^+$ satisfies conditions (iii) and (iv) as MP-inverse of $\boldsymbol{A}$) and thus $\boldsymbol{A}\boldsymbol{A}^+$ satisfies conditions (iii) and (iv) as inverse of MP-inverse of $\boldsymbol{A}\boldsymbol{A}^+$ and we conclude $(\boldsymbol{A}\boldsymbol{A}^+)^+ = \boldsymbol{A}\boldsymbol{A}^+$.

(6) Show that $(\boldsymbol{A}^+\boldsymbol{A})^+ = \boldsymbol{A}^+\boldsymbol{A}$.

This follows by near identical arguments to those used in Exercise (5).

(7) If A is symmetric (i.e., if $A' = A$) show that $AA^+ = A^+A$.

$AA^+ = (AA^+)'$ (because AA^+ is symmetric by the third MP-condition) $= (A^+)'A' = (A')^+A = A^+A$.

(8) Show that $A'AA^+ = A' = A^+AA'$.

$A'AA^+ = A'(AA^+)' = A'(A^+)'A' = (A^+A)'A' = A^+AA'$, noting that AA^+ and A^+A are symmetric by MP-conditions (iii) and (iv). Also $A'(A^+)'A' = (AA^+A)' = A'$ by MP-condition (i).

(9) If $A = \begin{pmatrix} A_1 & 0 \\ 0 & A_2 \end{pmatrix}$ show $A^+ = \begin{pmatrix} A_1^+ & 0 \\ 0 & A_2^+ \end{pmatrix}$.

This follows by direct verification of the four Moore–Penrose conditions.

(10) If $x = \begin{pmatrix} 1 \\ 6 \\ 3 \end{pmatrix}$ find x^+ without using any of the **R** functions for finding Moore–Penrose inverses and check the result with `ginv(.)`.

Note that x is of rank 1 so $x^+ = \text{tr}(xx')^{-1}x'$:

```
> x<-matrix(c(1,6,3)) ; t(x)/sum(x*x)
      [,1] [,2]  [,3]
[1,] 0.022 0.13 0.065
> #CHECK
> ginv(x)
      [,1] [,2]  [,3]
[1,] 0.022 0.13 0.065
```

(11) If $X = \begin{pmatrix} 4 & 12 & 8 \\ 6 & 18 & 12 \\ 5 & 15 & 10 \end{pmatrix}$ find X^+ without using any of the **R** functions for finding Moore–Penrose inverses and check the result with `ginv(.)`.

Note that X is of rank 1 so $X^+ = \text{tr}(XX')^{-1}X'$:

```
> X<-matrix(c(4,12,8,6,18,12,5,15,10),3,3,byrow=T)

> X
     [,1] [,2] [,3]
[1,]    4   12    8
[2,]    6   18   12
[3,]    5   15   10

> t(X)/sum(diag(X%*%t(X)))
       [,1]   [,2]   [,3]
[1,] 0.0037 0.0056 0.0046
[2,] 0.0111 0.0167 0.0139
[3,] 0.0074 0.0111 0.0093

> #CHECK
> ginv(X)

>        [,1]   [,2]   [,3]
[1,] 0.0037 0.0056 0.0046
[2,] 0.0111 0.0167 0.0139
[3,] 0.0074 0.0111 0.0093
```

(12) If $X = \begin{pmatrix} 6 & 2 & 8 \\ 5 & 1 & 6 \\ 1 & 7 & 8 \end{pmatrix}$, find X^+ without using any of the **R** functions for finding Moore–Penrose inverses and check the result with `ginv(.)`.

It was shown in Exercises 3.1, Exercise (1) (vi) that X has rank 2 and so is singular. We cannot use the result $X^+ = \text{tr}(XX')^{-1}X'$ because this only applies to matrices of rank 1. Instead we can use the result based on the singular value decomposition but note that since X is of rank 2 one of the singular values is zero and so we need to extract just those eigenvectors corresponding to non-zero singular values.

```
> X<-matrix(c(6,5,1,2,1,7,8,6,8),3,3)
```

```
> U<-svd(X)$u[,1:2]
> V<-svd(X)$v[,1:2]
> D<-svd(X)$d[1:2]
> M<-V%*%solve(diag(D))%*%t(U)
> M
       [,1]   [,2]   [,3]
[1,]  0.069  0.066 -0.077
[2,] -0.038 -0.044  0.112
[3,]  0.031  0.023  0.036
```

```
> #CHECK
> ginv(X)
       [,1]   [,2]   [,3]
[1,]  0.069  0.066 -0.077
[2,] -0.038 -0.044  0.112
[3,]  0.031  0.023  0.036
>
```

(13) Let $A = \begin{pmatrix} 1 & 1 \\ 2 & 2 \\ 3 & 4 \end{pmatrix}$ and $y = \begin{pmatrix} 1 \\ 1 \\ 1 \end{pmatrix}$.

(a) Show that $AA^+y = \begin{pmatrix} 0.6 \\ 1.2 \\ 1.0 \end{pmatrix} \neq y$.

```
> library(MASS)
> A<-matrix(c(1,2,3,1,2,4),
+3,2)
> A
     [,1] [,2]
[1,]    1    1
[2,]    2    2
[3,]    3    4
```

```
> y<-matrix(c(1,1,1),3,1)
> y
     [,1]
[1,]    1
[2,]    1
[3,]    1
```

```
> G<-ginv(A)
> G
     [,1] [,2] [,3]
[1,]  0.8    2   -1
[2,] -0.6   -1    1
```

```
> A%*%G%*%y
     [,1]
[1,]  0.6
[2,]  1.2
[3,]  1.0
```

(b) What is the least squares solution to the equation $Ax = y$?

The least squares solution is given by $x = A^+y$.

```
> G%*%y
      [,1]
[1,]   1.4
[2,]  -0.8
```

so $\boldsymbol{x} = \boldsymbol{A}^{+}\boldsymbol{y} = \begin{pmatrix} 1.4 \\ -0.8 \end{pmatrix}$ is the least squares solution.

(c) If instead $\boldsymbol{y} = \begin{pmatrix} 1 \\ 2 \\ 7 \end{pmatrix}$ or $\boldsymbol{y} = \begin{pmatrix} 3 \\ 6 \\ 11 \end{pmatrix}$ show that in both cases the equation is consistent and find solutions. Are these solutions unique in either or both cases?

```
> y<-matrix(c(1,2,7),
+3,1)
> A%*%G%*%y
      [,1]
[1,]     1
[2,]     2
[3,]     7
> G%*%y
      [,1]
[1,]    -3
[2,]     4
```

```
> y<-matrix(c(3,6,11),
+3,1)
> A%*%G%*%y
      [,1]
[1,]     3
[2,]     6
[3,]    11
> G%*%y
      [,1]
[1,]     1
[2,]     2
```

In both cases $\boldsymbol{AA}^{+}\boldsymbol{y} = \boldsymbol{y}$ and so the equations are consistent and have solutions given by $\boldsymbol{x} = \boldsymbol{A}^{+}\boldsymbol{y}$ of $\boldsymbol{x} = \begin{pmatrix} -3 \\ 4 \end{pmatrix}$ and $\boldsymbol{x} = \begin{pmatrix} 1 \\ 2 \end{pmatrix}$.
In each case the solution is unique because $\boldsymbol{A}$ has full column rank.

(14) Let $\boldsymbol{A} = \begin{pmatrix} 2 & 3 & 3 & 1 \\ 3 & 4 & 5 & 1 \\ 1 & 2 & 1 & 1 \end{pmatrix}$ and $\boldsymbol{y} = \begin{pmatrix} 14 \\ 22 \\ 6 \end{pmatrix}$.

(a) Show that the equation $\boldsymbol{Ax} = \boldsymbol{y}$ is consistent.

```
> library(MASS)
> options(digits=1)
> A<-matrix(c(2,3,1,3,4,2,
+ 3,5,1,1,1,1),3,4)
> y<-matrix(c(14,22,6),
+ 3,1)
```

```
> A
     [,1] [,2] [,3] [,4]
[1,]    2    3    3    1
[2,]    3    4    5    1
[3,]    1    2    1    1
```

```
> y
     [,1]
[1,]   14
[2,]   22
[3,]    6
```

```
> G<-ginv(A)
> A%*%G%*%y
     [,1]
[1,]   14
[2,]   22
[3,]    6
```

So $AA^{+}y = y$ and thus the equation is consistent.

(b) Show that $x = \begin{pmatrix} 1.3 \\ 1.1 \\ 2.8 \\ -0.2 \end{pmatrix}$ and $x = \begin{pmatrix} 1.5 \\ 0.7 \\ 2.9 \\ 0.2 \end{pmatrix}$ are both solutions of the equation.

```
> x<-matrix(c(1.3,1.2,
+ 2.8,-0.2),4,1)
> x; A%*%x
     [,1]
[1,]  1.3
[2,]  1.2
[3,]  2.8
[4,] -0.2
     [,1]
[1,]   14
[2,]   22
[3,]    6
```

```
> x<-matrix(c(1.5,0.7,
+ 2.9,0.2),4,1)
> x; A%*%x
     [,1]
[1,]  1.5
[2,]  0.7
[3,]  2.9
[4,]  0.2
     [,1]
[1,]   14
[2,]   22
[3,]    6
```

Thus both values of x satisfy the equation.

(c) Find a different solution to the equation.

All solutions to the equation are of the form $x = A^{+}y + (I_4 - A^{+}A)q = \begin{pmatrix} 1.3 \\ 1.1 \\ 2.8 \\ -0.2 \end{pmatrix} + q - \begin{pmatrix} 0.18 & 0.24 & 0.29 & 0.06 \\ 0.24 & 0.65 & 0.06 & 0.41 \\ 0.29 & 0.06 & 0.82 & -0.24 \\ 0.06 & 0.41 & -0.24 & 0.35 \end{pmatrix} q$, where q is any 4×1 vector. Taking q as a random vector generated by choosing elements randomly from the first ten integers (using 137 for the seed of the random number generator) gives:

```
> set.seed(137)
> q<-matrix(c(sample
+ (c(1:10),4)),4,1)
```

```
> q
     [,1]
[1,]    7
[2,]    4
[3,]    8
[4,]    6
```

```
> x<-G%*%y+q-G%*%A%*%q
> x
     [,1]
[1,]    3
[2,]   -2
[3,]    3
[4,]    3
```

```
> #### verify solution:
> A%*%x
     [,1]
[1,]   14
[2,]   22
[3,]    6
```

(15) Show that if $\boldsymbol{x}$ is a $n\times 1$ vector then $\mathrm{tr}(\boldsymbol{xx}') = \mathbf{1}_{\mathbf{n}}'\boldsymbol{x}\odot\boldsymbol{x}$, where $\iota_{\mathbf{n}}$ is the $n\times 1$ vector with all entries equal to one.

The $(i,i)^{th}$ element of $\boldsymbol{xx}'$ is x_i^2 and so $\mathrm{tr}(\boldsymbol{xx}') = \sum_i^n x_i^2$ and the ith element of $\boldsymbol{x}\odot\boldsymbol{x}$ is x_i^2 and the result follows.

(16) If $\boldsymbol{A}$ is a square $n\times n$ matrix then show that $\boldsymbol{A}\odot\mathbf{I_n} = \mathrm{diag}(\mathrm{diag}(\boldsymbol{A}))$.

Since $\mathbf{I_n}$ has zero entries everywhere except along the diagonal elements which are all 1, $\boldsymbol{A}\odot\mathbf{I_n}$ is a diagonal matrix with elements a_{ii}, i.e., $\mathrm{diag}(\mathrm{diag}(\boldsymbol{A}))$.

(17) If $\boldsymbol{u}$ and $\boldsymbol{x}$ are $m\times 1$ vectors and $\boldsymbol{v}$ and $\boldsymbol{y}$ are $n\times 1$ vectors then show that $(\boldsymbol{uv}')\odot(\boldsymbol{xy}') = (\boldsymbol{u}\odot\boldsymbol{x})(\boldsymbol{v}\odot\boldsymbol{y})'$.

The $(i,j)^{th}$ element of $(\boldsymbol{u}\odot\boldsymbol{x})(\boldsymbol{v}\odot\boldsymbol{y})'$ is $u_ix_iv_jy_j$ and the $(i,j)^{th}$ element of $(\boldsymbol{uv}')\odot(\boldsymbol{xy}')$ is $u_iv_jx_iy_j = u_ix_iv_jy_j$, so $(\boldsymbol{uv}')\odot(\boldsymbol{xy}') = (\boldsymbol{u}\odot\boldsymbol{x})(\boldsymbol{v}\odot\boldsymbol{y})'$.

(18) If $\rho(\boldsymbol{A}) = \rho(\boldsymbol{B}) = 1$ then show that $\rho(\boldsymbol{A}\odot\boldsymbol{B}) \leq 1$.

Since $\boldsymbol{A}$ and $\boldsymbol{B}$ are of rank 1 they can be expressed as $\boldsymbol{A} = \boldsymbol{uv}'$ and $\boldsymbol{B} = \boldsymbol{xy}'$ (see §3.2.1, Page 53) and so using the previous exercise $\rho(\boldsymbol{A}\odot\boldsymbol{B})$ can be expressed as the product of two rank 1 matrices and so has rank ≤ 1 (see §3.3.2).

(19) Show that $\rho(\boldsymbol{A}\odot\boldsymbol{B}) \leq \rho(\boldsymbol{A}).\rho(\boldsymbol{B})$.

Suppose $\rho(\boldsymbol{A}) = r$ and $\rho(\boldsymbol{B}) = s$ then $\boldsymbol{A}$ and $\boldsymbol{B}$ can be expressed as the sum of r and s matrices each of rank 1, so $\rho(\boldsymbol{A}\odot\boldsymbol{B})$ can be expressed as the sum of rs matrices each of rank 1 (the products of the individual rank 1 matrices) so $\rho(\boldsymbol{A}\odot\boldsymbol{B}) \leq rs$, using the result of the previous exercise and §3.3.1.

(20) If $\boldsymbol{A}$ and $\boldsymbol{B}$ are both $m\times n$ and $\boldsymbol{x}$ is $n\times 1$ show that $\mathrm{diag}(\boldsymbol{A}\mathrm{diag}(\mathrm{diag}(\boldsymbol{x}))\boldsymbol{B}') = (\boldsymbol{A}\odot\boldsymbol{B})\boldsymbol{x}$.

Suppose $\boldsymbol{A} = (\boldsymbol{a}_1, \boldsymbol{a}_2, \ldots, \boldsymbol{a}_n$ and $\boldsymbol{B} = (\boldsymbol{b}_1, \boldsymbol{b}_2, \ldots, \boldsymbol{b}_n)$, i.e., $\boldsymbol{a}_i$ and $\boldsymbol{b}_i$ are the column vectors of $\boldsymbol{A}$ and $\boldsymbol{B}$, then

$$\mathrm{diag}(\boldsymbol{A}\mathrm{diag}(\mathrm{diag}(\boldsymbol{x}))\boldsymbol{B}') = \left(\sum_{i=1}^{n} x_i\boldsymbol{a}_i\boldsymbol{b}_i'\right) = \sum_{i=1}^{n} x_i\mathrm{diag}(\boldsymbol{a}_i\boldsymbol{b}_i') = \sum_{i=1}^{n} x_i\boldsymbol{c}_i = (\boldsymbol{A}\odot\boldsymbol{B})\boldsymbol{x}$$

where $\boldsymbol{c}_i$ is the ith column of $(\boldsymbol{A}\odot\boldsymbol{B})$.

(21) Show that $\boldsymbol{x}'(\boldsymbol{A}\odot\boldsymbol{B})\boldsymbol{y} = \mathrm{tr}(\mathrm{diag}(\mathrm{diag}(\boldsymbol{x}))\boldsymbol{A}\mathrm{diag}(\mathrm{diag}(\boldsymbol{y}))\boldsymbol{B}')$, where $\boldsymbol{x}$, $\boldsymbol{y}$, $\boldsymbol{A}$ and $\boldsymbol{B}$ are conformable so that the various products are well-defined.

From the previous exercise we have $(\boldsymbol{A}\odot\boldsymbol{B})\boldsymbol{y} = \mathrm{diag}(\boldsymbol{A}\mathrm{diag}(\mathrm{diag}(\boldsymbol{y}))\boldsymbol{B}')$ and noting that if $\boldsymbol{x}$ is $n\times 1$ and $\boldsymbol{X}$ is $n\times n$ then $\boldsymbol{x}'\mathrm{diag}(X) = \mathrm{tr}(\mathrm{diag}(\mathrm{diag}(\boldsymbol{x}))\boldsymbol{X})$ we have $\boldsymbol{x}'(\boldsymbol{A}\odot\boldsymbol{B})\boldsymbol{y} = \mathrm{tr}(\mathrm{diag}(\mathrm{diag}(\boldsymbol{x}))\boldsymbol{A}\mathrm{diag}(\mathrm{diag}(\boldsymbol{y}))\boldsymbol{B}')$.

(22) If $\boldsymbol{A}$ and $\boldsymbol{B}$ are $n \times n$ matrices and Λ is a $n \times n$ matrix with diagonal elements λ_i then show that $\text{diag}(\boldsymbol{A}\Lambda\boldsymbol{B}) = (\boldsymbol{A} \odot \boldsymbol{B})\text{diag}(\Lambda)$.

Let $\boldsymbol{a}_{\star i}$ be the ith column vector of $\boldsymbol{A}$ and $\boldsymbol{b}'_{i\star}$ be the ith row vector of $\boldsymbol{B}$ then

$$\text{diag}(\boldsymbol{A}\Lambda\boldsymbol{B} = \left(\sum_{i=1}^{n} \lambda_i \boldsymbol{a}_{\star i}\boldsymbol{b}'_{i\star}\right) = \sum_{i=1}^{n} \lambda_i \text{diag}(\boldsymbol{a}_{\star i}\boldsymbol{b}'_{i\star}) = \sum_{i=1}^{n} \lambda_i \boldsymbol{c}_i = (\boldsymbol{A} \odot \boldsymbol{B}')diag(\Lambda),$$

where $\boldsymbol{c}_i$ is the ith column of $(\boldsymbol{A} \odot \boldsymbol{B}')$.

(23) If $\boldsymbol{A}$ is $m \times n$ and $\boldsymbol{B}$ is $p \times q$ then show that $(\boldsymbol{A} \otimes \boldsymbol{B})^- = \boldsymbol{A}^- \otimes \boldsymbol{B}^-$.

$(\boldsymbol{A} \otimes \boldsymbol{B})(\boldsymbol{A}^- \otimes \boldsymbol{B}^-)(\boldsymbol{A} \otimes \boldsymbol{B}) = (\boldsymbol{A}\boldsymbol{A}^-\boldsymbol{A}) \otimes (\boldsymbol{B}\boldsymbol{B}^-\boldsymbol{B})$ (applying the mixed product rule, §8.5.2, twice in succession) $= (\boldsymbol{A} \otimes \boldsymbol{B})$.

(24) If $\boldsymbol{A}$ is $m \times n$ and $\boldsymbol{B}$ is $p \times q$ then show that $(\boldsymbol{A} \otimes \boldsymbol{B})^+ = \boldsymbol{A}^+ \otimes \boldsymbol{B}^+$.

This follows by checking each of the four Moore–Penrose conditions in a similar way to the solution to the previous exercise.

(25) If $\boldsymbol{A} = \left(\begin{smallmatrix} 0 & 0 \\ 1 & 0 \end{smallmatrix}\right)$ find the eignevectors of $\boldsymbol{A}$ and $\boldsymbol{A} \otimes \boldsymbol{A}$.

```
> A<-matrix(c(0,1,0,0),2)
> eigen(A)
$values
[1] 0 0
$vectors
     [,1]    [,2]
[1,]    0  2e-292
[2,]    1  -1e+00
```

```
> eigen(A%x%A)
$values
[1] 0 0 0 0
$vectors
     [,1] [,2] [,3] [,4]
[1,]    0    0    0    0
[2,]    0    0    1    0
[3,]    0    0    0    1
[4,]    1   -1    0    0
```

```
> eigen(A)$vec%x%eigen(A)$vec
     [,1] [,2] [,3] [,4]
[1,]    0    0    0    0
[2,]    0    0    0    0
[3,]    0    0    0    0
[4,]    1   -1   -1    1
```

So the eigenvectors of $\boldsymbol{A} \otimes \boldsymbol{A}$ are not given by the Kronecker products of the eigenvectors of $\boldsymbol{A}$.

Chapter 9: Key Applications to Statistics

(1) Suppose the random variable $\boldsymbol{x}$ has variance Σ_0, a known $p \times p$ positive definite symmetric matrix, and $\boldsymbol{X}'$ is a $n \times p$ data matrix of independent observations of $\boldsymbol{x}$ with sample variance $\boldsymbol{S}$. Find a matrix $\boldsymbol{A}$ such that the data matrix $\boldsymbol{Y}' = \boldsymbol{X}'\boldsymbol{A}$ has sample variance matrix Σ_0.

If $\boldsymbol{Z}' = \boldsymbol{X}'\boldsymbol{S}^{-1/2}$ then the sample variance of the observations of $\boldsymbol{z} = \boldsymbol{S}'\boldsymbol{x}$ is $\boldsymbol{S}^{-1/2}\boldsymbol{S}\boldsymbol{S}^{-1/2} = \mathbf{I_p}$ so if $\boldsymbol{Y}' = \boldsymbol{Z}'\Sigma_0^{1/2}$ then $\boldsymbol{Y}'$ has sample variance $\Sigma_0^{1/2}.1.\Sigma_0^{1/2} = \Sigma_0$. Thus we need to take $\boldsymbol{A} = \Sigma_0^{1/2}\boldsymbol{S}^{-1/2}$.

(2) Suppose $\Sigma_0 = \begin{pmatrix} 4.031 & 3.027 \\ 3.027 & 3.021 \end{pmatrix}$. Using the **R** function `mvrnorm(.)` in the `MASS` library and the previous exercise, generate a sample of 47 two-dimensional observations which have a sample variance of Σ_0.

First define function for matrix square roots used in Exercises 6, Exercise (8) on Page 198:

```
> SQR<-function(S){
+ Z<-svd(S)
+ SQR<-Z$u%*%sqrt(diag(Z$d))%*%t(Z$v)
+ return(SQR)}
```

Next, generate 47 two-dimensional multivariate normal observations with any mean and variance (taken here as $\mathbf{0}$ and $\mathbf{I_2}$) and find covariance $\boldsymbol{S}$:

```
> library(MASS)
> V<-matrix(c(1,0,0,1),2,2)
> X<-t(mvrnorm(47,c(0,0),V))
> S<-var(t(X))
```

Finally set up Σ_0 and using previous exercise find required matrix $\boldsymbol{A}$ to multiply $\boldsymbol{X}$ to obtain required data matrix $\boldsymbol{Y}'$ and check its sample variance:

```
> Sig<-matrix(c(4.031,3.027,3.027,3.021 ),2,2)
> Y<-SQR(Sig)%*%solve(SQR(S))%*%X
> ### Require observations are in data matrix Y'
> ### Check:
> var(t(Y))
      [,1]  [,2]
[1,] 4.031 3.027
[2,] 3.027 3.021
```

(3) With Σ_0 as given in the previous exercise generate 27 observations whose sample mean is $(20.25, 29.83)'$ and sample variance Σ_0 and a further 20 observations with sample mean

$(18.95, 28.63)'$ and sample variance Σ_0 (Rao's paradox.)

Continuing the **R** session from the previous exercise the extra steps needed are to ensure that the sample mean is as required, so initially the random numbers are generated with an arbitrary mean, then this is corrected to that specified for each sample. The first sample will be in data matrix $\boldsymbol{Y}'$ and the second in $\boldsymbol{Z}'$.

```
> X<-t(mvrnorm(27,c(0,0),V))
> S<-var(t(X))
> Y<-SQR(Sig)%*%solve(SQR(S))%*%X
> ybar<-matrix(apply(Y,1,mean),2)
> Ybar<-(ybar-c(20.25, 29.83))%*%t(matrix(rep(1,27),27))
> Y<-Y-Ybar
>
> X<-t(mvrnorm(20,c(0,0),V))
> S<-var(t(X))
> Z<-SQR(Sig)%*%solve(SQR(S))%*%X
> zbar<-matrix(apply(Z,1,mean),2)
> Zbar<-(zbar-c(18.95, 28.63))%*%t(matrix(rep(1,20),20))
> Z<-Z-Zbar
```

NB: This example exhibits ***Rao's paradox***, named by Healy (1969) and first discussed by Rao (1966), where the difference between the two sample means of each individual component would be assessed as significant at the 5% level by a t-test but taking the multivariate Hotelling's T^2-test of the two components together the difference is not significant at the 5% level. Demonstration of this is left to the reader.

(4) Using the **R** function `runif(.)`, `matrix(.)`, `var(.)` and `eigen(.)` generate a random 5×5 orthogonal matrix.

This is one possible method but there are many other ways of generating random orthogonal matrices.

```
> options(digits=2)
> set.seed(137)
> X<-matrix(runif(100,0,1),20,5)
> S<-var(X)
> A<-eigen(S)$vectors
> A
      [,1]   [,2]  [,3]   [,4]   [,5]
[1,] -0.37  0.210  0.19 -0.326  0.821
[2,] -0.32  0.571  0.32 -0.428 -0.536
[3,] -0.25 -0.477 -0.51 -0.658 -0.132
[4,]  0.83  0.081  0.17 -0.523  0.107
[5,] -0.12 -0.629  0.76 -0.066 -0.099
> ## A is orthogonal.  Check:
> A%*%t(A); t(A)%*%A
```

```
        [,1]     [,2]     [,3]     [,4]     [,5]
[1,]  1.0e+00  5.6e-17  0.0e+00  1.2e-16 -1.2e-16
[2,]  5.6e-17  1.0e+00 -2.5e-16 -1.7e-16 -1.9e-16
[3,]  0.0e+00 -2.5e-16  1.0e+00 -1.4e-16 -3.6e-16
[4,]  1.2e-16 -1.7e-16 -1.4e-16  1.0e+00 -1.5e-16
[5,] -1.2e-16 -1.9e-16 -3.6e-16 -1.5e-16  1.0e+00
        [,1]     [,2]     [,3]     [,4]     [,5]
[1,]  1.0e+00 -1.2e-16 -1.4e-17 -7.1e-17  5.4e-17
[2,] -1.2e-16  1.0e+00  5.6e-17 -4.1e-16 -6.9e-17
[3,] -1.4e-17  5.6e-17  1.0e+00  1.7e-16 -1.4e-17
[4,] -7.1e-17 -4.1e-16  1.7e-16  1.0e+00 -2.5e-16
[5,]  5.4e-17 -6.9e-17 -1.4e-17 -2.5e-16  1.0e+00
```

(5) Suppose $\boldsymbol{x}_1, \boldsymbol{x}_2, \ldots, \boldsymbol{x}_n$ are independent observations of $N_p(\lambda\mu_0, \Sigma_0)$ where μ_0 and Σ_0 are known and λ is an unknown scalar.

(i) Show that the mle of λ is given by $\hat{\lambda} = \mu_0'\Sigma_0^{-1}\bar{\boldsymbol{x}}/\mu_0'\Sigma_0^{-1}\mu_0$.

The log-likelihood is

$$\ell(\lambda; \boldsymbol{X}) = -{}^1\!/\!_2\sum_{i=1}^{n}(\boldsymbol{x}_i - \bar{\boldsymbol{x}})'\Sigma_0^{-1}(\boldsymbol{x}_i - \bar{\boldsymbol{x}}) - {}^1\!/\!_2 n(\bar{\boldsymbol{x}} - \lambda\mu_0)'\Sigma_0^{-1}(\bar{\boldsymbol{x}} - \lambda\mu_0)$$
$$- {}^1\!/\!_2 np\log(2\pi) - {}^1\!/\!_2 n\log(|\Sigma_0|)$$

so $$\frac{\partial \ell}{\partial \lambda} = n\mu_0'\Sigma_0^{-1}\bar{\boldsymbol{x}} - n\lambda\mu_0'\Sigma_0^{-1}\mu_0.$$

Setting this equal to zero gives $\hat{\lambda} = \mu_0'\Sigma_0^{-1}\bar{\boldsymbol{x}}/\mu_0'\Sigma_0^{-1}\mu_0$.

(ii) Find the mean and variance of $\hat{\lambda}$ and hence give its distribution.

$\mathrm{E}[\hat{\lambda}] = \mu_0'\Sigma_0^{-1}\mathrm{E}[\bar{\boldsymbol{x}}]/\mu_0'\Sigma_0^{-1}\mu_0 = \mu_0'\Sigma_0^{-1}\lambda\mu_0/\mu_0'\Sigma_0^{-1}\mu_0 = \lambda.$

$$\begin{aligned}\mathrm{var}(\hat{\lambda}) &= \mu_0'\Sigma_0^{-1}\mathrm{var}(\bar{\boldsymbol{x}})(\mu_0'\Sigma_0^{-1})'/(\mu_0'\Sigma_0^{-1}\mu_0)^2 \\ &= \mu_0'\Sigma_0^{-1}(\Sigma_0/n)(\Sigma_0^{-1}\mu_0)/(\mu_0'\Sigma_0^{-1}\mu_0)^2 = 1/\mu_0'\Sigma_0^{-1}\mu_0.\end{aligned}$$

Hence $\hat{\lambda} \sim N(\lambda, 1/\mu_0'\Sigma_0^{-1}\mu_0)$.

(iii) Show that the LRT statistic for testing $H_0: \ \mu = \lambda\mu_0$ for some scalar λ (where μ_0 and Σ_0 are known) is $n(\bar{\boldsymbol{x}} - \hat{\lambda}\mu_0)'\Sigma_0^{-1}(\bar{\boldsymbol{x}} - \hat{\lambda}\mu_0)$ which under H_0 follows a χ_p^2-distribution.

Under H_0 $\hat{\mu} = \hat{\lambda}\mu_0$ and under $\overline{H}_0$ $\hat{\mu} = \bar{\boldsymbol{x}}$ so the LRT statistic is $2\{\ell_{\max}(\overline{H}_0) - \ell_{\max}(H_0)\} = n(\bar{\boldsymbol{x}} - \hat{\lambda}\mu_0)'\Sigma_0^{-1}(\bar{\boldsymbol{x}} - \hat{\lambda}\mu_0)$.

(iv) In a standard feeding experiment on greenfinches, four types of sunflower seeds were placed in identical quadruple compartment bird feeders in each of 27 suburban gardens. The mean weights consumed of the four types after 120 minutes were 47 g, 45 g, 39 g and 42 g. Experience from a long series of such standard experiments suggests that the standard deviations of the amounts consumed of any type of sunflower seed in a single garden can be taken to be 10 g and the pairwise correlations between the weights consumed are -0.1. Do these data suggest that greenfinches have unequal preferences for the various types of sunflower seeds?

We have $\mu_0 = (1,1,1,1)' = \iota_4$, $\bar{x} = (47,45,39,42)'$ and (noting that a standard deviation of 10 implies a variance of 100 and a correlation 0.1 implies a covariance of 10) $\Sigma_0 = \begin{pmatrix} 100 & -10 & -10 & -10 \\ -10 & 100 & -10 & -10 \\ -10 & -10 & 100 & -10 \\ -10 & -10 & -10 & 100 \end{pmatrix}$.

Although it is possible to find the inverse of Σ_0 explicitly using the result in §5.4.1 there is little advantage in doing so if the calculations are going to be performed in **R**.

```
 options(digits=3)
> xbar<-matrix(c(47,45,39,42),4)
> mu0<-matrix(c(1,1,1,1),4)
> Sig<-110*diag(c(1,1,1,1))-10*mu0%*%t(mu0)
> Z<-solve(Sig)
> lamhat<-(t(mu0)%*%Z%*%xbar/t(mu0)%*%Z%*%mu0)[1,1]
> LRT<-27*t(xbar-lamhat*mu0)%*%Z%*%(xbar-lamhat*mu0)
> lamhat; LRT; 1-pchisq(LRT,4)
[1] 43.2
     [,1]
[1,] 9.02
       [,1]
[1,] 0.0606
```

Note that although the LRT statistic is a scalar **R** calculates it as a 1×1 matrix and so it is necessary to extract the $(1,1)^{st}$ element. Thus we have $\hat{\lambda} = 43.2$ and the value of the LRT statistic is 9.02 which under the null hypothesis that greenfinches have equal preferences is an observation from a χ_4^2-distribution, yielding a p-value of 0.061 and so there is only slight evidence that greenfinches have unequal preferences for the types of sunflower seeds.

(6) Suppose $\boldsymbol{x}_1, \boldsymbol{x}_2, \ldots, \boldsymbol{x}_n$ are independent observations of $N_p(\mu, \lambda\Sigma_0)$ where Σ_0 is a known positive definite matrix and λ is an unknown scalar and μ is not assumed to be known.

(i) Show that the mle of λ is $\hat{\lambda} = \frac{n-1}{np}\mathrm{tr}(\Sigma_0^{-1}\boldsymbol{S})$ where $\boldsymbol{S}$ is the sample variance matrix.

The log-likelihood is (see §9.2.4)

$$\begin{aligned} \ell(\mu,\lambda;\boldsymbol{X}) &= -\tfrac{1}{2}(n-1)\mathrm{tr}(\Sigma_0^{-1}\boldsymbol{S})/\lambda - \tfrac{1}{2}n\big(\mathrm{tr}\big(\Sigma_0^{-1}(\bar{\boldsymbol{x}}-\mu)(\bar{\boldsymbol{x}}-\mu)'\big)\big)/\lambda \\ &\quad - \tfrac{1}{2}np\log(2\pi) - \tfrac{1}{2}n\log(|\Sigma_0|) - \tfrac{1}{2}np\log(\lambda), \end{aligned}$$

$$\begin{aligned} \text{so } \frac{\partial\ell}{\partial\lambda} &= \tfrac{1}{2}(n-1)\mathrm{tr}(\Sigma_0^{-1}\boldsymbol{S})/\lambda^2 + \tfrac{1}{2}n\big(\mathrm{tr}\big(\Sigma_0^{-1}(\bar{\boldsymbol{x}}-\mu)(\bar{\boldsymbol{x}}-\mu)'\big)\big)/\lambda^2 \\ &\quad -\tfrac{1}{2}np/\lambda. \end{aligned}$$

Differentiating ℓ wrt μ gives $\hat{\mu} = \bar{\boldsymbol{x}}$ and so setting the derivative of ℓ wrt λ equal to zero gives $\hat{\lambda} = \frac{n-1}{np}\text{tr}(\Sigma_0^{-1}\boldsymbol{S})$.

Under $\overline{H}_0$ $\hat{\mu} = \bar{\boldsymbol{x}}$ and $\hat{\Sigma} = \frac{n-1}{n}\boldsymbol{S}$ so $\ell_{\max}(\overline{H}_0) = {}^1\!/\!_2 np - {}^1\!/\!_2 np\log(2\pi) - {}^1\!/\!_2 n\log(|\frac{n-1}{n}\boldsymbol{S}|)$.

Under H_0 $\hat{\mu} = \bar{\boldsymbol{x}}$ and $\hat{\Sigma} = \hat{\lambda}\Sigma_0$ so $\ell_{\max}(H_0) = {}^1\!/\!_2 np - {}^1\!/\!_2 np\log(2\pi) - {}^1\!/\!_2 n\log(|\Sigma_0|) - {}^1\!/\!_2 np\log(\frac{n-1}{np}\text{tr}(\Sigma_0^{-1}\boldsymbol{S}))$.

Thus the LRT statistic is

$2\{\ell_{\max}(\overline{H}_0) - \ell_{\max}(H_0)\} = np\log(\frac{n-1}{np}\text{tr}(\Sigma_0^{-1}\boldsymbol{S})) - n\log(|\frac{n-1}{n}\Sigma_0^{-1}\boldsymbol{S}|)$.

(Mardia et al. (1979), Page 134). Note that this statistic can be written as $np\log(a/g)$ where a and g are the arithmetic and geometric means of the eigenvalues of $\frac{n-1}{n}\Sigma_0^{-1}\boldsymbol{S}$; see key results (1) and (2) of §6.9).

(ii) In a feeding experiment on sea urchins, equal amounts of three types of algæ were placed in 27 tanks each containing a single sea urchin. After 24 hours the mean weight losses over the 27 samples of the three types of algæ were 4.7 g, 3.9 g and 4.2 g with sample variance matrix $\boldsymbol{S} = \begin{pmatrix} 1.1 & 0.0 & 0.1 \\ 0.0 & 0.9 & 0.0 \\ 0.1 & 0.0 & 0.8 \end{pmatrix}$. Are these data consistent with the theory that the amounts of algæ have equal variances with pairwise correlations of 0.1?

Note that the LRT does not involve the actual amounts of algæ consumed.

We have $\boldsymbol{S} = \begin{pmatrix} 1.1 & 0.0 & 0.1 \\ 0.0 & 0.9 & 0.0 \\ 0.1 & 0.0 & 0.8 \end{pmatrix}$ and take $\Sigma_0 = \begin{pmatrix} 1.0 & 0.1 & 0.1 \\ 0.1 & 1.0 & 0.1 \\ 0.1 & 0.1 & 1.0 \end{pmatrix}$.

```
> S<-matrix(c(1.1,0,0.1,0,0.9,0,0.1,.0,0.8),3,3)
> one<-matrix(c(1,1,1),3)
> Sig<-0.9*diag(c(1,1,1))+0.1*one%*%t(one)
> Z<-26/27*solve(Sig)%*%S
> LRT<-27*(3*log(sum(diag(Z))/3)-log(det(Z)))
> LRT;1-pchisq(LRT,5)
[1] 1.21
[1] 0.944
```

The approximate null distribution of the LRT statistic is χ_5^2 (5 is the difference in numbers of independent parameters estimated under H_0 and $\overline{H}_0$). Thus the value of the LRT statistic of 1.21 provides little evidence against the hypothesis that the variance is a scalar multiple of Σ_0. Note that *only* high values of the LRT statistic provide evidence against the null hypothesis.

Bibliography

Abadir, K. M. and J. R. Magnus (2005). *Matrix algebra*, Volume 1. Cambridge: Cambridge University Press.

Adler, J. (2010). *R in a nutshell: a desktop quick reference*. Sebastopol, CA: O'Reilly Media, Inc.

Anderson, T. (2003). *An introduction to multivariate statistical analysis*. Hoboken, NJ: John Wiley & Sons.

Banerjee, S. and A. Roy (2014). *Linear algebra and matrix analysis for statistics*. Boca Raton, Fl: CRC Press.

Basilevsky, A. (2013). *Applied matrix algebra in the statistical sciences*. Mineola, NY: Courier Dover Publications.

Bates, D. and M. Maechler (2014). *Matrix: sparse and dense matrix classes and methods*. R package version 1.1-4.

Cox, T. (2005). *An introduction to multivariate data analysis*. London: Arnold.

Crawley, M. J. (2005). *Statistics: An introduction using R*. Chichester: John Wiley & Sons.

Crawley, M. J. (2012). *The R book*. Chichester: John Wiley & Sons.

Dalgaard, P. (2008). *Introductory statistics with R*. New York: Springer.

Draper, N. R. and H. Smith (1998). *Applied regression analysis 3rd ed.* Hoboken, NJ: John Wiley & Sons.

Faraway, J. J. (2014). *Linear models with R*. Boca Raton, Fl: CRC Press.

Gnanadesikan, R. (1997). *Methods for statistical data analysis of multivariate observations*. Hoboken, NJ: John Wiley & Sons.

Gonzlez, I. and S. Djean (2012). *CCA: Canonical correlation analysis*. R package version 1.2.

Gower, J. C. and D. J. Hand (1995). *Biplots*, Volume 54. Boca Raton, Fl: CRC Press.

Greenacre, M. (2010). *Correspondence analysis in practice*. Boca Raton, Fl: CRC Press.

Guttman, I. (1982). *Linear models: an introduction*. New York: John Wiley & Sons.

Harville, D. A. (2008). *Matrix algebra from a statistician's perspective*. New York: Springer.

Healy, M. (1969). 259 note: Rao's paradox concerning multivariate tests of significance. *Biometrics*, 411–413.

Jones, B. and M. G. Kenward (2003). *Design and analysis of cross-over trials*. Boca Raton, Fl: CRC Press.

Magnus, J. R. and H. Neudecker (1988). *Matrix differential calculus*. Chichester: John Wiley & Sons.

Mardia, K. V., J. T. Kent, and J. M. Bibby (1979). *Multivariate analysis*. San Diego, CA: Academic Press.

Nordhausen, K., S. Sirkia, H. Oja, and D. E. Tyler (2012). *ICSNP: Tools for multivariate nonparametrics*. R package version 1.0-9.

Novomestky, F. (2012). *matrixcalc: Collection of functions for matrix calculations*. R package version 1.0-3.

Puntanen, S., G. P. Styan, and J. Isotalo (2011). *Matrix tricks for linear statistical models: our personal top twenty*. Heidelberg: Springer.

R Core Team (2014). *R: A language and environment for statistical computing*. Vienna: R Foundation for Statistical Computing.

Rao, C. R. (1966). Covariance adjustment and related problems in multivariate analysis. *Multivariate analysis 1*, 87–103.

Ringrose, T. J. (1996). Alternative confidence regions for canonical variate analysis. *Biometrika 83*(3), 575–587.

Ringrose, T. J. (2012). Bootstrap confidence regions for correspondence analysis. *Journal of Statistical Computation and Simulation 82*(10), 1397–1413.

Searle, S. R. (1982). *Matrix algebra useful for statistics*. Hoboken, NJ: John Wiley & Sons.

Seber, G. A. and A. J. Lee (2012). *Linear regression analysis*. Hoboken, NJ: John Wiley & Sons.

Styan, G. P. (1973). Hadamard products and multivariate statistical analysis. *Linear Algebra and its Applications 6*, 217–240.

Venables, W. N. and B. D. Ripley (2002). *Modern applied statistics with S* (4th ed.). New York: Springer.

Venables, W. N., D. M. Smith, R Development Core Team, et al. (2014). An introduction to R.

Verzani, J. (2011). *Getting started with RStudio*. Sebastopol, CA: O'Reilly Media.

Vinod, H. D. (2011). *Hands-on matrix algebra using R: Active and motivated learning with applications*. Singapore: World Scientific.

Wuertz, D. and Rmetrics Core Team Members (2013). *fBasics: Rmetrics markets and basic statistics*. R package version 3010.86.

Zuur, A., E. N. Ieno, and E. Meesters (2009). *A beginner's guide to R*. New York: Springer.

Index